Presentism

Presentism

Essential Readings

Edited with Introductions by
Ernâni Magalhães and L. Nathan Oaklander

LEXINGTON BOOKS
A division of
ROWMAN & LITTLEFIELD PUBLISHERS, INC.
Lanham • Boulder • New York • Toronto • Plymouth, UK

Published by Lexington Books
A division of Rowman & Littlefield Publishers, Inc.
A wholly owned subsidiary of The Rowman & Littlefield Publishing Group, Inc.
4501 Forbes Boulevard, Suite 200, Lanham, Maryland 20706
http://www.lexingtonbooks.com

Estover Road, Plymouth PL6 7PY, United Kingdom

British Library Cataloguing in Publication Information Available

Library of Congress Cataloging-in-Publication Data

Presentism : essential readings / edited with introductions by Ernâni Magalhães and L. Nathan Oaklander.
p. cm.
Includes index.
ISBN 978-0-7391-4500-5 (cloth : alk. paper)
1. Presentism (Philosophy) I. Magalhães, Ernâni, 1975- II. Oaklander, L. Nathan, 1945-
BD642.P74 2010
115—dc22

2010031006

For my wife and greatest supporter, Lisa Wellinghoff, and for the wonderful philosophers who taught me at the University of Iowa

Ernâni Magalhães

For my granddaughters Ayonna and Symphony

L. Nathan Oaklander

Contents

Preface xi
Acknowledgments xiii

General Introduction 1
Ernâni Magalhães and L. Nathan Oaklander

Part I: Presentism and Temporal Becoming

Introduction: Presentism and Temporal Becoming 11
L. Nathan Oaklander

1 What Is Time? 27
St. Augustine

2 The Unreality of Time 35
J. M. E. McTaggart

3 Changes in Events and Changes in Things 45
Arthur Prior

4 Is Presentness a Property? 57
William Lane Craig

5 Presentism: A Critique 75
L. Nathan Oaklander

Part II: Defining Presentism: The Triviality Objection

Introduction: Defining Presentism: The Triviality Objection 93
Ernâni Magalhães

6 The Presentist's Dilemma 99
Ulrich Meyer

7 On Presentism and Triviality 109
Thomas Crisp

Part III: Overviews of Presentism

Introduction: Overviews of Presentism 117
Ernâni Magalhães

8 Presentism and Properties 125
John Bigelow

9 A Theory of Presentism 141
Craig Bourne

10 A Defense of Transient Presentism 161
M. Oreste Fiocco

Part IV: The Case against Presentism

Section 1: Presentism and Inter-Temporal Relations

Introduction: Presentism and Inter-Temporal Relations 191
Ernâni Magalhães

11 Presentism and the Non-Present 195
Matthew Davidson

12 Presentism and the Problem of Cross-Time Relations 207
Rafael De Clercq

Section 2: Presentism and Relativity

Introduction: Presentism and Relativity 225
Ernâni Magalhães

13 Time and Physical Geometry 229
Hilary Putnam

14 A Defense of Presentism in a Relativistic Setting 237
Mark Hinchliff

Section 3: Presentism and Truthmaking

Introduction: Presentism and Truthmaking 251
L. Nathan Oaklander

15 Presentism and Truthmaking 259
Simon Keller

16 Presentism and the Grounding Objection 277
Thomas Crisp

Part V: Presentism and Persistence

Introduction: Presentism and Persistence 299
L. Nathan Oaklander

17 Temporary Intrinsics and Presentism 305
Dean Zimmerman

18 On the Incompatibility of Enduring and Perduring Entities 319
Trenton Merricks

Index 327
About the Editors 333

Preface

Over the past two decades the philosophy of time known as "presentism," the view that only present things exist, has become the subject of numerous expositions, critiques and defenses, and there has not been an anthology that contains the essential readings on the topic. The aim of our book is to fill that lacuna. The book is intended to be used in upper level courses on metaphysics and seminars on the philosophy of time. It should be of use to students and scholars working in the area of time, and the additional references should give the reader the resources to delve more deeply into the philosophy of time than the essays alone could provide. In addition, the substantial introductions should aid those who are new to the philosophy of time by providing useful critical summaries of the readings to follow.

Both authors developed an interest in the philosophy of time while attending the University of Iowa, and although their stay there was separated by roughly 30 years they are grateful for the education received at Iowa. They subsequently met in 2006 at the University of Iowa during a conference celebrating the one hundredth anniversary of the birth of the ontologist par excellence, Gustav Bergmann, whose thoughts and writings influenced students at Iowa for many generations. We met again when Ernâni Magalhães organized a Conference in 2008 on McTaggart and One Hundred Years of the Unreality of Time. Within a few months after that the plan for this book was hatched.

We wish to thank the Office of the Vice President for Research at the University of Michigan for a subvention grant, and the Office of Research and the Department of Philosophy at the University of Michigan-Flint for additional funding to help defray to cost of producing this book. We also thank Linda G. Oaklander for preparing the index and Daniel Grafton for his assistance in editing the chapters.

L. Nathan Oaklander
Flint, MI 48502
July 2010

Ernâni Magalhães
Morgantown, WV
July 2010

Acknowledgments

Augustine. Selections from Book XI, 230-243, *Confessions* (1998) edited by Henry Chadwick. Reprinted by permission of Oxford University Press.

Bigelow, John. "Presentism and Properties," *Philosophical Perspectives* 10 (1996), pp. 35-52. © 1996 Blackwell Publishing Ltd. Reproduced by permission of Blackwell Publishing Ltd.

Bourne, Craig. "A Theory of Presentism," *Canadian Journal of Philosophy* 36 (2006): 1-23. Reprinted by permission of *Canadian Journal of Philosophy* and Craig Bourne.

Craig, William Lane, "Is Presentness a Property?" *American Philosophical Quarterly* (1997): 27-40. Reprinted by permission of the *American Philosophical Quarterly* and William Lane Craig.

Crisp, Thomas. "On Presentism and Triviality," *Oxford Studies in Metaphysics, volume 1*, D. Zimmerman (ed.) (Oxford: Oxford University Press, 2004), pp. 15-20. Reprinted by permission of Oxford University Press.

Crisp, Thomas. "Presentism and the Grounding Objection," *Noûs* 41 (2007): 90-109. Copyright © 2007 Blackwell Publishing Ltd. Reproduced by permission of Blackwell Publishing Ltd.

Davidson, Matthew. "Presentism and the Non-Present," *Philosophical Studies* 113 (2003): 77-92. © Springer 2003. Reprinted with kind permission from Springer Science and Business Media.

De Clercq, Rafael. "Presentism and the Problem of Cross-Time Relations," *Philosophy and Phenomenological Research*72 (2006): 386-402. © 2006 by Blackwell Publishing Ltd. Reproduced by permission of Blackwell Publishing Ltd.

Fiocco, M. Oreste. "A Defense of Transient Presentism," *American Philosophical Quarterly* 44 (2007): 191-212. Reprinted by permission of the *American Philosophical Quarterly* and M. Oreste Fiocco.

Hinchliff, Mark. "A Defense of Presentism in a Relativistic Setting," *Philosophy of Science* 67 (2000): S575-S586. © 2000 The University of Chicago Press. Reprinted by permission of The University of Chicago Press.

Keller, Simon. "Presentism and Truthmaking," *Oxford Studies in Metaphysics, volume 1*, D. Zimmerman (ed.) (Oxford: Oxford University Press, 2004), pp. 83-104. Reprinted by permission of Oxford University Press.

McTaggart, J. M. E. "Time," from *The Nature of Existence,* vol. 2 (Cambridge: Cambridge University Press, 1927), pp. 9-22. © Cambridge University Press. Reprinted with the permission of Cambridge University Press.

Merricks, Trenton. "On the Incompatibility of Enduring and Perduring Entities," *Mind* 104 (1995): 523-531. Reprinted by permission of Oxford University Press.

Meyer, Ulrich. "The Presentist's Dilemma," *Philosophical Studies* 122 (2005): 213-225. © Springer 2005. Reprinted with kind permission from Springer Science and Business Media.

Oaklander, L. Nathan. "Presentism: A Critique," in Hallvard Lillehammer and G. Rodriguez Pereyra (eds.), *Real Metaphysics: Essays in Honour of D. H. Mellor, With His Replies* (London: Routledge Press, 2002), pp. 196–211. Reprinted by permission of Taylor & Francis.

Prior, Arthur. "Changes in Events and Changes in Things," the 1962 Lindley Lecture at the University of Kansas. Reprinted by permission of the Department of Philosophy, the University of Kansas.

Putnam, Hilary. "Time and Physical Geometry," *Journal of Philosophy* 64 (1967): 240-247. Reprinted by permission of the *Journal of Philosophy* and Hilary Putnam.

Zimmermann, Dean. "Temporary Intrinsics and Presentism," *Metaphysics: The Big Questions*, D. Zimmerman and P. van Inwagen (eds.) (Oxford: Blackwell Publishers, 1998), pp. 206-220. © 1998 Blackwell Publishing Ltd. Reproduced by permission of Blackwell Publishing Ltd.

General Introduction

Ernâni Magalhães and L. Nathan Oaklander

Presentism is the view that everything is present. Equivalently, it is the view that nothing exists which is not present. Reality is exhausted by the present: future things such as tomorrow and the inauguration of the forty-ninth president are not real. Past things such as yesterday and the inauguration of the forty-third president are not real. (Notice these consequences assume that nothing can be both past and present or future and present.) To put it most jarringly: there is no past and there is no future. Yet presentists do not deny that tomorrow *will* exist and that yesterday *did* exist. To understand presentism we need to take a step back.

As we ordinarily speak and think about and experience time, it is understood in two seemingly different ways. On the one hand, we understand time in terms of the notions of *past*, *present*, and *future*, and, on the other hand, we think of time in terms of the notions of *earlier* and *later*. Thus, the inauguration of the forty-fifth president is in the future, the presidency of Barack Obama is in the present, and the presidency of Bill Clinton is in the past. Also, the inauguration of the forty-eighth president *is later than* Obama's presidency, and *earlier than* the inauguration of the forty-ninth president. Past, present, and future are transitory aspects of time: future events will cease to be future and become present and then past; present events used to be future and will be past; past events used to be present and future. The relational facts—such as that the inauguration of the forty-eighth president is later than Obama's presidency—seem not to change regardless of what time it is. In other words, there is a sense in which *time changes* and a sense in which *time does not change*. One way of understanding the philosophical problem of time and the different theories that attempt to answer that problem is to ask: How are these two seemingly incompatible aspects of time to be understood? Are our ordinary ways of thinking about time, in terms of the so-called "A-series" (after McTaggart's series of terms that are past, present and future (see chapter 2)) and the so called "B-series" (after McTaggart's series of terms that are earlier/later and simultaneous) equally fundamental to the nature of time, or does a rational understanding of temporal reality, one which includes both logical coherence and fidelity to the latest scientific evidence, require us to choose one of these seemingly different ways of understanding time, or is there some third alternative?

To get a clearer understanding of the issues involved and to understand some of the motives and interpretation of presentism, it will be useful to distinguish two broad camps into which most philosophers of time fall, the A- and B-theories. According to the A-theory, our ordinary language of time with the past, present, and future tenses reflects something fundamental about the metaphysical nature of time, what the essence of time consists in. On the A-theory, in its purist form, events or times (if there be such) have the non-relational properties of *pastness*, *presentness*, and *futurity* (known as "A-properties") and change with respect to them as they pass from the future to the present and into the past. Temporal relations either do not exist at all, or if they do exist they are analyzable in terms of the non-relational temporal properties of events or times or in some other way. For the B-theorist, on the other hand, it is the B-series with its unchanging temporal or B-relations of *earlier/later than* and *simultaneous with* between events or times that is fundamental, and the A-series is analyzed away in terms of it. There have been so-called A/B theorists who grant ontological status to both A-determinations and B-relations, but for our purposes the distinction between the pure A-theorist and the B-theorist will suffice.

The distinction between different theories of time is a metaphysical distinction; it concerns what exist, or what is real or what has ontological status. Generally, and in fact virtually invariably, ontological questions concern what *kind* of entities there are and how they are connected—it does not concern how many of that kind exist. Although A-theorists agree in identifying time with its transitory aspect, they disagree about which A-properties are exemplified. According to the presentist, only the present exists: President Obama exists, but neither World War II nor his grandchildren exist. According to the open future or growing block theory, the present and past exist, but the future does not exist: World War II and President Obama, but not his grandchildren, exist. According to the full A-theory, past, present and future events, things or times all exist. On the full A-theory, World War II, President Obama and his grandchildren all are real. In a way superficially resembling the full A-theory (for reasons to be explained shortly), B-theorists—or eternalists, as they are also called—also believe in the equal existence of past, present, and future entities. There is, however, no disagreement—among A-theorists or B-theorists—over what events did exist or currently exist or will exist. Whether one is a presentist, a growing block theorist, a full A-theorist or a B-theorist it is agreed that World War II *did* exist, President Obama *now* exists, and so *will* the first president of the twenty-second century.

How can there be such disagreement among the various philosophies of time when there is basic agreement about what things do, did or will exist? What is the fundamental basis of the disagreement between presentists and A-theorists on the one hand and presentists and B-theorists on the other? Although almost all philosophers of time agree that it is true that World War II occurred, that President Obama exists, and that there will a first twenty-second century president, they disagree about what ontological account should be given of these truths. They disagree, in other words, about what in reality it is that makes these truths true. For the full A-theorist, for example, it is the real event of World War II exemplifying the property of *pastness* that makes it true that World War II occurred, and it is the real future individual exemplifying *futurity* that makes it true that there will be a first president of the twenty-second century. The presentist cannot say the same. Since neither World War II nor the first twenty-second century president are present, and there are no such properties as *pastness* and *futurity*, those entities cannot make it true that World War II occurred and that there will be a first twenty-second century president. What does the presentist say makes those statements true? Different presentists say different things—the most prominent answers are reflected in the chapters below—but given their view it must be something present.

The superficial resemblance between the B-theory and full A-theory—both are committed to equally existing past, present, and future things—masks a deeper disagreement, which can be seen in terms of their differing accounts of the truthmakers for tensed statements about the past, present and future. For the full A-theory it cannot be World War II by itself that makes it true that World War II occurred. Its mere existence, after all, in no way implies whether it is past, present, or future. It is World War II's *pastness* that makes it true that World War II occurred, rather than that it is occurring or will occur. Pastness as we have noticed is a transient characteristic—an event may lack and then acquire this characteristic with the passage of time. The B-theorist has no truck with such transient truthmakers. For the B-theorist what makes it true that World War II occurred is the real event together with its standing in the *earlier than* relation to some things that are occurring now, such as the writing of this sentence. And "occurring now" denotes the time at which the phrase is written, uttered, or thought. Thus, for the B-theorist, events are not really past, present or future at all, since those terms simply indicate the perspective or temporal location from which events are viewed, not intrinsic characteristics of the events themselves. Indeed the disagreement over the dependence on transient facts to account for tensed truths similarly separates presentists and B-theorists. Presentists, believing that time fundamentally passes, also account for tensed truths in terms of the possession of a transient property.

Thus, at bottom, what separates presentism and eternalism are their differing views of whether a transient property or the *earlier/later than* relation should be taken as fundamental to time. Each view faces challenges arising from its commitment. On the one hand, given that our ordinary conception of time and our experience and language of time involve temporal relations between and among events, the onus is on the presentist to give us an account of how that is possible given that on that view nothing really is earlier or later than anything else. On the other hand, the eternalist has the equally pressing challenge of giving some explanation of the so-called "transitory aspect of time," that feature of time contained in our ordinary conception of time as well as our experience and language of time that reflects the passage of time or the movement of events in time from the future through the present and into the past. How can time have these features if in reality nothing literally is past, present, and future and time does not move?

The collection we have compiled contains selections that are concerned exclusively with presentism. Intuitively, there is much to be said for presentism. For what it's worth, it may even be the most faithful expression of the commonsense ontology of time. Pre-theoretically, the catalogue of the real will surely include any present things, but it will not include such past things as dinosaurs or French kings, or such future things as flying cars and Martian colonies. Dinosaurs and Martian colonies do not exist, even if they are past and future. Indeed, the fact that they are past and future seems to guarantee that they do not exist. How can something past or future be real? To become past, as Augustine says, is to "pas[s] away into non-existence" (chapter 1, p. 27).

There is also a deeper philosophical motivation for presentism, namely, its alleged way out of McTaggart's conundrum concerning the nature of time. Recall that what we referred to as the transitory aspect of time consists in the fact that events which are once future become present and then recede into the more and more distant past. This *sui generis* or unique type of change was understood by McTaggart to involve a single event changing its non-relational temporal properties: first having *futurity*, then having *presentness* and then having *pastness*. McTaggart argued, however, that a single event cannot have these incompatible properties as it moves through time since either a contradiction would result (every event *being past, being present*, and *being future*) or a vicious infinite regress of time dimensions would result (since the different times at which events had these incompatible A-properties would themselves have incompatible A-properties). Presentists who took McTaggart's paradox seriously argued that this dilemma can easily be avoided by doing away with the properties of *pastness* and *futurity*. For if *presentness* is the only non-relational temporal property that events possess, then temporal becoming, however it is to be understood, would not involve a single event having incompatible A-properties. McTaggart's paradox, temporal becoming and presentism are discussed in Part I, "Presentism and Temporal Becoming."

Another challenge to presentism consists in rendering it a significant position. One such challenge, and initially the most pressing, is to state the view in a way that enables it to avoid being trivially true on the one hand, or wildly implausible on the other. A critic may argue that if the word "exists" means "exist now" then the thesis that only the present exists reduces to the tautology that only the present exists now or only what is happening at present is present. On the other hand, if "exists" is taken to mean "is past, present or future," then the claim that only the present exists would imply that only what is present is past, present, or future, and that is absurd. If only the present exists in this second sense of the term, then the past and the future are present, which is tantamount to saying that there is no future or past; that the entire span of time is limited to what exists now, and that is absurd. As Lewis once put it, to say that only the present exists is to say that we have no past or future, which is obviously false. The selections by Meyer and Crisp in Part II, "Defining Presentism: The Triviality Objection," discuss that issue.

As we have indicated in previous paragraphs, there are a whole host of issues raised by presentism. Part III, "Overviews of Presentism," contains wide-ranging articles that attempt in different ways to expound a version of presentism that can withstand the numerous objections that have been raised against it. Fiocco, Bigelow, and Bourne attempt to formulate views that can handle the problems addressed in more detail by the articles in Part IV, "The Case against Presentism." Part IV is divided into three sections. Together, these sections address the major objections that have been brought against presentism. The articles in Section 1 consider whether presentism is compatible with the possibility of relations between things that do not exist at the same time. Present things seem to be related to past things; they exist after them, for example. It seems, however, that in order for two or more things to be related, there must *be* the relevant things. But past things, according to presentism, *are* not.

Section 2 considers whether presentism is compatible with relativity theory. Since presentism holds that only the present is real, it seems to require that there is a privileged frame of reference, the one that uniquely determines which things are present. Relativity, however, denies that there is any such special frame.

Section 3 considers whether presentism is compatible with the doctrine that every truth has a truthmaker. When it is true that so-and-so it must be because some real things make it the case that so-and-so. It is true that Caesar crossed the Rubicon. Barring a crude identification of past facts with present evidence, what can the presentist identify as the truthmaker for this truth and others of its kind? Each section includes an article arguing that presentism is not up to the challenge and another arguing that it is.

Part V, "Presentism and Persistence," considers the relationship between presentism and persistence or duration. The remarkable feature about a persisting thing is that it can exemplify (over time) incompatible characteristics. In other words, it can change. Presentism is thought by some—such as Zimmerman—to provide an elegant explanation of this remarkable feature. A thing can be bent and straight because it (say) *is* bent and *was* straight. Merricks argues that presentism is indeed only compatible with the view that things persist by being wholly present at every time that they exist—the so-called "endurantist" theory of persistence. It is thus incompatible with the view that persisting things are sums of temporal "stages."

Notes

1. These claims are not quite universally agreed. Some open future theorists—see Tooley (1997)—deny that any future-tensed sentences are true. Some B-theorists—see Oaklander (2003)—deny that any tensed sentences are true.

References and Further Readings

Augustine. *Confessions*. Translated by H. Chadwick. Oxford University Press, 1991. Chapter one in this volume.

McTaggart, J. M. E. *The Nature of Existence*, 2 vols., edited by C. D. Broad. Cambridge: Cambridge University Press, 1927; rep. ed.: 1968. Chapter two in this volume.

Oaklander, L. N. "Two Versions of the New Theory of B-Language." Pp. 271-303 in *Time, Tense, and Reference*, edited by A. Jokic and Q. Smith. Cambridge, Mass.: MIT Press, 2003.

Tooley, M. *Time, Tense, and Causation*. Oxford: Oxford University Press, 1997.

Full A-Theory

Gale, R. *The Language of Time*. New York: Humanities Press, 1968.

Schlesinger, G. *Aspects of Time*. Indianapolis: Hackett Publishing Company, 1980.

Smith, Q. *Language and Time*. Oxford: Oxford University Press, 1993.
———. "Time and Degrees of Existence: A Theory of 'Degree Presentism.'" Pp. 119-36 in *Time, Reality & Experience*, edited by C. Callender. Cambridge: Cambridge University Press, 2002.

B-Theory

Christensen, F. *Space-like Time*. Toronto: University of Toronto Press, 1993.
Craig, W. L. *The Tenseless Theory of Time*. Dordrecht: Kluwer Academic Publisher, 2000.
Goodman, N. *The Structure of Appearance*. Cambridge, Mass: Harvard University Press, 1951.
Grünbaum, A. "The Exclusion of Becoming from the Physical World." Pp. 471-500 in *The Concepts of Space and Time*, edited by M. Capek. Boston: D. Reidel Publishing Co., 1976.
Le Poidevin, R. *Change, Cause and Contradiction*. Basingstoke: Macmillan, 1991.
Mellor, D. H. *Real Time*. Cambridge: Cambridge University Press, 1981.
Quine, W. V. O. *Word and Object*. Cambridge, Mass.: MIT Press, 1960.
Reichenbach, H. *Elements of Symbolic Logic*. London: Macmillan, 1947.
Russell, B. "On the Experience of Time," *Monist* 25 (1915): 212-33.
———. "Lectures on the Philosophy of Logical Atomism." Pp. 175-281 in *Logic and Knowledge*, edited by R. C. Marsh. London: Routledge, 1956.
Sider, T. *Four-Dimensionalism: An Ontology of Persistence and Time*. Oxford: Oxford University Press, 2001.
Smart, J. J. C. *Philosophy and Scientific Realism*. London: Routledge & Kegan Paul, 1963.
———. "Time and Becoming." Pp. 3-16 in *Time and Cause: Essays Presented to Richard Taylor*, edited by P. v. Inwagen. Dordrecht: D. Reidel, 1980.

A-Theory vs. B-Theory

Broad, C. D. *Examination of McTaggart's Philosophy*. Cambridge: Cambridge University Press, 1938.
Callender, C., ed. *Time, Reality and Experience*. Cambridge: Cambridge University Press, 2002.
Craig, W. L. *The Tensed Theory of Time: A Critical Examination*. Dordrecht: Kluwer, 2000.
Gale, Richard. *The Language of Time*. New York: Humanities Press, 1968.
———, ed. *The Philosophy of Time*. New York: Doubleday, 1967.
Le Poidevin, R. *Time, Cause and Contradiction*. Basingstoke: Macmillan, 1991.
———. *Questions of Time and Tense*. Oxford: Clarendon Press, 1998.
Mellor, D. H. *Real Time II*. London: Routledge, 1998.
Oaklander, L. N. *Temporal Relations and Temporal Becoming*. Lanham, MD: University Press of America, 1984.
Oaklander, L. N., and Q. Smith, eds. *The New Theory of Time*. New Haven, CT: Yale University Press, 1994.
Smith, Q. *Language and Time*. Oxford: Oxford University Press, 1993.
Smith, Q., and A. Jokic, eds. *Time, Tense and Reference*. London: MIT Press, 2003.
Tichý, P. "The Transiency of Truth," *Theoria* 46 (1980): 165-82.

Open Future or Growing Block Theory

Bourne, C. "When Am I? A Tense Time for Some Tense Theorists?" *Australasian Journal of Philosophy* 55 (2002): 359-71.

———. *A Future for Presentism*. Oxford: Clarendon Press, 2006.

Broad, C. D. *Scientific Thought*. London: Routledge, 1923.

Cameron, R., and E. Barnes. "The Open Future: Bivalence, Determinism, and Ontology," *Philosophical Studies* 146 (2009): 291-309.

Earman, J. "Reassessing the Prospects for a Growing Block Model of the Universe," *International Studies in the Philosophy of Science* 22 (2008): 135-64.

Forrest, P. "The Real but Dead Past," *Analysis* 64 (2004): 358-63.

———. "Uniform Grounding of Truth and the Growing Block Theory: A Reply to Heathwood," *Analysis* 66 (2006): 161-63.

Heathwood, C. "The Real Price of the Dead Past: A Reply to Forrest and Braddon-Mitchell," *Analysis* 65 (2005): 249-51.

Markosian, N. "The Open Past," *Philosophical Studies* 79 (2005): 95-105.

McCall, S. "Objective Time Flow," *Philosophy of Science* 43 (1976): 337-62.

———. *A Model of the Universe*. Oxford: Clarendon Press, 1994.

Tooley, M. *Time, Tense, and Causation*. Oxford: Oxford University Press, 1997.

Williams, D. C. "The Sea-Fight Tomorrow." Pp. 282-306 in *Structure, Method, and Meaning*, edited by P. Henle, H. M. Kallen, and S. K. Langer. New York: The Liberal Arts Press, 1951.

Part I

Presentism and Temporal Becoming

Introduction: Presentism and Temporal Becoming

L. Nathan Oaklander

Introduction

One of the most fascinating and puzzling aspects of our ordinary language, thought, and experience of time is the way they reflect its passage. In some sense, time seems to move from the future to the present and then from the present into the more and more distant past. We use tensed language to reflect this allegedly *sui generis* form of change. For example, it is now true to say that "I *will* retire within the next 10 years," and in a few years it will be true to say that "I *am now* retiring," and some years later it will be true that "I *have been* retired for x number of years." Furthermore, we think differently about events and experiences that appear to be moving toward or away from us. Thus, an unpleasant future event is thought of with *dread* whereas the same event when it becomes past is thought of with *relief*. Finally, there is some plausibility in the claim that we experience the passage of time since we seem to experience the present as having a reality that the past and future do not have. How, then, are we to understand this enigmatic feature of reality—temporal becoming or passage—in all its various manifestations?

To give an adequate answer to that question is one of the most fundamental problems in the philosophy of time. Minimally an adequate answer to the question "What is temporal becoming?" must be able to specify what there is in reality that provides an ontological ground for those aspects of language, thought, and experience that purportedly reflect the passage of time. B-theorists maintain that a logically coherent account of passage cannot be gleaned from a superficial examination of ordinary language, thought, or experience. For, to suppose that our use of tensed *language* represents the non-relational temporal A-properties of *pastness*, *presentness*, and *futurity*, or that our *thoughts* (and attitudes) about events reflect their movement through time, or that our *experience* implies that the present has some special ontological status not bestowed on the past or future is, B-theorists allege, fraught with insurmountable dialectical difficulties, such as McTaggart's paradox (see chapter 2). A-theorists, on the other hand, maintain that a careful examination of our ordinary language, thought, and experience *is* a sure path to the truth about time, and that B-theorists are mistaken in thinking that all A-theoretical accounts of temporal becoming are internally inconsistent. Although A-theorists universally agree that difficulties, such as McTaggart's paradox, pose no lasting threat to our so-called "intuitive" conception of time, they do debate amongst themselves over which version of the A-theories is best capable of avoiding the charge of incoherence leveled against them by B-theorists.

In this introduction I will consider two presentist responses to McTaggart by first discussing Augustine's discussion of time in his *Confessions* (chapter 1) where he raises problems with understanding temporal becoming and the presentist response to them. His essay sets the stage for the more general attack on the reality of time found in McTaggart's essay in chapter 2 that I will then discuss. I shall turn next to two recent defenders of presentism, Arthur Prior (chapter 3) and William Lane Craig (chapter 4), who each, in their own distinctive ways, attempt to respond to both McTaggart and Augustine. Finally, I will turn to the essay by Oaklander (chapter 5) and will explain the main reasons why Oaklander believes that neither Prior nor Craig avoid the problems posed by Augustine and McTaggart, and that, consequently, they do not give an adequate account of temporal becoming.

Saint Augustine

Augustine famously asks: "What then is time? Provided that no one asks me, I know. If I want to explain it to an inquirer, I do not know" (chapter 1, p. 27). The answer to the question "What is time?" is mysterious. It seems that we know what is meant when we speak and think about it. We say and understand utterances such as "I had my breakfast *before* I had my lunch" and "My breakfast is *past*." These familiar ways of speaking about time are well-known and seemingly understood in everyday life, but they raise a variety of philosophical perplexities and paradoxes, as Augustine makes clear. For example, how can my breakfast *be* past if the past does not exist? And how can my breakfast *be* or exist before my lunch if only the present exists?

The first question that puzzles Augustine concerns temporal passage or temporal becoming. We ordinarily conceive of there being three times—past, present, and future—and as time passes future times or events become present and then present times or events become past, but how is this possible? In order for something to move through time (much like if things are to move through space) they must exist, but "How can they "be" when the past is not now present and the future is not yet present?" (p. 27). In other words, in order for time to exist it must pass. But how can time pass if the past and future are not? In order for time to pass the future must become present and the present must pass into the past. However, we here face a problem, according to Augustine. If the future is to become present then it must "be" but since the future is not yet present, it does not exist, and if the past has ceased to be present then it does not exist. The problem then is how can what is not become (or be) present and how can what no longer is become past since only what is can become?

Thus, Augustine finds the notion of temporal becoming puzzling. In order for time to pass the future must change into or become present and the present must change by becoming past, and thus the past, present, and future must exist. However, the future and the past do not exist so can no more move through time and change their temporal location than objects that do not exist can move through space and change their spatial location. It would appear, therefore, that if our ordinary conception of time which contains the notion of temporal passage is to be rendered intelligible the past and future must exist. Yet that same ordinary conception of time involves the idea that the past and future do not exist; thus time cannot pass.

We have in this paradoxical state of affairs two seemingly different ways of conceiving of time. According to one conception only the present exists, the past and future do not exist, and according to the other, the past, present, and future are ontologically on a par, being equally existing parts of reality. Although Augustine's initial intuition that the present exists and the past and future do not exist, leads to presentism, he recognizes that presentism (although he does not call it such) also leads to a related difficulty, namely, that the present can have no temporal extension since if it did it would have parts that are past and future and so do not exist. If, however, it has no temporal extension or as Augustine puts it, "the present occupies no space" (p. 28), how then can the future pass through it, or how can there even *be* an unextended present time?

Augustine struggles with this paradox and briefly attempts to resolve it by denying one of its horns by entertaining the possibility that the past and future do exist. Indeed, there is, according to Augustine, an important reason to believe in the reality of the past and future, namely, what has come to be called "the intentionality of thought." Whenever we conceive, imagine, perceive, remember, or more generally engage in a so-called "mental act" of thinking, there is always some object distinct from the mental act itself that the thought is about. Thus, whenever I conceive, there is also something that I am conceiving, and so on for all other mental acts. Given the intentionality of thought and the truisms that in memory we are thinking about the past and in anticipation we are thinking about the future, it follows that the past and the future must in some sense "be" in order to be thought about.

Augustine is sympathetic with this line of thought and indeed subscribes to it as is clear from the following passages:

> Where did those who sang prophecies see these events if they do not yet exist? To see what has no existence is impossible. And those who narrate past history would surely not be telling a true story if they did not discern events by their

> soul's insight. If the past were non-existent, it could not be discerned at all. Therefore both the future and past events exist. (p. 29)

> Who will tell me that there are not three times, past, present, and future, as we learnt when children and as we have taught children, but only the present, because the other two have no existence? Or do they exist in the sense that, when the present emerges from the future, times comes out of some secret store, and then recedes into some secret place when the past comes out of the present? (p. 29)

As we shall see in the essays by Bigelow, Bourne, Fiocco, and Crisp (8-10, and 16), presentists are concerned with how we can make judgments about the past and the future if the past and the future do not exist. Moreover, presentists worry not only about what such judgments are about, but how they can be true or, if they are true, what is the truth ground of past and future tense judgments. Augustine hints at answers to these questions which reveal his commitment to a presentist ontology since he maintains,

> Whatever may be the way in which the hidden presentiment of the future is known, nothing can be seen if it does not exist. Now that which already exists is not future but present. When therefore people speak of knowing the future, what is seen is not events which do not yet exist (that is, they really are future), but perhaps their causes or signs which already exist. In this way, to those who see them they are not future but present, and that is the basis on which the future can be conceived in the mind and made the subject of predication. (p. 29)

Clearly then, the present is the ground of things that are thought to be past or future, and in that sense Augustine is a presentist: neither the future nor the past is now present, and since only the present exists, it follows that neither the past nor the future exist, except insofar as they have a ground in the present. Yet presentism, paradoxically, raises seemingly insurmountable problems for our ordinary conception of time insofar as our commonsense view of time requires temporal becoming: how can what does not exist pass from the future to the present, and how can what does exist (the present) continue its journey into the past when, if it is past, it does not exist? Augustine never really answers those two questions.

J. M. E. McTaggart

McTaggart also struggles with trying to make sense of our ordinary notion of temporal becoming, but he sees that any attempt to do it justice will require a richer account of time than is found in Augustine. It is richer because in addition to the three times of past, present, and future there are also the times, or better, the relations between the times or events that are past, present, and future. Thus, according to McTaggart, as we ordinarily conceive of time it involves both temporal or B-relations such as *simultaneity* and *earlier/later than* (so-called after McTaggart's B-series) and temporal or A-properties of *pastness*, *presentness*, and *futurity* (after which the A-series is named). If we recognize both these features as having a foothold in our ordinary conception of time, and we provide some foundation for them in our metaphysical account of time, then we can provide a model of temporal becoming that does not face the challenges that are found in Augustine's theory, since the past, present, and future exist, although it does, as we shall see, have insurmountable difficulties of its own.

Before explaining McTaggart's account of temporal becoming and his argument for the paradoxes he finds in it, it will be instructive to see what motivates his view of becoming. McTaggart arrives at his positive theory of time, not through a consideration of intentionality (although his view provides a possible response to that problem), but through a consideration of change, ordinary change. He argues that ordinary change requires temporal change or temporal becoming, and it requires it in a certain form.

McTaggart maintains that it is "universally admitted that time involves change" (chapter 2, p. 36). What then is change? Change involves a single thing having a property and then losing it. Clearly, change also involves time, for otherwise a contradiction ensues: if an apple is green and red it must be so at different times. What then is time? If time is composed of a series of moments so that "the apple is green at t_1" and "the apple is red at t_2," then there is no contradiction, but for McTaggart there is no change, since there is nothing in either of those facts that change. In other words, it is always the case that the apple is red at t_2 and green at t_1, and thus it is not the case that the apple or anything else changes. The object of McTaggart's attack is Russell's account of change as stated in his *The Principles of Mathematics*, Section 442:

> Change is the difference, in respect of truth or falsehood, between a proposition concerning an entity and the time *T*, and a proposition concerning the same entity and the time *T'*, provided that these proposition differ only by the fact that *T* occurs in the one where *T'* occurs in the other. (Russell 1938, p. 467)

McTaggart expresses his reasoning against Russell in the following passage:

> It will be noticed that Mr. Russell looks for change, not in the events in the time-series, but in the entity to which that events happen, or which they are states. If my poker, for example, is hot on a particular Monday, and never before or since, the event of the poker being hot does not change. But the poker changes, because there is a time when this event is happening to it, and a time when it is not happening to it.
>
> But this makes no change in the qualities of the poker. It is always a quality of that poker that it is one which is hot on that particular Monday. And it is always a quality of that poker that it is one which is not hot at any other time. Both these qualities are true of it at any time—the time when it is hot and the time

> when it is cold. And therefore it seems to be erroneous to say that there is any change in the poker. The fact that it is hot at one point in a series and cold at other points cannot give change, if neither of these facts change—and neither of them does. Nor does any other fact about the poker change, unless its presentness, pastness, or futurity change. (pp. 38-39)

Of course, one may here question McTaggart's assumption that change requires that *facts* (such as the apple's being red at t_2) or *events* (such as the apple's being red) must change and instead insist that it is *things* that change. However, a more detailed discussion of that issue will have to wait until Prior (chapter 3) and part V—"Presentism and Persistence"—where it will reemerge.

Essentially the same type of argument is given against the attempt to analyze the time involved in change in terms of B-relations. Thus, McTaggart argues that we cannot account for or analyze the changing apple by saying that the apple is green *before* or *earlier than* the apple is red because there is nothing about either of those facts that change. No event in the relational fact changes by ceasing to be or coming to be, since it is always the case that, say, the apple is green before it is red. The relational fact itself does not change, nor can we say that the apple's being green "merges itself into another event" (p. 37)—the apple's being red—since in order to merge into another event the apple's being green must cease to exist and the apple's being red must come into existence, but that, as we have just seen, is impossible. Thus, the B-series of earlier, later, and simultaneous events (or moments of absolute time) is not the essence of time because time involves change and the B-series alone cannot contain or constitute change.

McTaggart concludes, in his positive account of time, that the A-series is necessary for time and change. He argues that ordinary change requires temporal change or temporal becoming, and it requires it in a certain form. In order for an apple to change from green to red it must be the case that the time at which it is red is future when the apple is green and that when the apple is red the apple's being green is past. Thus, there are events such as the apple's being green and the apple's being red, and these events change their A-characteristics of *pastness*, *presentness*, and *futurity*. It is this form of change known as temporal becoming that accounts for and is the ultimate foundation of the apple changing from green to red, and without it time and change would not exist.

Thus, McTaggart recognizes that there is a difference between events which are past and future from those which are present, but it does not concern their *existence* as events; rather it concerns their *temporal properties*. For that reason McTaggart does provide, perhaps unwittingly, a response to the two problems that Augustine found to exist when it comes to explaining the nature of time. Recall, the first problem that perplexed Augustine was to understand how events in the future could pass or flow into the present and then recede into the past given that the future does not exist and the past does not exist. How can what is not—the future—become what is—the present—and then change again to what is not—the past, and thereafter continually change while receding further and further into the past? McTaggart's response is to deny that the past and the future do not exist. The past and the future *do exist* and thus they don't change by coming into or going out of existence, but rather by their taking on different and incompatible temporal properties. This is called the full A-theory since it recognizes the existence of, and gives ontological status to all events in the history of the world regardless of whether they are past, present, or future.

The second problem that worried Augustine concerned the intentionality of thought, in particular, how can we think about the future and the past if they do not exist? Given that there is nothing in the future or in the past, what are such thoughts about? Again, the answer, although McTaggart does not consider this problem in the reading, is that the past and future exist so that when I am truly thinking about a past or future event what I am thinking about exists.

McTaggart's positive account of time and change emphasizes the A-series and temporal becoming and relegates the B-series and B-relations to a subservient position. Indeed, he gives a reductive analysis of B-relations in terms of A-determinations when he says that "a term P is earlier than a term Q if it is ever the case that P is future while Q is present or P is present while Q is past" (McTaggart 1927, p. 271). But does McTaggart's positive treatment of temporal becoming and his view that it constitutes the essence of time adequately answer Augustine's question, "What is time?" McTaggart's own answer to that question is "no" since he argues that temporal becoming is contradictory and that therefore the distinctions between past, present, and future events are illusory. Since, however, time and change depend on A-series distinctions, he concludes that nothing that exists can be temporal and that therefore time is unreal. I want to turn next to the argument in support of the paradox that he is so justly famous (from a B-theorist perspective) or infamous (from an A-theorist perspective) for.

McTaggart begins his negative attack on time by asking whether past, present, and future are relations to a term outside the series of events or non-relational properties of events. Since the difference makes no difference to his argument we will treat A-determinations as qualities. The problem of temporal becoming, as McTaggart understands it, can be understood as follows. Temporal becoming involves events changing from future to present, and from present to past. Yet, as McTaggart correctly says,

> Past, present, and future are incompatible determinations. Every event must be one or the other, but no event can be more than one. If I say that any event is past, that implies that it is neither present nor future, and so with the others. And this exclusiveness is essential to change, and therefore to time. (p. 42)

McTaggart claims that events must have one and only one of the three incompatible A-determinations. The reason for this exclusiveness is that if an event had all three such characteristics then the same event would have incompatible A-properties which is impossible. Moreover, if an event had all three A-determinations then there would be no basis for determining whether an event was first future then present and then past or the other way around and thus no basis for determining the direction of time and change.

The problem, however, is that change requires temporal becoming and becoming requires that every event have each of the three incompatible properties of *pastness*, *presentness*, and *futurity*. That is, in order for a thing to change from, say, green to red, the event of its being green must be present when the event of its being red is future and past when its being red is present and so on. For that reason McTaggart says that

> The characteristics, therefore, are incompatible. But every event has them all. . . . How is this consistent with their being incompatible? (p. 42)

It would appear, then, that temporal becoming gives rise to exactly the same problem that haunts the problem of change. For now, instead of the problem of understanding how an apple can have incompatible properties, for example, being green and being red, we are faced with the problem of understanding how an event, such as the death of Queen Anne, can have the incompatible properties of *being past, being present*, and *being future*?

According to McTaggart, the obvious solutions to that problem will not work. For example, one could say that events undergo temporal becoming by having different and incompatible A-characteristics in *succession* or at *different times*, but what is "succession" and what is "time"? If succession is analyzed in terms of the B-series relations of *earlier/later than* so that Queen Anne's death is future *earlier than* her death is present and her death is present *earlier than* her death is past, then we avoid the contradiction, but do away with the fact of change (and hence with time) since there is no change in those B-series facts, according to McTaggart.

Similarly, if the incompatible A-properties are explained in terms of the different times at which it has them then the same dilemma presents itself. For if "times" are analyzed in terms of absolute moments in a B-series then a contradiction apparently can be avoided by claiming that Queen Anne's death is future at t_1, present at t_2, and past and t_3, but only at the cost of doing away with change since there is nothing in any of those facts that changes. It is always the case that, for example, Queen Anne's death is future at t_1. On the other hand, if we reintroduce change by claiming, as according to McTaggart we must, that the times (t_1, t_2, t_3, . . . t_i) at which Queen Anne's death has incompatible A-properties themselves undergo temporal becoming, then moments must have incompatible A-properties successively or at different higher order times that are themselves past, present, and future and so on *ad infinitum*.

Finally, if we analyze "succession" in terms of the A-series, as McTaggart does, so that it is never the case that an event e is past, present, and future, but that "It *is* present, *will be* past, and *has been* future. Or it *is* past, and *has been* future and present, or again *is* future, and *will be* present and past" then a vicious infinite regress is unavoidable. For, once again we are understanding temporal becoming in terms of events having incompatible A-determinations at different moments of A-time, which themselves have incompatible A-determinations at different moments of A-time, and so on. As McTaggart puts it:

> Such an infinity is vicious. The attribution of the characteristics past, present, and future to the terms of any series leads to a contradiction, unless it is specified that they have them successively. This means, as we have seen, that they have them in relation to terms specified as past, present, and future. These again, to avoid a like contradiction, must in turn be specified as past, present, and future. And, since this continues infinitely, the first set of terms never escapes from contradiction at all. (p. 43)

Since time implies change and change implies the A-series and temporal becoming (which is contradictory) it follows that change and hence time is unreal. What then is to be done?I shall turn next to two prominent presentists who offer different responses to McTaggart.

Arthur Prior

Whether we think of the future and past as existent and temporal becoming as involving the change of A-properties that events undergo as they move from the future through the present and into the past as McTaggart did, or maintain that the past and the future do not exist and thus that temporal becoming is the passing of the non-existent future through the present into the non-existent past, as Augustine did, there are problems, as we have seen. What then is to be done?

In the article reprinted here, "Changes in Events and Changes in Things," A. N. Prior rejects McTaggart's notion of passage thereby avoiding the contradictions in McTaggart's theory of time while at the same time developing a theory that is sympathetic with Augustine's presentism, without its failings. Thus, Prior denies the A-properties of *pastness* and *futurity* since there are no past and future existents to exemplify them, and he rejects that temporal passage or change must involve events moving through time from future to present to past. What he replaces those notions of temporal becoming with while remaining true to a presentist view of time are interesting questions that I shall briefly explain.

Prior believes that in order to solve the metaphysical problem of time and avoid its alleged paradoxes, we must look at grammar. To simplify his point we could say that in ordinary language we say things like "My having breakfast is past," or "I was eating my breakfast" and "I will be eating my lunch." These sentences mistakenly suggest that there are the events, "my eating my lunch" and "my having breakfast" that have the non-relational properties of *being past* and *being future*. Prior claims, however, that "it is only an historical accident that we generally form the past tense by modifying the present tense, e.g., by changing 'am' to 'was,' rather than by tacking on an adverb" (chapter 3, p. 50).

In a language constructed to reflect the true nature of reality, "we could form the past tense by prefixing to a given sentence the phrase 'It was the case that' . . . and the future tense by prefixing 'It will be the case that'" (p. 50). For example, instead of saying that "I was eating my breakfast," we could say that "It was the case that I am eating my breakfast," and in that sentence "it was the case that" is an adverb that no more implies the existence of *pastness* (or a past tense fact) than the sentence "It is not the case" in the sentence, "It is not the case that I am eating my breakfast" implies the existence of *notness* (or a negative fact).

For Prior it is misleading grammatical expressions that give rise to some of the problems about temporal passage. Thus,

> (3) It is now six years since it was the case that I am falling out of a punt,
>
> could be re-written as
>
> (4) My falling out of a punt has receded six years into the past.
>
> This suggests the something called an event, my falling out of a punt, has gone through a performance called receding into the past, and moreover has been going through this performance even after it has ceased to exist, i.e., after it has stopped happening. But of course, (4) is just a paraphrase of (3), and like (3) is not about any objects except me and that punt—there is no real reason to believe in the existence either now or six years ago of a further object called "my falling out of a punt."
>
> What I am suggesting is that what looks like talk about events is really at bottom talk about things, and that what looks like talk about changes in events is really just slightly more complicated talk about changes in things. (p. 52)

On Prior's version of presentism, events do not literally flow from the future into the present and then recede into the past. To say that something is present is to say that it is real, and to say that it is past or future is to say that it was present or will be present. McTaggart's paradox is thus easily answered. It is not the case that every time t is past, present, and future, but rather, either it was the case that (t_1 is present) or it will be the case that (t_1 is present) or t_1 is present, and there is no contradiction in that disjunctive state of affairs.

Thus, the metaphorical statement that Queen Anne's death is receding more and more into the distant past can therefore be rephrased as:

> It was the case that it was that case only 250 years ago that Queen Anne is dying, and is not now the case that it was the case only 250 years ago that Queen Anne is dying. (p. 53)

This fits the formula for change, but it does not entail that something called an "event" endures through time and thus it is not susceptible to the paradoxes that are generated by McTaggart and Augustine's view of becoming.

There is, however, a problem with Prior's restriction of the real to the present, namely, how can we think about the past if the past does not exist. In the case of the statement "Queen Anne died," it would appear that this is a statement *about* Queen Anne, but how is this possible if Queen Anne does not exist?

Augustine raises this problem (p. 29) and Prior attempts to resolve it by means of a distinction between particular and general facts. When I am talking about or thinking about a past thing my statement or thought is about a general fact, whereas when I am perceiving what exists the object of my perception is a particular fact. Since there are no past or future things they cannot be the object of thought, perception, language, or experience. Prior continues, "We might paraphrase 'Queen Anne has died' as 'Once there was a person named "Anne" who reigned over England, etc., but there is not now any such person'" (p. 53).

Prior's point seems to be that "On this view, the fact that Queen Anne has been dead for some years is not, in the strict sense of "about," a fact about Queen Anne; it is not a fact about anyone or anything—it is a *general* fact" (p. 54). The problem with this account of the intentionality of thought about the past and the future is that it is a theory made to fit his view of time, but not to fit the phenomenology of the situation. For clearly when I think that Queen Anne died, I am not thinking that someone in general died or someone called "Queen Anne" died, but rather that a particular individual died and thus that that person entered into the fact I am thinking about, especially since the thought is true. In other words, it is not a fact about someone called "Anne" at all. The question of what, for the presentist, judgments about past individuals are about is dealt with in detail in later selections, (see for example, De Clercq, chapter 12), and we shall return to it when we consider them.

William Lane Craig

Although Prior's version of presentism provides a *prima facie* solution to McTaggart's and Augustine's problems of time, it has left many, including some presentists, unsatisfied. On the one hand they agree that he has it right when he rejects the A-properties of *futurity* and *pastness* since without those properties McTaggart's paradox cannot get off the ground. According to Craig, McTaggart's paradox is generated by the attempt to combine a B-theoretic ontology with A-theoretic becoming. That is, events are construed as strung out along a tenseless B-series and temporal becoming consists in the terms of that series remaining the same through a change of their A-properties. In that case, however, events are enduring continuants that have incompatible characteristics leading to an endless series of time dimensions in which the incompatible temporal properties problem is never resolved.

For Craig there are no B-relations and as we shall see there are no A-determinations (not even the premier A-determination of *presentness*), and he treats temporal becoming as consisting "in things' coming to be absolutely" (chapter 4, p. 59). He is, however, sensitive to the "lack of ontology" objection raised against Prior's presentism and seeks to avoid it. Recall that for Prior the tenses should not be construed as properties and temporal qualifications should not be construed as predicates. He achieves that goal by rephrasings which replace them with propositional prefixes analogous to negation or as operators in much the same way as negation is an operator. Thus instead of saying that "My eating breakfast is past," we should say that "It was the case that I am eating my breakfast" and there is no more reason to reify the operator "it was the case" in that sentence than there is to reify "It is not the case" in the sentence "It is not the case that the moon is made of Swiss cheese."

Yet questions still remain, what then are the truth grounds for propositions about the past and the future? What are such judgments about? What do past and future tenses represent? How should we understand what appear to be grammatical ascriptions of pastness and futurity? What are their truth makers? Furthermore, if we have a pure A-theory without a B-theoretic ontology, how are we to understand or what is the ground of B-relations?

While Craig is sensitive to these questions his answers to them are not satisfactorily explored in the essay in this collection, although in fairness that is not his aim in this article. He claims that grammatical ascriptions of pastness and futurity should not be construed as attributing pastness and futurity to events/things/times, but "Rather such ascriptions should be parsed as asserting that the entity in question did or will exist" (p. 60). Of course, to construe A-predicates in terms of ascriptions of what was or will be the case is equally susceptible to the lack of ontology charge raised against Prior since it still leaves unaccounted for the ontological correlates of "was" and "will be" in the propositions in which they occur. Craig does not deal with these questions in the article found in this volume although he does elsewhere as we shall see when I discuss Oaklander's paper.

The main focus of Craig's paper is the question: Is presentness a property? Having eliminated the other A-properties, can this one be eliminated too? Since there is obviously on his view a close connection between being present and existing (since only the present exists), Craig pursues the question of whether or not existence is a property to see if the results can be applied to the question of whether or not presentness is a property. Craig considers several arguments in favor of the view that existence is not a property and expresses sympathy for the Kantian view (versions of which are advocated by Broad and Moore, see pp. 63-64). "On this view, for a thing to exist is for its whole concept (or, alternatively, its individual essence) to be instantiated. In either case, to exist is to exemplify properties" (pp. 65-66).

Since it is natural for the presentist to identify a thing's being present with that thing's existing, then presentness is no more a property than existence, and this is in fact Craig's view. Craig initially claims that "*To be present* seems to be identical with *to exist*" (p. 68) although that is not his final view. He recognizes the possibility of timeless existence. Thus, Craig concludes that presentness is properly construed as a mode of existence. There are two modes of existing, timeless existence and temporal existence, and presentness is identical with this second mode. There are two ways of exemplifying properties: timelessly or presently. On a presentist ontology, to exist temporally is to be present. Since presentness is identical with temporal existence (or occurrence) and existence is not a property, neither is presentness a property. Presentness is the act of temporal being. (p. 68)

There are, it seems to me, difficult questions with this analysis. First, it explains what it is for a thing to exist, but it does not explain what it is for a property to exist. At a certain level the instantiation of properties comes to an end, but that higher order property still exists. So, what is it for a higher order property to exist? Furthermore, what is the difference between a thing timelessly exemplifying a property and its presently exemplifying a property? That is, what is the difference between timeless existence and temporal existence? In what sense do both timeless and temporal entities exist? These are some of the questions the reader should consider in reading Craig's essay.

L. Nathan Oaklander

We have seen that Prior attempts to do away with McTaggart's paradox by first rejecting the notion of becoming that involves events changing their A-properties as they flow from the future to the present and into the past. On Prior's ontology there are no events and there are no A-properties, with the possible exception of what Craig calls the "premier A-property," presentness. Moreover, temporal becoming is not a literal change in that it does not involve something that has a property and then loses it through the period of time during which it exists. Neither the future nor the present exist so no event/thing/moment could endure while it moves through time. Nevertheless, they do undergo what looks like a change since the flow of time or the receding of things in time to the more and more distant past is captured by a formula that represents change, namely,

It was the case that *p*, and it is not the case that *p*. (p. 51)

Thus, for example, the "recession of events into the past that we are really talking about when we say that time flows or passes," (p. 51) can be captured by the following sentence:

It was the case 5 months ago that (it was the case only 47 years ago that (I am being born)), and it is not now the case that (it was the case only 47 years ago that (I am being born)),

that is, I am not as young as I used to be. (p. 51)

We have seen that Craig echoes the sentiment of many who accuse Prior of providing a mere verbal solution to the ontological problems surrounding time and becoming. What does the past tense stand for in these sentences, what does it represent, what are the truth makers for these sentences? Moreover, how is the presentist to ground temporal relations given that the terms of temporal relations exist at different times, but for the presentist, there is only one time, the present time? Finally, how, once presentism becomes more ontologically perspicuous, are we still going to be able to say McTaggart's paradox is avoided?

Craig's critique of McTaggart is based on his view that McTaggart attempts to combine a B-theoretic ontology with the A-theory according to which pastness, presentness, and futurity are monadic properties of events which therefore must exist, and that temporal becoming consists as a species of qualitative change. The questions that haunt Prior's analysis of time reemerge in a consideration of Craig's response to McTaggart:

(1) If temporal becoming is not to be understood as a species of qualitative change, then how is it to be understood?
(2) If McTaggart mistakenly combines a B-theoretic ontology with the A-theory, how then does Craig attempt to analyze temporal relations between and among items existing in the B-series?
(3) If the past and the future do not exist, then what are the truth makers of past- and future-tense statements?

In his article, Oaklander argues that Craig's answers to these questions are unacceptable and that, therefore, his A-theoretic, presentist account of time, change, and becoming is subject to McTaggart's paradox and must therefore be rejected. More specifically, Craig understands becoming as "the successive actualization of the history of the actual world" (2000b, p. 207). This *prima facie* appeal to temporal relations ("successive actualization") raises questions about the consistency of his presentism since it seems to require that non-present items exist. Moreover, his understanding of past and future tense "facts" in terms of present tense facts that have *elapsed* by the present moment (that is, are *earlier than* the present moment) or present tense facts that have not *elapsed* by the present moment (that is, are *later than* the present moment) also require B-relations, unless he can reduce B-relations to A-determinations and becoming.

The key question, therefore, is how is succession to be analyzed? Recall that McTaggart argued that the incompatible A-properties problem can be resolved by saying that events have their A-properties successively, but he went on to argue that given an A-theoretic analysis of succession, the original paradox is not avoided, but just shifted to another level and so on *ad infinitum*. Craig disagrees. He argues that one can have an ontological foundation in the metaphysics of presentism for affirming the existence of temporal relations without construing them as B-theoretic and unanalyzable. Oaklander considers three attempts that Craig gives to analyze temporal relations, but argues that they fail, and since Craig's account of past and future tense facts depends upon his analysis of temporal relations (since they assume the notion of what has or has not elapsed by the present moment) it fails too. Without an adequate account of the past and future and B-relations, Craig does not have adequate resources to escape McTaggart's conundrum and therefore his version of presentism should be rejected, or so Oaklander argues.

References

Augustine. *Confessions*. Translated by H. Chadwick. Oxford University Press, 1991. Chapter 1 in this volume.

Bigelow, J. "Presentism and Properties," *Philosophical Perspectives* 10 (1996): 35-52. Chapter 8 in this volume.

———. "Worlds Enough for Time," *Noûs* 25 (1991): 1-19.

Broad, C. D. *Examination of McTaggart's Philosophy*. Cambridge: Cambridge University Press, 1938.

———. *Scientific Thought*. London: Routledge, 1923.

Craig, W. L. "The Extent of the Present," *International Studies in the Philosophy of Science* 14 (2000a): 165-85.

———. *The Tensed Theory of Time: A Critical Examination*. Dordrecht: Kluwer, 2000b.

Dummett, M. "A Defence of McTaggart's Proof of the Unreality of Time," *Philosophical Review* 69 (1960): 497-504.

Dyke, H. "McTaggart and the Truth about Time," *Philosophy* 50 (2002): 137-52.

Hestervold, H. S. "Presentism through Thick and Thin," *Pacific Philosophical Quarterly* 89 (2008): 325-47.

McTaggart, J. M. E. *The Nature of Existence*, 2 vols., edited by C. D. Broad. Cambridge: Cambridge University Press, 1927; rep. ed.: 1968. Chapter 2 in this volume.

———. "The Unreality of Time," *Mind* 18 (1908): 457-74; reprinted in *Philosophical Studies*, edited by S. V. Keeling. London: Edward & Arnold & Co., 1934.

Mellor, D. H. *Real Time II*. London: Routledge, 1998.

Oaklander, L. N. "McTaggart's Paradox Defended." Pp. 51-62 in *The Ontology of Time*, by L. N. Oaklander. Amherst, N. Y.: Prometheseus Books, 2004.

Prior, A. "Changes in Events and Changes in Things." Pp. 1-14 in *Papers on Time and Tense*, by A. Prior. Oxford: Clarendon Press, 1968. Chapter 3 in this volume.

———. "The Notion of the Present," *Studium Generale* 23 (1970): 245-48.

Russell, B. *The Principles of Mathematics*. New York: W. W. Norton & Co., 1938. First published in 1903.

Smart, J. J. C. "Time and Becoming." Pp. 3-16 in *Time and Cause: Essays Presented to Richard Taylor*, edited by P. v. Inwagen. Dordrecht: D. Reidel, 1980.

Smith, Q. "Reference to the Past and Future." Pp. 357-90 in *Time, Tense, and Reference*, edited by A. Jokic and Q. Smith. Cambridge: MIT, 2003.

Sosa, E. "The Status of Becoming: What Is Happening Now?" *The Journal of Philosophy* 76 (1979): 26-42.

1
What Is Time?

Saint Augustine

What is time? Who can explain this easily and briefly? Who can comprehend this even in thought so as to articulate the answer in words? Yet what do we speak of, in our familiar everyday conversation, more than of time? We surely know what we mean when we speak of it. We also know what is meant when we hear someone else speak about it. What then is time? Provided that no one asks me, I know. If I want to explain it to an inquirer, I do not know. But I confidently affirm myself to know that if nothing passes away, there is no past time, and if nothing arrives, there is no future time, and if nothing existed there would be no present time. Take the two tenses, past and future. How can they "be" when the past is not now present and the future is not yet present? Yet if the present were always present, it would not pass into the past: it would not be time but eternity. If then, in order to be time at all, the present is so made that it passes into the past, how can we say that this present also "is"? The cause of its being is that it will cease to be. So indeed we cannot truly say that time exists except in the sense that it tends towards non-existence.

Nevertheless we speak of "a long time" and "a short time," and it is only of the past or the future that we say this. Of the past we speak of "a long time," when, for example, it is more than a hundred years ago. "A long time" in the future may mean a hundred years ahead. By "a short time ago" we would mean, say, ten days back, and "a short time ahead" might mean "in ten days time". But how can something be long or short which does not exist? For the past now has no existence and the future is not yet. So we ought not to say of the past "It is long," but "it was long," and of the future "it will be long." My Lord, my light, does not your truth mock humanity at this point? This time past which was long, was it long when it was past or when it was still present? It could be long only when it existed to be long. Once past, it no longer was. Therefore it could not be long if it had entirely ceased to exist.

Therefore let us not say "The time past was long." For we cannot discover anything to be long when, after it has become past, it has ceased to be. But let us say "That time once present was long" because it was long at the time when it was present. For it had not yet passed away into non-existence. It existed so as to be able to be long. But after it had passed away, it simultaneously ceased to be long because it ceased to be.

Human soul, let us see whether present time can be long. To you the power is granted to be aware of intervals of time, and to measure them. What answer will you give me? Are a hundred years in the present a long time? Consider first whether a hundred years can be present. For if the first year of the series is current, it is present, but ninety-nine are future, and so do not yet exist. If the second year is current, one is already past, the second is present, the remainder lie in the future. And so between the extremes, whatever year of this century we assume to be present, there will be some years before it which lie in the past, some in the future to come after it. It follows that a century could never be present.

Consider then whether if a single year is current, that can be present. If in this year the first month is current, the others lie in the future; if the second, then the first lies in the past and the rest do not yet. Therefore even a current year is not entirely present; and if it is not entirely present, it is not a year which is present. A year is twelve months, of which any month which is current is present; the others are either past or future. Moreover, not a month which is current is present, but one day. If the first day, the others are future; if the last day, the others are past; any intermediary day falls between past and future.

See—present time, which alone we find capable of being called long, is contracted to the space of hardly a single day. But let us examine that also; for not even one day is entirely present. All the hours of night and day add up to twenty-four. The first of them has the others in the future, the last has them in the past. Any hour between these has past hours before it, future hours after it. One hour is itself constituted of fugitive moments. Whatever part of it has flown away is past. What remains to it is future. If we can think of some bit of time which cannot be divided into even the smallest instantaneous moments, that alone is what we can call "present." And this time flies so quickly from future into past that it is an interval with no duration. If it has duration, it is divisible into past and future. But the present occupies no space.

Where then is the time which we call long? Is it future? We do not really mean "It is long," since it does not yet exist to be long, but we mean it will be long. When will it be long? If it will then still lie in the future, it will not be long, since it will not yet exist to be long. But if it will be long at the time when, out of the future which does not yet exist, it begins to have being and will become present fact, so that it has the potentiality to be long, the present cries out in words already used that it cannot be long.

Nevertheless, Lord, we are conscious of intervals of time, and compare them with each other, and call some longer, others shorter. We also measure how much longer or shorter one period is than another, and answer that the one is twice or three times as much as the other, or that the two periods are equal. Moreover, we are measuring times which are past when our perception is the basis of measurement. But who can measure the past which does not now exist or the future which does not yet exist, unless perhaps someone dares to assert that he can measure what has no existence? At the moment when time is passing, it can be perceived and measured. But when it has passed and is not present, it cannot be.

I am investigating, Father, not making assertions. My God, protect me and rule me (Ps. 22: I; 27: 9). Who will tell me that there are not three times, past, present, and future, as we learnt when children and as we have taught children, but only the present, because the other two have no existence? Or do they exist in the sense that, when the present emerges from the future, time comes out of some secret store, and then recedes into some secret place when the past comes out of the present? Where did those who sang prophecies see these events if they do not yet exist? To see what has no existence is impossible. And those who narrate past history would surely not be telling a true story if they did not discern events by their soul's insight. If the past were non-existent, it could not be discerned at all. Therefore both future and past events exist.

Allow me, Lord, to take my investigation further. My hope, let not my attention be distracted. If future and past events exist, I want to know where they are. If I have not the strength to discover the answer, at least I know that wherever they are, they are not there as future or past, but as present. For if there also they are future, they will not yet be there. If there also they are past, they are no longer there. Therefore, wherever they are, whatever they are, they do not exist except in the present. When a true narrative of the past is related, the memory produces not the actual events which have passed away but words conceived from images of them, which they fixed in the mind like imprints as they passed through the senses. Thus my boyhood, which is no longer, lies in past time which is no longer. But when I am recollecting and telling my story, I am looking on its image in present time, since it is still in my memory. Whether a similar cause is operative in predictions of the future, in the sense that images of realities which do not yet exist are presented as already in existence, I confess, my God, I do not know. At least I know this much: we frequently think out in advance our future actions, and that premeditation is in the present; but the action which we premeditate is not yet in being because it lies in the future. But when we have embarked on the action and what we were premeditating begins to be put into effect, then that action will have existence, since then it will be not future but present.

Whatever may be the way in which the hidden presentiment of the future is known, nothing can be seen if it does not exist. Now that which already exists is not future but present. When therefore people speak of knowing the future, what is seen is not events which do not yet exist (that is, they really are future), but perhaps their causes or signs which already exist. In this way, to those who see them they are not future but present, and that is the basis on which the future can be conceived in the mind and made the subject of prediction.

Again, these concepts already exist, and those who predict the future see these concepts as if already present to their minds.

Among a great mass of examples, let me mention one instance. I look at the dawn. I forecast that the sun will rise. What I am looking at is present, what I am forecasting is future. It is not the sun which lies in the future (it already exists) but its rise, which has not yet arrived. Yet unless I were mentally imagining its rise, as now when I am speaking about it, I could not predict it. But the dawn glow which I see in the sky is not sunrise, which it precedes, nor is the imagining of sunrise in my mind the actuality. These are both discerned as present so that the coming sunrise may be foretold.

So future events do not yet exist, and if they are not yet present, they do not exist; and if they have no being, they cannot be seen at all. But they can be predicted from present events which are already present and can be seen.

Governor of your creation, what is the way by which you inform souls what lies in the future? For you instructed your prophets. By what method then do you give information about the future—you to whom nothing is future? Is it rather that you inform how to read the future in the light of the present? What does not exist, certainly cannot be the subject of information. This method is far beyond my power of vision. "It is too mighty for me, I cannot attain it." (Ps. 138: 6). But it would be in my power with your help if you granted it, sweet light of my uncomprehending eyes.

What is by now evident and clear is that neither future nor past exists, and it is inexact language to speak of three times—past, present, and future. Perhaps it would be exact to say: there are three times, a present of things past, a present of things present, a present of things to come. In the soul there are these three aspects of time, and I do not see them anywhere else. The present considering the past is the memory, the present considering the present is immediate awareness, the present considering the future is expectation. If we are allowed to use such language, I see three times, and I admit they are three. Moreover, we may say, There are three times, past, present, and future. This customary way of speaking is incorrect, but it is common usage. Let us accept the usage. I do not object and offer no opposition or criticism, as long as what is said is being understood, namely that neither the future nor the past is now present. There are few usages of everyday speech which are exact, and most of our language is inexact. Yet what we mean is communicated.

A little earlier I observed that we measure past periods of time so that we can say that one period is twice as long as another or equal to it, and likewise of other periods of time which we are capable of measuring and reporting. Therefore, as I was saying, we measure periods of time as they are passing, and if anyone says to me "How do you know?" I reply: I know it because we do measure time and cannot measure what has no being; and past and future have none. But how do we measure present time when it has no extension? It is measured when it passes, but not when it has passed, because then there will be nothing there to measure.

When time is measured, where does it come from, by what route does it pass, and where does it go? It must come out of the future, pass by the present, and go into the past; so it comes from what as yet does not exist, passes through that which lacks extension, and goes into that which is now non-existent. Yet what do we measure but time over some extension? When we speak of lengths of time as single, duple, triple, and equal, or any other temporal relation of this kind, we must be speaking of periods of time possessing extension. In what extension then do we measure time as it is passing? Is it in the out it comes to pass by? No, for we do not measure what does not yet exist. Is it in the present through which it passes? No, for we cannot measure that which has no extension. Is it in the past into which it is moving? No, for we cannot measure what now does not exist.

My mind is on fire to solve this very intricate enigma. Do not shut the door, Lord my God. Good Father, through Christ I beg you, do not shut the door on my longing to understand these things which are both familiar and obscure. Do not prevent me, Lord, from penetrating them and seeing them illuminated by the light of your mercy. Whom do I ask about them? And to whom but you I more profitably confess my incompetence? You are not irritated by the burning zeal with which I study your scriptures. Grant what I love. For I love, and this love was your gift. Grant it, Father. You truly know how to give good gifts to your children (Matt 7: II). Grant it, since I have undertaken to acquire understanding and "the labour is too much for me" (Ps. 72: 16) until you open the way. Through Christ I beg you, in the name of him who is the holy of holy ones, let no one obstruct my inquiry. "I also have believed, and therefore speak" (Ps. 115: I; 2 Cor. 4: 13). This is my hope. For this I live "that I may contemplate the delight of the Lord" (Ps. 26: 4). "Behold you have made my days subject to ageing" (Ps. 38: 6). They pass away, and how I do not know. And we repeatedly speak of time and time, of times and times: "How long ago did he say this?" "How long ago did he do this?" "For how long a time did I fail to see that?" And "These syllables take twice the time of that single, short syllable." We speak in this way, and hear people saying this, and we are understood and we understand. These usages are utterly commonplace and everyday. Yet they are deeply obscure and the discovery of the solution is new.

I have heard a learned person say that the movements of sun, moon, and stars in themselves constitute time. But I could not agree. Why should not time consist rather of the movement of all physical objects? If the heavenly bodies were to cease and a potter's wheel were revolving, would there be no time by which we could measure its gyrations, and say that its revolutions were equal; or if at one time it moved more slowly and at another time faster, that some rotations took longer, others less? And when we utter these words do not we also speak in time? In our words some syllables are long, others short, in that the sounding of the former requires a longer time, whereas the latter are shorter.

God grant to human minds to discern in a small thing universal truths valid for both small and great matters. There are stars and heavenly luminaries to be "for signs and for times, and for days and for years" (Gen. I: 14). But I would not say that a revolution of that wooden wheel is a day; and that learned friend could not assert that its rotation was not a period of time.

I desire to understand the power and the nature of time, which enables us to measure the motions of bodies and to say that, for instance, this movement requires twice as long as that. I have this question to raise: the word "day" is used not only of the interval of time when the sun is up over the earth, so that day is one thing, night another, but also of the sun's entire circuit from east to west, as when we say "so many days have passed" where "so many days" includes the nights, and the periods of night-time are not counted separately. So a complete day is marked by the movement and circuit of the sun from east to west. My question then is whether the sun's movement itself constitutes the day? or the actual interval of time during which it is accomplished? or both?

In the first instance, it would still be a day even if the sun completed its course in the space of a single hour. In the second case, it would not be a day if from one sunrise to the next so short an interval as one hour passed, but only if the sun completed a day of twenty-four hours. In the third case—a day being both the circuit and the time taken—it could not be called a day if the sun completed its entire circuit in an hour, nor if the sun ceased to move and the length of time passed were the twenty-four hours normally taken by the sun in completing its entire course from sunrise to sunrise. I will not, therefore, now investigate what it is which we call a day, but the nature of time by which we can measure the sun's circuit and by which we might say that, if all was accomplished in twelve hours, the sun had completed its course in half the usual time. I ask what time is when we make a comparison and say that one interval is single and another double, even if the sun were to make its transit from east to west sometimes in single time, sometimes in twice the time.

Let no one tell me then that time is the movements of heavenly bodies. At a man's prayer the sun stood still, so that a battle could be carried through to victory (Josh. 10: 12 ff.): the sun stopped but time went on. That battle was fought and completed in its own space of time such as was sufficient for it. I therefore see that time is some kind of extension. But do I really see that? Or do I imagine that I see? You, light and truth, will show me.

Do you command me to concur if someone says time is the movement of a physical entity? You do not. For I learn that no body can be moved except in time. You tell me so, but I do not learn that the actual movement of a body constitutes time. That is not what you tell me. For when a body is moved, it is by time that I measure the duration of the movement, from the moment it begins until it ends. Unless I have observed the point when it begins, and if its movement is continuous so that I cannot observe when it ceases, I am unable to measure except for the period from the beginning to the end of my observation. If my observing lasts for a considerable time, I can only report that a long time passed, but not precisely how much. When we say how much, we are making a comparison—as, for example, "This period was of the same length as that," or "This period was twice as long as that," or some such relationship.

If, however, we have been able to note the points in space from which and to which a moving body passes, or the parts of a body when it is spinning on its axis, then we can say how much time the movement of the body or its parts required to move from one point to another. It follows that a body's movement is one thing, the period by which we measure is another. It is self-evident which of these is to be described as time. Moreover, a body may sometimes be moving, sometimes be at rest. We measure by time and say "It was standing still for the same time that it was in movement," or "It was still for two or three times as long as it was in movement," or any other measurement we may make, either by precise observation or by a rough estimate (we customarily say "more or less"). Therefore time is not the movement of a body.

I confess to you, Lord, that I still do not know what time is, and I further confess to you, Lord, that as I say this I know myself to be conditioned by time. For a long period I have been speaking about time, and that long period can only be an interval of time. So how do I know this, when I do not know what time is? Perhaps what I do not know is how to articulate what I do know. My condition is not good if I do not even know what it is I do not know. See, my God, "before you I do not lie" (Gal. I: 21). As I speak, so is my heart. You, Lord, "will light my lamp." Lord, my God, "you will lighten my darknesses" (Ps. 17: 29).

My confession to you is surely truthful when my soul declares that times are measured by me. So my God, I measure, and do not know what I am measuring. I measure the motion of a body by time. Then am I not measuring time itself? I could not measure the movement of a body, its period of transit and how long it takes to go from A to B, unless I were measuring the time in which this movement occurs. How then do I measure time itself? Or do we use a shorter time to measure a longer time, as when, for example, we measure a transom by using a cubit length? So we can be seen to use the length of a short syllable as a measure when we say that a long syllable is twice its length. By this method we measure poems by the number of lines, lines by the number of feet, feet by the number of syllables, and long vowels by short, not by the number of pages (for that would give us a measure of space, not of time). The criterion is the time words occupy in recitation, so that we say "That is a long poem, for it consists of so many lines. The lines are long, for they consist of so many feet. The feet are long for they extend over so many syllables. The syllable is long, for it is double the length of a short one."

Nevertheless, even so we have not reached a reliable measure of time. It may happen that a short line, if pronounced slowly, takes longer to read aloud than a longer line taken faster. The same principle applies to a poem or a foot or a syllable. That is why I have come to think that time is simply a distension. But of what is it a distension? I do not know, but it would be surprising if it is not that of the mind itself. What do I measure, I beg you, my God, when I say without precision "This period is longer than that,"or with precision "This is twice as long as that?" That I am measuring time I know. But I am not measuring the future which does not yet exist, nor the present which has no extension, nor the past which is no longer in being. What then am I measuring? Time as it passes but not time past? That is what I affirmed earlier.

Stand firm, my mind, concentrate with resolution. "God is our help, he has made us and not we ourselves" (Ps. 61: 9; 99: 3). Concentrate on the point where truth is beginning to dawn. For example, a physical voice begins to sound. It sounds. It continues to sound, and then ceases. Silence has now come, and the voice is past. There is now no sound. Before it sounded it lay in the future. It could not be measured because it did not exist; and now it cannot be measured because it has ceased to be. At the time when it was sounding, it was possible because at that time it existed to be measured. Yet even then it had no permanence. It came and went. Did this make it more possible to measure? In process of passing away it was extended through a certain space of time by which it could be measured, since the present occupies no length of time. Therefore during that transient process it could be measured. But take, for example, another voice. It begins to sound and continues to do so unflaggingly without any interruption. Let us measure it while it is sounding; when it has ceased to sound, it will be past and will not exist to be measurable. Evidently we may at that stage measure it by saying how long it lasted. But if it is still sounding, it cannot be measured except from the starting moment when it began to sound to the finish when it ceased. What we measure is the actual interval from the beginning to the end. That is why a sound which has not yet ended cannot be measured: one cannot say how long or short it is, nor that it is equal to some other length of time or that in relation to another it is single or double or any such proportion. But when it has come to an end, then it will already have ceased to be. By what method then can it be measured?

Nevertheless we do measure periods of time. And yet the times we measure are not those which do not yet exist, nor those which already have no existence, nor those which extend over no interval of time, nor those which reach no conclusions. So the times we measure are not future nor past nor present nor those in process of passing away. Yet we measure periods of time.

"God, Creator of all things"—*Deus Creator omnium*—the line consists of eight syllables, in which short and long syllables alternate. So the four which are short (the first, third, fifth, and seventh) are single in relation to the four long syllables (the second, fourth, sixth, and eighth). Each of the long syllables has twice the time of the short. As I recite the words, I also observe that this is so, for it is evident to sense-perception. To the degree that the sense-perception is unambiguous, I measure the long syllable by the short one, and perceive it to be twice the length. But when one syllable sounds after another, the short first, the long after it, how shall I keep my hold on the short, and how use it to apply a measure to the long, so as to verify that the long is twice as much? The long does not begin to sound unless the short has ceased to sound. I can hardly measure the long during the presence of its sound, as measuring becomes possible only after it has ended. When it is finished, it has gone into the past. What then is it which I measure? Where is the short syllable with which I am making my measurement? Where is the long which I am measuring? Both have sounded; they have flown away; they belong to the past. They now do not exist. And I offer my measurement and declare as confidently as a practised sense-perception will allow, that the short is single, the long double—I mean in the time they occupy. I can do this only because they are past and gone. Therefore it is not the syllables which I am measuring, but something in my memory which stays fixed there.

2
The Unreality of Time

J. M. E. McTaggart

303. It will be convenient to begin our enquiry by asking whether anything existent can possess the characteristic of being in time. I shall endeavor to prove that it cannot.

It seems highly paradoxical to assert that time is unreal, and that all statements which involve its reality are erroneous. Such an assertion involves a departure from the natural position of mankind which is far greater than that involved in the assertion of the unreality of space or the unreality of matter. For in each man's experience there is a part—his own states as known to him by introspection—which does not even appear to be spatial or material. But we have no experience which does not appear to be temporal. Even our judgments that time is unreal appear to be themselves in time.

304. Yet in all ages and in all parts of the world the belief in the unreality of time has shown itself to be singularly persistent. In the philosophy and religion of the West—and still more, I suppose, in the philosophy and religion of the East—we find that the doctrine of the unreality of time continually recurs. Neither philosophy nor religion ever hold themselves apart from mysticism for any long period, and almost all mysticism denies the reality of time. In philosophy, time is treated as unreal by Spinoza, by Kant, and by Hegel. Among more modern thinkers, the same view is taken by Mr. Bradley. Such a concurrence of opinion is highly significant, and is not the less significant because the doctrine takes such different forms, and is supported by such different arguments.

I believe that nothing that exists can be temporal, and that therefore time is unreal. But I believe it for reasons which are not put forward by any of the philosophers I have just mentioned.

305. Positions in time, as time appears to us *primâ facie*, are distinguished in two ways. Each position is Earlier than some and Later than some of the other positions. To constitute such a series there is required a transitive asymmetrical relation, and a collection of terms such that, of any two of them, either the first is in this relation to the second, or the second is in this relation to the first. We may take here either the relation of "earlier than" or the relation of "later than," both of

which, of course, are transitive and asymmetrical. If we take the first, then the terms have to be such that, of any two of them, either the first is earlier than the second, or the second is earlier than the first.

In the second place, each position is either Past, Present, or Future. The distinctions of the former class are permanent, while those of the latter are not. If *M* is ever earlier than *N*, it is always earlier. But an event, which is now present, was future, and will be past.

306. Since distinctions of the first class are permanent, it might be thought that they were more objective, and more essential to the nature of time, than those of the second class. I believe, however, that this would be a mistake, and that the distinction of past, present, and future is as *essential* to time as the distinction of earlier and later, while in a certain sense it may, as we shall see, be regarded as more *fundamental* than the distinction of earlier and later. And it is because the distinctions of past, present, and future seem to me to be essential for time, that I regard time as unreal.

For the sake of brevity I shall give the name of the *A* series to that series of positions which runs from the far past through the near past to the present, and then from the present through the near future to the far future, or conversely. The series of positions which runs from earlier to later, or conversely, I shall call the *B* series. The contents of any position in time form an event. The varied simultaneous contents of a single position are, of course, a plurality of events. But, like any other substance, they form a group, and this group is a compound substance. And a compound substance consisting of simultaneous events may properly be spoken of as itself an event.[1]

307. The first question which we must consider is whether it is essential to the reality of time that its events should form an *A* series as well as a *B* series. It is clear, to begin with, that, in present experience, we never *observe* events in time except as forming both these series. We perceive events in time as being present, and those are the only events which we actually perceive. And all other events which, by memory or by inference, we believe to be real, we regard as present, past, or future. Thus the events of time as observed by us form an *A* series.

308. It might be said, however, that this is merely subjective. It might be the case that the distinction of positions in time into past, present, and future, is only a constant illusion of our minds, and that the real nature of time contains only the distinctions of the *B* series—the distinctions of earlier and later. In that case we should not perceive time as it really is, though we might be able to *think* of it as it really is.

This is not a very common view, but it requires careful consideration. I believe it to be untenable, because, as I said above, it seems to me that the *A* series is essential to the nature of time, and that any difficulty in the way of regarding the *A* series as real is equally a difficulty in the way of regarding time as real.

309. It would, I suppose, be universally admitted that time involves change. In ordinary language, indeed, we say that something can remain unchanged through time. But there could be no time if nothing changed. And if anything changes, then all other things change with it. For its change must change some of their relations to

it, and so their relational qualities. The fall of a sand-castle on the English coast changes the nature of the Great Pyramid.

If, then, a *B* series without an *A* series can constitute time, change must be possible without an *A* series. Let us suppose that the distinctions of past, present, and future do not apply to reality. In that case, can change apply to reality?

310. What, on this supposition, could it be that changes? Can we say that, in a time which formed a *B* series but not an *A* series, the change consisted in the fact that the event ceased to be an event, while another event began to be an event? If this were the case, we should certainly have got a change.

But this is impossible. If *N* is ever earlier than *O* and later than *M*, it will always be, and has always been, earlier than *O* and later than *M*, since the relations of earlier and later are permanent. *N* will thus always be in a *B* series. And as, by our present hypothesis, a *B* series by itself constitutes time, *N* will always have a position in a time-series, and always has had one. That is, it always has been an event, and always will be one, and cannot begin or cease to be an event.

Or shall we say that one event *M* merges itself into another event *N*, while still preserving a certain identity by means of an unchanged element, so that it can be said, not merely that *M* has ceased and *N* begun, but that it is *M* which has become *N*? Still the same difficulty recurs. *M* and *N* may have a common element, but they are not the same event, or there would be no change. If, therefore, *M* changed into *N* at a certain moment, then at that moment, *M* would have ceased to be *M*, and *N* would have begun to be *N*. This involves that, at that moment, *M* would have ceased to be an event, and *N* would have begun to be an event. And we saw, in the last paragraph, that, on our present hypothesis, this is impossible.

Nor can such change be looked for in the different moments of absolute time, even if such moments should exist. For the same argument will apply here. Each such moment will have its own place in the *B* series, since each would be earlier or later than each of the others. And, as the *B* series depends on permanent relations, no moment could ever cease to be, nor could it become another moment.

311. Change, then, cannot arise from an event ceasing to be an event, nor from one event changing into another. In what other way can it arise? If the characteristics of an event change, then there is certainly change. But what characteristics of an event can change? It seems to me that there is only one class of such characteristics. And that class consists of the determinations of the event in question by the terms of the *A* series.

Take any event—the death of Queen Anne, for example—and consider what changes can take place in its characteristics. That it is a death, that it is the death of Anne Stuart, that it has such causes, that it has such effects—every characteristic of this sort never changes. "Before the stars saw one another plain," the event in question was the death of a Queen. At the last moment of time—if time has a last moment—it will still be the death of a Queen. And in every respect but one, it is equally devoid of change. But in one respect it does change. It was once an event in the far future. It became every moment an event in the nearer future. At last it was present. Then it became past, and will always remain past, though every moment it becomes further and further past.[2]

Such characteristics as these are the only characteristics which can change. And, therefore, if there is any change, it must be looked for in the *A* series, and in the *A* series alone. If there is no real *A* series, there is no real change. The *B* series, therefore, is not by itself sufficient to constitute time, since time involves change.

312. The *B* series, however, cannot exist except as temporal, since earlier and later, which are the relations which connect its terms, are clearly time-relations. So it follows that there can be no *B* series when there is no *A* series, since without an *A* series there is no time.

313. We must now consider three objections which have been made to this position. The first is involved in the view of time which has been taken by Mr. Russell, according to which past, present, and future do not belong to time *per se*, but only in relation to a knowing subject. An assertion that *N* is present means that it is simultaneous with that assertion, an assertion that it is past or future means that it is earlier or later than that assertion. Thus it is only past, present, or future, in relation to some assertion. If there were no consciousness, there would be events which were earlier and later than others, but nothing would be in any sense past, present, or future. And if there were events earlier than any consciousness, those events would never be future or present, though they could be past.

If *N* were ever present, past, or future in relation to some assertion *V*, it would always be so, since whatever is ever simultaneous to, earlier than, or later than, *V*, will always be so. What, then, is change? We find Mr. Russell's views on this subject in his *Principles of Mathematics*, Section 442. "Change is the difference, in respect of truth or falsehood, between a proposition concerning an entity and the time *T*, and a proposition concerning the same entity and the time *T'*, provided that these propositions differ only by the fact that *T* occurs in the one where *T'* occurs in the other." That is to say, there is change, on Mr. Russell's view, if the proposition "at the time *T* my poker is hot" is true, and the proposition "at the time *T'* my poker is hot" is false.

314. I am unable to agree with Mr. Russell. I should, indeed, admit that, when two such propositions were respectively true and false, there would be change. But then I maintain that there can be no time without an *A* series. If, with Mr. Russell, we reject the *A* series, it seems to me that change goes with it, and that therefore time, for which change is essential, goes too. In other words, if the *A* series is rejected, no proposition of the type "at the time *T* my poker is hot" can ever be true, because there would be no time.

315. It will be noticed that Mr. Russell looks for change, not in the events in the time-series, but in the entity to which those events happen, or of which they are states. If my poker, for example, is hot on a particular Monday, and never before or since, the event of the poker being hot does not change. But the poker changes, because there is a time when this event is happening to it, and a time when it is not happening to it.

But this makes no change in the qualities of the poker. It is always a quality of that poker that it is one which is hot on that particular Monday. And it is always a quality of that poker that it is one which is not hot at any other time. Both these qualities are true of it at any time—the time when it is hot and the time when it is cold. And therefore it seems to be erroneous to say that there is any change in the poker. The fact that it is hot at one point in a series and cold at other points cannot give change, if neither of these facts change—and neither of them does. Nor does any other fact about the poker change, unless its presentness, pastness, or futurity change.

316. Let us consider the case of another sort of series. The meridian of Greenwich passes through a series of degrees of latitude. And we can find two points in this series, *S* and *S'*, such that the proposition "at *S* the meridian of Greenwich is within the United Kingdom" is true, while the proposition "at *S'* the meridian of Greenwich is within the United Kingdom" is false. But no one would say that this gave us change. Why should we say so in the case of the other series?

Of course there is a satisfactory answer to this question if we are correct in speaking of the other series as a time-series. For where there is time, there is change. But then the whole question is whether it is a time-series. My contention is that if we remove the *A* series from the *primâ facie* nature of time, we are left with a series which is not temporal, and which allows change no more than the series of latitude does.

317. If, as I have maintained, there can be no change unless facts change, then there can be no change without an *A* series. For, as we saw with the death of Queen Anne, and also in the case of the poker, no fact about anything can change, unless it is a fact about its place in the *A* series. Whatever other qualities it has, it has always. But that which is future will not always be future, and that which was past was not always past.

It follows from what we have said that there can be no change unless some propositions are sometimes true and sometimes false. This is the case of propositions which deal with the place of anything in the *A* series—"the battle of Waterloo is in the past," "it is now raining." But it is not the case with any other propositions.

318. Mr. Russell holds that such propositions are ambiguous, and that to make them definite we must substitute propositions which are always true or always false—"the battle of Waterloo is earlier than this judgment," "the fall of rain is simultaneous with this judgment." If he is right, all judgments are either always true, or always false. Then, I maintain, no facts change. And then, I maintain, there is no change at all.

I hold, as Mr. Russell does, that there is no *A* series. (My reasons for this will be given below, pp. 42-43.) And, as I shall explain on p. 31 [not included in this volume—*Editors*], I regard the reality lying behind the appearance of the *A* series in a manner not completely unlike that which Mr. Russell has adopted. The difference between us is that he thinks that, when the *A* series is rejected, change, time, and the *B* series can still be kept, while I maintain that its rejection involves the rejection of change, and, consequently, of time, and of the *B* series.

319. The second objection rests on the possibility of nonexistent time-series—such, for example, as the adventures of Don Quixote. This series, it is said, does not form part of the *A* series. I cannot at this moment judge it to be either past, present, or future. Indeed, I know that it is none of the three. Yet, it is said, it is certainly a *B*

series. The adventure of the galley-slaves, for example, is later than the adventure of the windmills. And a *B* series involves time. The conclusion drawn is that an *A* series is not essential to time.

320. I should reply to this objection as follows. Time only belongs to the existent. If any reality is in time, that involves that the reality in question exists. This, I think, would be universally admitted. It may be questioned whether all of what exists is in time, or even whether anything really existent is in time, but it would not be denied that, if anything is in time, it must exist.

Now what is existent in the adventures of Don Quixote? Nothing. For the story is imaginary. The states of Cervantes' mind when he invented the story, the states of my mind when I think of the story—these exist. But then these form part of an *A* series. Cervantes' invention of the story is in the past. My thought of the story is in the past, the present, and—I trust—the future.

321. But the adventures of Don Quixote may be believed by a child to be historical. And in reading them I may, by an effort of my imagination, contemplate them as if they really happened. In this case, the adventures are believed to be existent, or are contemplated as existent. But then they are believed to be in the *A* series, or are contemplated as being in the *A* series. The child who believes them to be historical will believe that they happened in the past. If I contemplate them as existent, I shall contemplate them as happening in the past. In the same way, if I believed the events described in Jefferies' *After London* to exist, or contemplated them as existent, I should believe them to exist in the future, or contemplate them as existing in the future. Whether we place the object of our belief or of our contemplation in the present, the past, or the future, will depend upon the characteristics of that object. But somewhere in the *A* series it will be placed.

Thus the answer to the objection is that, just as far as a thing is in time, it is in the *A* series. If it is really in time, it is really in the *A* series. If it is believed to be in time, it is believed to be in the *A* series. If it is contemplated as being in time, it is contemplated as being in the *A* series.

322. The third objection is based on the possibility that, if time were real at all, there might be in reality several real and independent time-series. The objection, if I understand it rightly, is that every time-series would be real, while the distinctions of past, present, and future would only have a meaning within each series, and would not, therefore, be taken as absolutely real. There would be, for example, many presents. Now, of course, many points of time can be present. In each time-series many points are present, but they must be present successively. And the presents of the different time-series would not be successive, since they are not in the same time.[3] And different presents, it would be said, cannot be real unless they are successive. So the different time-series, which are real, must be able to exist independently of the distinction between past, present, and future.

323. I cannot, however, regard this objection as valid. No doubt in such a case, no present would be *the* present—it would only be the present of a certain aspect of the universe. But then no time would be *the* time—it would only be the time of a certain aspect of the universe. It would be a real time-series, but I do not see that the present would be less real than the time.

I am not, of course, maintaining that there is no difficulty in the existence of several distinct *A* series. In the second part of this chapter I shall endeavor to show that the existence of *any A* series is impossible. What I assert here is that, if there could be an *A* series at all, and if there were any reason to suppose that there were several distinct *B* series, there would be no additional difficulty in supposing that there should be a distinct *A* series for each *B* series.

324. We conclude, then, that the distinctions of past, present, and future are essential to time, and that, if the distinctions are never true of reality, then no reality is in time. This view, whether true or false, has nothing surprising in it. It was pointed out above that we always perceive time as having these distinctions. And it has generally been held that their connection with time is a real characteristic of time, and not an illusion due to the way in which we perceive it. Most philosophers, whether they did or did not believe time to be true of reality, have regarded the distinctions of the *A* series as essential to time.

When the opposite view has been maintained it has generally been, I believe, because it was held (rightly, as I shall try to show) that the distinctions of past, present, and future cannot be true of reality, and that consequently, if the reality of time is to be saved, the distinction in question must be shown to be unessential to time. The presumption, it was held, was for the reality of time, and this would give us a reason for rejecting the *A* series as unessential to time. But, of course, this could only give a presumption. If the analysis of the nature of time has shown that, by removing the *A* series, time is destroyed, this line of argument is no longer open.

325. I now pass to the second part of my task. Having, as it seems to me, succeeded in proving that there can be no time without an *A* series, it remains to prove that an *A* series cannot exist, and that therefore time cannot exist. This would involve that time is not real at all, since it is admitted that the only way in which time can be real is by existing.

326. Past, present, and future are characteristics which we ascribe to events, and also to moments of time, if these are taken as separate realities. What do we mean by past, present, and future? In the first place, are they relations or qualities? It seems quite clear to me that they are not qualities but relations, though, of course, like other relations, they will generate relational qualities in each of their terms.[4] But even if this view should be wrong, and they should in reality be qualities and not relations, it will not affect the result which we shall reach. For the reasons for rejecting the reality of past, present, and future, which we are about to consider, would apply to qualities as much as to relations.

327. If, then, anything is to be rightly called past, present, or future, it must be because it is in relation to something else. And this something else to which it is in relation must be something outside the time-series. For the relations of the *A* series are changing relations, and no relations which are exclusively between members of the time-series can ever change. Two events are exactly in the same places in the time-series, relatively to one another, a million years before they take place, while each of them is taking place, and when they are a million years in the past. The same is true of the relation of moments to one another, if moments are taken as separate realities. And the same would be true of the relations of events to

moments. The changing relation must be to something which is not in the time-series.

Past, present, and future, then, are relations in which events stand to something outside the time-series. Are these relations simple, or can they be defined? I think that they are clearly simple and indefinable. But, on the other hand, I do not think that they are isolated and independent. It does not seem that we can know, for example, the meaning of pastness, if we do not know the meaning of presentness or of futurity.

328. We must begin with the *A* series, rather than with past, present, and future, as separate terms. And we must say that a series is an *A* series when each of its terms has, to an entity *X* outside the series, one, and only one, of three indefinable relations, pastness, presentness, and futurity, which are such that all the terms which have the relation of presentness to *X* fall between all the terms which have the relation of pastness to *X*, on the one hand, and all the terms which have the relation of futurity to *X*, on the other hand.

We have come to the conclusion that an *A* series depends on relations to a term outside the *A* series. This term, then, could not itself be in time, and yet must be such that different relations to it determine the other terms of those relations, as being past, present, or future. To find such a term would not be easy, and yet such a term must be found, if the *A* series is to be real. But there is a more positive difficulty in the way of the reality of the *A* series.

329. Past, present, and future are incompatible determinations. Every event must be one or the other, but no event can be more than one. If I say that any event is past, that implies that it is neither present nor future, and so with the others. And this exclusiveness is essential to change, and therefore to time. For the only change we can get is from future to present, and from present to past.

The characteristics, therefore, are incompatible. But every event has them all.[5] If *M* is past, it has been present and future. If it is future, it will be present and past. If it is present, it has been future and will be past. Thus all the three characteristics belong to each event. How is this consistent with their being incompatible?

330. It may seem that this can easily be explained. Indeed, it has been impossible to state the difficulty without almost giving the explanation, since our language has verb-forms for the past, present, and future, but no form that is common to all three. It is never true, the answer will run, that *M is* present, past, and future. It *is* present, *will be* past, and *has been* future. Or it *is* past, and *has been* future and present, or again *is* future, and *will be* present and past. The characteristics are only incompatible when they are simultaneous, and there is no contradiction to this in the fact that each term has all of them successively.

331. But what is meant by "has been" and "will be"? And what is meant by "is," when, as here, it is used with a temporal meaning, and not simply for predication? When we say that *X* has been *Y*, we are asserting *X* to be *Y* at a moment of past time. When we say that *X* will be *Y*, we are asserting *X* to be *Y* at a moment of future time. When we say that *X* is *Y* (in the temporal sense of "is"), we are asserting *X* to be *Y* at a moment of present time.

Thus our first statement about *M*—that it is present, will be past, and has been future—means that *M* is present at a moment of present time, past at some moment of future time, and future at some moment of past time. But every moment, like every event, is both past, present, and future. And so a similar difficulty arises. If *M* is present, there is no moment of past time at which it is past. But the moments of future time, in which it is past, are equally moments of past time, in which it cannot be past. Again, that *M* is future and will be present and past means that *M* is future at a moment of present time, and present and past at different moments of future time. In that case it cannot be present or past at any moments of past time. But all the moments of future time, in which *M* will be present or past, are equally moments of past time.

332. And thus again we get a contradiction, since the moments at which *M* has any one of the three determinations of the *A* series are also moments at which it cannot have that determination. If we try to avoid this by saying of these moments what had been previously said of *M* itself—that some moment, for example, is future, and will be present and past—then "is" and "will be" have the same meaning as before. Our statement, then, means that the moment in question is future at a present moment, and will be present and past at different moments of future time. This, of course, is the same difficulty over again. And so on infinitely.

Such an infinity is vicious. The attribution of the characteristics past, present, and future to the terms of any series leads to a contradiction, unless it is specified that they have them successively. This means, as we have seen, that they have them in relation to terms specified as past, present, and future. These again, to avoid a like contradiction, must in turn be specified as past, present, and future. And, since this continues infinitely, the first set of terms never escapes from contradiction at all.[6]

The contradiction, it will be seen, would arise in the same way supposing that pastness, presentness, and futurity were original qualities, and not, as we have decided that they are, relations. For it would still be the case that they were characteristics which were incompatible with one another, and that whichever had one of them would also have the other. And it is from this that the contradiction arises.

Notes

1. It is very usual to contemplate time by the help of a metaphor of spatial movement. But spatial movement in which direction? The movement of time consists in the fact that later and later terms pass into the present, or—which is the same fact expressed in another way—that presentness passes to later and later terms. If we take it the first way, we are taking the B series as sliding along a fixed A series. If we take it the second way, we are taking the A series as sliding along a fixed B series. In the first case time presents itself as a movement from future to past. In the second case it presents itself as a movement from earlier to later. And this explains why we say that events come out of the future, while we say that we ourselves move towards the future. For each man identifies himself especially with his present state, as against his future or his past, since it is the only one which he is

directly perceiving. And this leads him to say that he is moving with the present towards later events. And as those events are now future, he says that he is moving towards the future.

Thus the question as to the movement of time is ambiguous. But if we ask what is the movement of either series, the question is not ambiguous. The movement of the A series along the B series is from earlier to later. The movement of the B series along the A series is from future to past.

2. The past, therefore, is always changing, if the A series is real at all, since at each moment a past event is further in the past than it was before. This result follows from the reality of the A series, and is independent of the truth of our view that all change depends exclusively on the A series. It is worthwhile to notice this, since most people combine the view that the A series is real with the view that the past cannot change—a combination which is inconsistent.

3. Neither would they be simultaneous, since that equally involves being in the same time. They would stand in no time-relation to one another.

4. It is true, no doubt, that my anticipation of an experience M, the experience itself, and the memory of the experience are three states which have different original qualities. But it is not the future M, the present M, and the past M, which have these three different qualities. The qualities are possessed by three different events—the anticipation of M, M itself, and the memory of M—each of which in its turn is future, present, and past. Thus this gives no support to the view that the changes of the A series are changes of original qualities.

5. If the time-series has a first term, that term will never be future, and if it has a last term, that term will never be past. But the first term, in that case, will be present and past, and the last term will be future and present. And the possession of two incompatible characteristics raises the same difficulty as the possession of three.

6. It may be worthwhile to point out that the vicious infinite has not arisen from the impossibility of defining past, present, and future, without using the terms in their own definitions. On the contrary, we have admitted these terms to be indefinable. It arises from the fact that the nature of the terms involves a contradiction, and that the attempt to remove the contradiction involves the employment of the terms, and the generation of a similar contradiction.

3

Changes in Events and Changes in Things

Arthur Prior

The basic question to which I wish to address myself in this lecture is simply the old one, does time really flow or pass? The problem, of course, is that genuine flowing or passage is something which occurs *in* time, and *takes* time to occur. If time itself flows or passes, must there not be some "super-time" in which it does so? Again, whatever flows or passes does so at some *rate,* but a rate of flow is just the amount of movement in a given *time,* so how could there be a rate of flow of time itself? And if time does not flow at any rate, how can it flow at all?

A natural first move towards extricating ourselves from these perplexities is to admit that talk of the flow or passage of time is just a metaphor. Time may be, as Isaac Watts says, *like* an ever-rolling stream, but it isn't really and literally an ever-rolling stream. But *how* is it like an ever-rolling stream? What is the literal truth behind this metaphor? The answer to this is not, at first sight, difficult. Generally when we make such remarks as, "Time does fly, doesn't it?—why, it's already the 16th," we mean that some date or moment which we have been looking forward to as future has ceased to be future and is now present and on its way into the past. Or more fundamentally, perhaps, some future *event* to which we have been looking forward with hope or dread is now at last occurring, and soon will have occurred, and will have occurred a longer and longer time *ago*. We might say, for example, "Time does fly, I'm already 47," that is, my birth is already that much past, "and soon I shall be 48," i.e., it will be more past still. Suppose we speak about something "becoming more past" not only when it moves from the comparatively near past to the comparatively distant past, but also when it moves from the present to the past, from the future to the present, and from the comparatively distant future to the comparatively near future. Then whatever is happening, has happened, or will happen is all the time "becoming more past" in this extended sense; and just this is what we mean by the flow or passage of time. And if we want to give the *rate* of this flow or passage, it is surely very simple—it takes one exactly a year to get a year older, i.e., events become more past at the rate of a year per year, an hour per hour, a second per second.

Does this remove the difficulty? It is far from obvious that it does. It's not just that an hour per hour is a queer sort of rate—*this* queerness, I think, has been exaggerated, and I shall say more about it in a minute—but the whole idea of events changing is at first sight a little strange, even if we abandon the admittedly figurative description of this change as a *movement.* By and large, to judge by the way that we ordinarily talk, it's *things* that change, and events don't change but *happen.* Chairs, tables, horses, people change—chairs get worn out and then mended, tables get dirty and then clean again, horses get tired and then refreshed, people learn things and forget them, or are happy and then miserable, active and then sleepy, and so on, and all these are changes, and chairs, tables, horses, and people are all what I mean by things as opposed to events. An accident, a coronation, a death, a prizegiving, are examples of what we'd call events, and it does seem unnatural to describe these as changing—what these do, one is inclined to say, is not to change but to happen or occur.

One of the things that make us inclined to deny that events undergo changes is that events *are* changes—to say that such and such an event has occurred is generally to say that some thing has, or some things have, changed in some way. To say, for instance, that the retirement of Sir Anthony Eden occurred in such and such a year is just to say that Sir Anthony then retired and so suffered the change or changes that retirement consists in—he had been Prime Minister, and then was not Prime Minister. Sir Anthony's retirement is or was a change concerning Sir Anthony; to say that it itself changes or has changed sounds queer because it sounds queer to talk of a change changing.

This queerness, however, is superficial. When we reflect further we realize that changes do change, especially if they go on for any length of time. (In this case we generally, though not always, call the change a *process* rather than an event, and there are other important differences between events and processes besides the length of time they take, but these differences are not relevant to the present discussion, so I shall ignore them and discuss changes generally, events and processes alike.) Changes do change—a movement, for example, may be slow at first and then rapid, a prizegiving or a lecture may be at first dull and afterwards interesting, or vice versa, and so on. It would hardly be too much to say that modem science began when people became accustomed to the idea of changes changing, e.g., to the idea of acceleration as opposed to simple motion. I've no doubt the ordinary measure of acceleration, so many feet per second per second, sounded queer when it was first used, and I think it still sounds queer to most students when they first encounter it. Ordinary speech is still resistant to it, and indeed to the expression of anything in the nature of a comparison of a comparison. We are taught at school that "more older," for example, is bad English, but why shouldn't I say that I am more older than my son than he is than my daughter? And if we have learned to talk of an acceleration of a foot per second per second without imagining that the second "second" must somehow be a different kind of "second" from the first one—without imagining that if motion takes place in ordinary time, acceleration must take place in some super-time—can we not accustom ourselves equally to a change of "a second per second" without any such imagining?

Changes do change, then, but this does not leave everything quite simple and solved. For there's still something odd about the change that we describe figuratively as the flow or passage of time—the change from an event's being future to its being present, and from its being present to its being more and more past. For the other changes in events which I have mentioned are ones which go on in the event *while it is occurring*; for example, if a lecture gets duller or a movement faster then this is something it does *as it goes on*; but the change from past to still further past isn't one that occurs while the event is occurring, for all the time that an event is occurring it isn't past but present, in fact the presentness of an event just *is* its happening, its occurring, as opposed to its merely having happened or being merely about to happen. We might put it this way: the things that change are *existing* things, and it's while they exist that they change, e.g., it's existing men, not non-existent men, that get tired and then pick up again; Julius Caesar, for example, isn't now getting tired and picking up again, unless the doctrine of immortality is true and he exists now as much as he ever did. And such changes as the change in the rate of movement are similarly changes that go on in events or processes while they exist, that is, while they exist in the only sense in which events and processes do exist, namely while they are occurring. But getting more and more past seems to be something an event does when it *doesn't* exist, and this seems very queer indeed.

We may retrace our steps to this point by looking at some of the literature of our subject. Professor C. D. Broad, in the second volume of his *Examination of McTaggart's Philosophy*, says that the ordinary view that an event, say the death of Queen Anne, is in the indefinitely distant future and then less and less future and then present and then goes into the more and more distant past—this ordinary story, Broad says, cannot possibly be true because it takes the death of Queen Anne to be at once a mere momentary thing and something with an indefinitely long history. We can make a first answer to this by distinguishing between the history that an event *has*, and the bit of history that it *is*. The bit of history that Queen Anne's death is, or was, is a very very short bit, but that doesn't prevent the history that it has from being indefinitely long. Queen Anne's death is part of the history of Queen Anne, and a very short part of it; what is long is not this part of the history of Queen Anne, but rather the history of this part of her history—the history of this part of her history is that first it was future, then it was present, and so on, and this can be a long history even if the bit of history that it is the history *of* is very short. There is not, therefore, the flat contradiction that Broad suggests here. There is, however, the difficulty that we generally think of the history of a thing as the sum of what it does and what happens to it *while it is there*—when it ceases to be, its history has ended—and this does make it seem odd that there should be an indefinitely long history of something which itself occupies a time which is indefinitely short.

But if there is a genuine puzzle here, it concerns what is actually going on also. For whatever goes on for any length of time—and that means: "whatever goes on"—will have future and past phases as well as the immediately present one; its going on is in fact a continual passage of one phase after another from being future through being present to being past. Augustine's reflections, in the eleventh chapter of his *Confessions,* on the notion of a "long time," are relevant here. Just when, he asks, is a long time long? Is it long when it is present, or when it is past or future? We need not, I think, attach much importance to the fact that Augustine concentrates on so abstract a thing as a "time" or an interval; his problems can be quite easily re-stated in terms of *what goes on* over the interval; in fact he himself slips into this, and talks about his childhood, a future sunrise, and so on. When, we may ask, does a process go on for a "long time"—while it is going on, or when it lies ahead of us, or is all over?

Augustine is at first driven to the view that it is when it is present that a time is long, for only what *is* can be long or short (paragraph 18) [see pp. 27-28 in this volume—*Editors*]. We can give the same answer with processes—it is when they are going on that they go on for a long time. But then, as Augustine points out, there are these phases. A hundred years is a long time, but it's not really present all at once; and even if we try to boil down the present to an hour, "that one hour passes away in flying particles." "The present hath no space" (20) [p. 28]. Augustine had apparently not heard of the "specious present," but even if he had it would not have helped him much—most of the happenings we are interested in take longer than that. He tries out the hypotheses that the past and the future, and past and future events, in some sense after all "are"—that there is some "secret place" where they exist all the time, and from which they come and to which they go. If there is no such place, then where do those who foresee the future and recall the past discern these things? "For that which is not, cannot be seen" (22) [p. 29].

Well, Augustine says, he doesn't know anything about that, but one thing that he does know is that wherever "time past and to come may be," "they are not there as future, or past, but present. For if there also they be future, they are not yet there; if there also they be past, they are no longer there. Wheresoever then is whatsoever is, it is only as present" (23) [p. 29]. Of course there are present "traces" or images of past things in our memories, and present signs and intentions on the basis of which we make our future forecasts (23, 24) [p. 29], and sometimes Augustine seems satisfied with this—past, present, and future, he says, "do exist in some sort, in the soul, but otherwhere do I not see them" (26) [p. 30]. But sometimes he seems far from content with this—*that which* we remember and anticipate, he says, is different from these signs, and is *not* present (23, 24) [p. 29] —and, one must surely add, is *not* "in the soul."

It is time now to be constructive, and as a preparation for this I shall indulge in what may seem a digression, on the subject of Grammar. English philosophers who visit the United States are always asked sooner or later whether they are "analysts." I'm not at all sure what the answer is in my own case, but there's another word that Professor Passmore once invented to describe some English philosophers who are often called "analysts," namely the word "grammaticist," and that's something I wouldn't at all mind calling myself. I don't deny that there are genuine metaphysical problems, but I think you have to talk about grammar at least a little bit in order to solve most of them. And in particular, I would want to maintain that most of the present group of problems about time and change, though not quite all of them, arise from the fact that many expressions which look like nouns, i.e., names of objects, are not really nouns at all but concealed verbs, and many expressions which look like verbs are not really verbs but concealed conjunctions and adverbs. That is a slight over-simplification, but before we can get it stated more accurately we must look more closely at verbs, conjunctions, and adverbs.

I shall assume that we are sufficiently clear for our present purposes as to what a noun or name is, and what a sentence is; and given these notions, we can define a verb or verb-phrase as an expression that constructs a sentence out of a name or names. For instance, if you tack the verb "died" on the name "Queen Anne" you get the sentence "Queen Anne died," and if you tack the phrase "is an undertaker" on the name "James Bowels" you get the sentence "James Bowels is an undertaker," so that this is a verb-phrase. I say "out of a name *or names*" because some verbs have to have an object as well as a subject. Thus if you put the verb "loves" between the names "Richard" and "Joan" you get the sentence "Richard loves Joan"; this verb constructs this sentence out of these two names; and the phrase "is taller than" would function similarly. Logicians call verbs and verb-phrases "predicates"; "died" and "is an undertaker" would be "one-place" predicates, and "loves" and "is taller than" are "two-place" predicates. There are also expressions which construct sentences, not out of names, but out of other sentences. If an expression constructs a sentence out of two or more other sentences it is a conjunction, or a phrase equivalent to a conjunction. For example "Either – or –" functions in this way in "Either it will rain or it will snow." If the expression constructs a sentence out of one other sentence it is an adverb or adverbial phrase, like "not" or "It is not the case that" or "allegedly" or "It is alleged that," or "possibly" or "It is possible that." Thus by attaching these expressions to "It is raining" we obtain the sentences

It is not raining;
It is not the case that it is raining;
It is allegedly raining;
It is alleged that it is raining;
It is possibly raining;
It is possible that it is raining.

One very important difference between conjunctions and adverbs, on the one hand, and verbs, on the other, is that because the former construct sentences out of sentences, i.e., the same sort of thing as they end up with, they can be applied again and again to build up more and more complicated sentences, like "It is allegedly possible that he will not come," which could be spread out as

It is said that (it is possible that (it is not the case that (he will come))).

You can also use the same adverb twice and obtain such things as double negation, alleged allegations, and so on. Verbs, because they do not end up with the same sort of expression as what they start with, cannot be piled up in this way.

Having constructed "Queen Anne died" by the verb "died" out of the name "Queen Anne," you cannot do it again—"Queen Anne died died" is not a sentence.

Turning now to our main subject, I want to suggest that putting a verb into the past or future tense is exactly the same sort of thing as adding an adverb to the sentence. "I *was* having my breakfast" is related to "I am having my breakfast" in exactly the same way as "I am *allegedly* having my breakfast" is related to it, and it is only an historical accident that we generally form the past tense by modifying the present tense, e.g., by changing "am" to "was," rather than by tacking on an adverb. In a rationalized language with uniform constructions for similar functions we could form the past tense by prefixing to a given sentence the phrase "It was the case that," or "It has been the case that" (depending on what sort of past we meant), and the future tense by prefixing "It will be the case that." For example, instead of "I will be eating my breakfast" we could say

> "It will be the case that I am eating my breakfast,"

and instead of "I was eating my breakfast" we could say

> "It was the case that I am eating my breakfast."

The nearest we get to the latter in ordinary English is "It was the case that I *was* eating my breakfast," but this is one of those anomalies like emphatic double negation. The construction I am sketching embodies the truth behind Augustine's suggestion of the "secret place" where past and future times "are," and his insistence that wherever they are, they are not there as past or future but as present. The past is not the present but it *is* the past present, and the future is not the present but it *is* the future present.

There is also, of course, the past future and the future past. For these adverbial phrases, like other adverbial phrases, can be applied repeatedly—the sentences to which they are attached do not have to be simple ones; it is enough that they be sentences, and they can be sentences which already have tense-adverbs, as we might call them, within them. Hence we can have such a construction as

> "It will be that case that (it has been the case that (I am taking off my coat)),"

or in plain English, "I will have taken off my coat." We can similarly apply repeatedly such *specific* tense-adverbs as "It was the case forty-eight years ago that." For example, we could have

> "It will be the case seven months hence that (it was the case forty-eight years ago that (I am being born)),"

that is, it will be my forty-eighth birthday in seven months' time.

To say that a change has occurred is to say at least this much: that something which was the case formerly is not the case now. That is, it is at least to say that for some sentence *p* we have

It was the case that *p*, and it is not the case that *p*.

This sentence *p* can be as complicated as you like, and can itself contain tense-adverbs, so that one example of our formula would be

It was the case 5 months ago that (it was the case only 47 years ago that (I am being born)), and it is not now the case that (it was the case only 47 years ago that (I am being born)),

that is, I am not as young as I used to be. This last change, of course, is a case of precisely that recession of events into the past that we are really talking about when we say that time flows or passes—and the piling of time-references on top of one another, with no suggestion that the time-words must be used in a different sense at each level, simply reflects the fact that tense-adverbs *are* adverbs, not verbs.

An important point to notice now is that while *I* have been talking about words—for example about verbs and adverbs—for quite a long time, the sentences that I have been using as examples have *not* been about words but about real things. When a sentence is formed out of another sentence or other sentences by means of an adverb or conjunction, it is not *about* those other sentences, but about whatever they are themselves about. For example, the compound sentence "Either I will wear my cap or I will wear my beret" is not about the sentences "I will wear my cap" and "I will wear my beret"; like them, it is about me and my headgear, though the information it conveys about these is a little less definite than what either of them would convey separately. Similarly, the sentence "It will be the case that I am having my tooth out" is not about the sentence "I am having my tooth out"; it is about me. A genuine sentence about the sentence "I am having my tooth out" would be one stating that it contained six words and nineteen letters, but "It will be the case that I am having my tooth out," i.e. ,"I will be having my tooth out," is quite obviously not a sentence of this sort at all.

Nor is it about some abstract entity named by the clause "that I am having my tooth out." It is about me and my tooth, and about nothing else whatever. The fact is that it is difficult for the human mind to get beyond the simple subject-predicate or noun-verb structure, and when a sentence or thought hasn't that structure but a more complex one we try in various ways to force it into the subject-predicate pattern. We thus invent new modes of speech in which the subordinate sentences are replaced by noun-phrases and the conjunctions or adverbs by verbs or verb-phrases. For example, instead of saying

(1) *If* you have oranges in your larder you have been to the greengrocer's,

we may say

(2) Your having oranges in your larder *implies* your having been to the greengrocer's,

which looks as if it has the same form as "Richard loves Joan" except that "Your having oranges in your larder" and "Your having been to the grocer" seem to name more abstract objects than Richard and Joan, and implying seems a more abstract activity than loving. We can rid ourselves of this suggestion if we reflect that (2) is nothing more than a paraphrase of (1). Similarly

(3) It is now six years since it was the case that I am falling out of a punt,

could be re-written as

(4) My falling out of a punt has receded six years into the past.

This suggests that something called an event, my falling out of a punt, has gone through a performance called receding into the past, and moreover has been going through this performance even after it has ceased to exist, i.e., after it has stopped happening. But of course, (4) is just a paraphrase of (3), and like (3) is not about any objects except me and that punt—there is no real reason to believe in the existence either now or six years ago of a further object called "my falling out of a punt."

What I am suggesting is that what looks like talk about events is really at bottom talk about things, and that what looks like talk about changes in events is really just slightly more complicated talk about changes in things. This applies *both* to the changes that we say occur in events when they are going on, like the change in speed of a movement ("movement" is a *façon de parler*; there is just the moving car, which moves more quickly than it did), *and* the changes that we say occur in events when they are not going on any longer, or not yet, e.g., my birth's receding into the past ("birth" is a *façon de parler*—there's just me being born, and then getting older).

It's not all quite as simple as this, however. This story works very well for me and my birth and my fall out of the punt, but what about Queen Anne? Does Queen Anne's death getting more past mean that *Queen Anne* has changed from having died 250 years ago to having died 251 years ago, or whatever the period is?—that *she* is still "getting older," though in a slightly extended sense? The trouble with this, of course, is just that Queen Anne doesn't exist now any more than her death does. There are at least two different ways in which we might deal with this one. We might, in the first place, say that our statement really is about Queen Anne (despite the fact that she "is no more"), and really is, or at least entails, a statement of the form

It was the case that p, and is not now the case that p,

namely

> It was the case that it was the case only 250 years ago that Queen Anne is dying, and is not now the case that it was the case only 250 years ago that Queen Anne is dying,

but we may add that this statement does not record a "change" in any natural sense of that word, and certainly not a change in Queen Anne. A genuine record of change, we could say, must not only be of the form above indicated but must meet certain further conditions which we might specify in various ways. And we could say that although what is here recorded *isn't* a change in the proper sense, it is *like* a change in fitting the above formula. The flow of time, we would then say, is merely metaphorical, not only because what is meant by it isn't a genuine change; but the force of the metaphor can still be explained—we use the metaphor because what we call the flow of time does fit the above formula. On this view it might be that not only the recession of Queen Anne's death but my own growing older will not count as a change in the strict sense, though growing older is normally *accompanied* by genuine changes, and the phrase is commonly extended to cover these—increasing wisdom, bald patches, and so on.

But can a statement really be *about* Queen Anne after she has ceased to be? I do not wish to dogmatize about this, but an alternative solution is worth mentioning. We might paraphrase "Queen Anne has died" as "Once there was a person named, 'Anne,' who reigned over England, etc., but there is not now any such person." This solution exploits a distinction which we may describe as one between *general facts* and *individual facts.* That someone has stolen my pencil is a general fact; that John Jones has stolen my pencil, if it is a fact at all, is an individual fact. It has often been said—for example, it was said by the Stoic logicians—that there are no general facts without there being the corresponding individual facts. It cannot, for example, be the case that "someone" has stolen my pencil, unless it is the case that some specific individual—if not John Jones, then somebody else—has stolen it. And in cases of this sort the principle is very plausible; indeed it is obviously true. I have read that some of the schoolmen described the subject of sentences like "someone has stolen my pencil" as an *individuum vagum,* but of course this is a makeshift—forcing things into a pattern again. There are no "vague individuals," and if a pencil has been stolen at all it has been stolen not by a vague individual but by some quite definite one, or else by a number of such. There are vague statements, however, and vague thoughts, and the existence of such statements and thoughts is as much a fact about the real world as any other; and when we describe the making of such statements and the entertaining of such thoughts, we do encounter at least partly general facts to which no wholly individual facts correspond. If I allege or believe that someone has stolen my pencil, there may be *no* specific individual with respect to whom I allege or believe that *he* stole my pencil. There is *alleged or believed to be* an individual who stole it, but there is *no individual who is alleged or believed* to have stolen it (not even a vague one). So while it is a fact that I allege or believe that someone stole it, there is no fact of the form "I allege (or believe) that X stole it." The one fact that there is, is no doubt an individual fact in so far as it concerns me, but is irreducibly general as far as the thief is concerned. (There may indeed be *no* thief—I am perhaps mistaken about the whole thing—but this is another question; our present point is that there may be no one who is even said or thought to be a thief, though it is said or thought *that there is* a thief.)

Returning now to Queen Anne, what I am suggesting is that the sort of thing that we unquestionably do have with "It is said that" and "It is thought that," we also have with "It will be the case that" and "It was the case that." It *was the case that someone* was called "Anne," reigned over England, etc., even though *there is not now anyone* of whom it was the case that *she* was called "Anne," reigned over England, etc. What we must be careful about here is simply getting our prefixes in the right order. Just as

(1) I think that (for some specific *X* (*X* stole my pencil))

does not imply

(2) For some specific *X* (I think that (*X* stole my pencil)),

so

(3) It was the case that (for some specific *X* (*X* is called "Anne," reigns over England, etc.))

does not imply

(4) For some specific *X* (it was the case that ex is called "Anne," reigns over England, etc.)).

On this view, the fact that Queen Anne has been dead for some years is not, in the strict sense of "about," a fact about Queen Anne; it is not a fact about anyone or anything—it is a *general* fact. Or if it is about anything, what it is about is not Queen Anne—it is about the earth, maybe, which has rolled around the sun so many times since there was a person who was called "Anne," reigned over England, etc. (It would then be a *partly* general fact—individual in so far as it concerns the earth, but irreducibly general as far as the dead queen is concerned. But if there are—as there undoubtedly are— irreducibly partly general facts, could there not be irreducibly wholly general ones?) Note, too, that the fact that this fact is not about Queen Anne cannot itself be a fact about Queen Anne—its statement needs rephrasing in some such way as "There is no person who was called "Anne," etc., and about whom it is a fact that, etc."

On this view, the recession of Queen Anne's death into the further past is quite decidedly not a change in Queen Anne, not because we are using "change" in so tight a sense that it is not a change at all, but because Queen Anne doesn't herself enter into this recession, or indeed, now, into any fact whatever. But the recession *is* still a change or quasi-change in the sense that it fits the formula "It was the case that *p*, but is not now the case that *p*"—this formula continues to express what is common to the flow of a literal river on the one hand (where it was the case that such-and-such drops were at a certain place, and this is the case no longer) and the flow of time on the other.

4
Is Presentness a Property?

William Lane Craig

Abstract

Ontological arguments suggest that presentness is not a property. Existence and presentness seem to be linked, with presentness being identical with the kind of existence that is bound by time. Various arguments strongly suggest that existence itself is not a property, such as the argument which states that a thing is prior to the properties it holds or exhibits. By this reasoning, if existence is a property, a thing would have to exist before it had the property of existence, which is illogical. If existence is not a property, then presentness, which is a type of existence, cannot be a property either.

I

McTaggart's infamous argument for the unreality of time crucially presupposes that the reality of time entails a B-theoretic event ontology to which objective temporal becoming is wedded. That McTaggart presupposes an event ontology is evident in his explanation of what an event is:

> The contents of any position in time form an event. The varied simultaneous contents of a single position are, of course, a plurality of events. But, like any other substance, they form a group, and this group is a compound substance. And a compound substance consisting of simultaneous events may properly be spoken of as itself an event.[1]

McTaggart's view of events as substances is underlined by his analysis of change. In arguing that change is impossible without an A-series, he considers whether events, moments of time, or characteristics of events change. He never even raises the question as to whether *things* change. For if time is real, the fundamental substance in time is events, on McTaggart's view, and so these must change in some way if change is to be real.

Moreover, it seems clear that McTaggart assumed that if time is real, then the substance-events are strung out in a B-series in which each member is equally real or existent. Events which are future or past are just as real as events which are present: "We perceive events in time as being present, and those are the only events which we actually perceive. All other events which, by memory or by inference, we believe to be real, we regard as present, past, or future."[2] The equal ontological status of past, present, and future events is reinforced by McTaggart's view that the B-series is derivable from the combination of the A-series with a C-series, which is a timelessly ordered series based on an atemporal ordering relation like the relation *greater than* which orders the natural numbers. The conjunction of A-determinations with the C-series renders its members earlier/later than one another.

An event changes only in its steady acquisition or loss of new A-determinations. Thus, for McTaggart *becoming present* does not imply *becoming existent*. Rather events exist tenselessly in a B-series and successively acquire the various A-determinations. Temporal becoming consists in the fact that one after another of these events acquires the A-determination of presentness.[3] Events first have the A-determination of distant futurity, then near futurity, then presentness, then near pastness, then distant pastness, all of these except presentness admitting of degrees.

McTaggart notes that pastness, presentness, and futurity may be construed as either relations or qualities of events (and also of moments of time, if these are taken as separate realities). Although McTaggart himself preferred the view that A-determinations are relations rather than qualities, his analysis of temporal becoming treats becoming as a sort of qualitative change in the terms of the B-series and so as a sort of motion of the present. As C. D. Broad commented, "The characteristic of presentness is then supposed to move along this series of event-particles, in the direction from earlier to later, as the light from a policeman's bullseye might move along a row of pilings."[4]

Broad denied that temporal becoming is a qualitative change. There are no events which first have the quality of futurity, which they then exchange successively for new qualities of presentness and pastness:

> When I utter the sentence "It has rained" I do not mean that, in some mysterious non-temporal sense of "is," there is a rainy event, which momentarily possessed the quality of presentness and has now lost it and acquired instead some determinate form of the quality of pastness. What I mean is that raininess has been, and no longer is being, manifested in my neighborhood. When I utter the sentence, "It will rain," I do not mean that in some mysterious, non-temporal sense of "is," there is a rainy event, which now possesses some determinate form of the quality of futurity and will in the course of time lose futurity and acquire instead the quality of presentness. What I mean is that raininess will be, but is not now being manifested in my neighborhood.[5]

Broad makes a very interesting comparison between McTaggart's argument and one version of Anselm's ontological argument. Anselm argued that God is greater if He exists than if He does not and, since He is the greatest conceivable being, He must therefore exist. The fallacy in this reasoning lies in the assumption that merely possible objects exist and so can possess properties and be compared to actual beings in respect to greatness.[6] But if *x* exists in some possible world *W* and does not exist in *W'*, *x* is not greater in *W* than it is in *W'*, nor does it lack some great-making property in *W'* in comparison to *W*. Rather there simply is no *x* in *W'*. Similarly, past and future things or events do not lack some quality of presentness or possess in its stead the qualities of pastness or futurity. Rather they simply do not exist at all. Temporal becoming consists in things coming to be absolutely. Thus, McTaggart's fallacy lay in his pursuing the "hopeless tack" of treating temporal becoming as a type of qualitative change.[7]

That an A-theoretical ontology of time is immune to McTaggart's argument is evident from the way in which McTaggart responds to Broad's critique in the latter's *Scientific Thought*. In this work Broad had defended a sort of hybrid A-B-theory according to which past and present events are real, but future events do not exist. McTaggart admits that Broad's rejection of the reality of the future does enable one to escape McTaggart's argument to a degree, since no event has the A-determination of futurity. But since on Broad's peculiar view the past is as real as the present, McTaggart's argument will still go through because it can be recast in terms of the determinations of pastness and presentness alone. Thus, McTaggart's handling of Broad's objection confirms that it is the attempted combination of a B-theoretical ontology with A-theoretical becoming that generates the contradiction.

By contrast, if we adopt a thorough-going A-theoretical ontology and deny the reality of both past and future events, then McTaggart's Paradox disappears completely. For then only present events/times exist, and all that remains of McTaggart's A-determinations is that an event has presentness at the present time. Since there is only one present, there is no sense in asking when the present moment is present. McTaggart's Paradox is thus ineffectual against pure A-theorists like Broad, Prior, Christensen, Levinson, and so forth; it has purchase only against hybrid A-B-theorists like McCall, Schlesinger, and Smith, who try to couple a B-theoretical ontology with objective, non-relational A-determinations.[8]

On a pure A-theoretic ontology, then, the only temporal items which exist exist presently. In other words, the pure A-theory entails a metaphysic of presentism. On a presentist ontology, past and future events/things/times are not real or existent and, hence, do not exemplify properties like pastness or futurity. Hence, there can be no question of an entity's trading in futurity for presentness or cashing in presentness for pastness. Rather entities come to be and pass away absolutely, so that the only temporal entities there are are the present ones.

It follows that on the A-theory of time, we should not understand grammatical ascriptions of pastness and futurity in sentences like "Past injustices must be forgotten" or "The future space station will further space research" as postulating the literal inherence of properties of pastness and futurity in the events/things/times so described.[9] On a presentist ontology, such entities do not exist and so possess no properties. Rather they did or will possess properties, and it is this tense distinction that is expressed by referring to such entities as past or future. The A-theorist thus agrees with the B-theorist that pastness and futurity are relational predicates, but he will differ in anchoring these relations in what is non-relationally present. The construal of pastness and futurity as relational predicates should not be taken to mean that these are relational properties inhering in events. Rather such ascriptions should be parsed as asserting that the entity in question did or will exist; for example, "One must forget the injustices which occurred" and "The space station will exist and further space research." Furthermore, odd locutions like "*E* is present in the past" or "*E* is present in the future" should be construed to mean that the statement "*E* is present" either was true or will be true.[10] Hence, only ascriptions of presentness may be taken literally as the possession of an A-determination by some temporal entity; in the case of non-present A-determinations, we are to understand ascriptions of them in terms of what was or will be the case.

II

Now, if pastness and futurity are not to be construed as properties, what about the premier A-determination of presentness itself? Is it a property? On this question there exists an intramural, if somewhat muted, debate among A-theorists. All agree that tense is an objective feature of reality; the dispute concerns the place of tense in one's ontology—what exactly is presentness? A good many A-theorists have followed A. N. Prior in denying that presentness should be considered a property possessed by events/things/times. Responding to Broad's point that an event does not possess properties except at the time at which it is present, Prior goes further to deny both that one's ontology ought to include events and that tense determinations are properties of either events or things. He asserts,

> "Is present," "is past," etc., are only quasi-predicates, and events only quasi-subjects. "*X*'s starting to be *Y* is past" just means "It has been that *X* is starting to be *Y*," and the subject here is not "*X*'s starting to be *Y*" but *X*. And in "It will always be that it has been that *X* is starting to be *Y*," the subject is still only *X*; there is just no need at all to think of another subject, *X*'s starting to be *Y* as momentarily doing something called "being present" and then doing something else called "being past" for much longer; and no need to argue as to whether *X*'s starting to be *Y* "is" only at the moment when it does the thing called being present, or also throughout the longer period when it does the other thing. It is *X* which comes to have started to be *Y*, and it is of *X* that it comes to be always the case that it once started to be *Y*; the other entities are superfluous, and we see how to do without them, how to stop treating them as subjects, when we see how to stop treating their temporal qualifications ("past," etc.) as predicates, by rephrasings which replace them with propositional prefixes ("It has been that," etc.) analogous to negation.[11]

Prior wants to excise events from his ontology; and the way to do this, he believes, is to deny that tense determinations are properties; rather they are propositional operators. We should no more reify tense operators than we do the operator of negation.

We may leave to the side Prior's rejection of an event ontology, since an A-theorist could agree with Prior on this score and still consider presentness to be a property of things or times. Rather Prior's relevant claim is that tense determinations are propositional operators rather than properties. On Prior's view, as Christensen explains, tenses are adverbial in nature and thus radically different from predicates: tenses do not tell *what* the properties of an individual are, but rather *when* an individual has its properties.[12] Thus, presentness is not a property of any event/thing/time.

Smith, however, has bitingly criticized what he calls the "no-property" view as obscure and unintelligible. Smith's fundamental complaint is that the construal of tenses as propositional operators is a semantical thesis only, which leaves unaddressed the metaphysical question "in what ontological category should we place *what is told*" by tenses?[13] Since A-theorists agree that tenses are not just egocentric features of human thought and language, there must be some feature of reality corresponding to the tenses we use. What in the propositional content of a present-tensed sentence corresponds to the tense in that sentence? Since any intelligible theory of tense attempts to classify the semantic relata of tenses in terms of some ontological category, Prior's theory, in failing to do so, is unintelligible.

Smith does not respond to Prior's comparison of tense operators to the operator of negation. It certainly seems a defensible theory to hold that there is no ontological correlate to the sentential operator of negation. One's ontology seems to become unnecessarily bloated if we countenance negative properties corresponding to every description which is not true of a thing. When we assert that something is not orange, we are not normally ascribing to it the property of being not-orange, but are just denying that it has the property of being orange. Similarly, perhaps presentness is not a property which events/things/times possess; perhaps our use of the present tense just indicates the reality of the events/things/time described. Indeed, in a later piece Prior claims that the present and the real are one and the same concept: "the present simply is the real considered in relation to two particular species of unreality, namely the past and the future."[14] Rejecting modal realism, Prior states that other worlds are made by prefixing "X is the case" with a phrase like "It could be that." What is real involves no such prefix. Similarly, "the present is *real* and the past and future are not. And I want to suggest that the reality of the present consists in what the reality of anything else consists in, namely the absence of a qualifying prefix."[15] This last remark illustrates the conflation of semantics and ontology that so exasperates Smith, for the reality of the lamp before me on my desk does not consist in the absence of a qualifying prefix, since prefixes do not operate on lamps. And while the sentence "The lamp is on my desk" lacks a tense logical operator, it is hard to see how the reality of my lamp or its location consists in this sentence's (or the proposition expressed by it) lacking an operator. Moreover, the analogy between tense operators and modal operators is much closer than that of the former with the negation operator, and it is undeniable that modal sentences or propositions raise a host of questions for ontology. The demonstrated inadequacy of Prior's views of modal assertions[16] makes one more than suspicious that the ontological questions concerning tense ascriptions cannot be so neatly circumvented as Prior hoped.

Still, the connection between presentness and reality which Prior affirmed is inescapable on a presentist ontology and suggests that we could perhaps make progress on the problem raised by Smith by exploring the question of existence. If to be present is simply to exist, as Prior suggests, then our ontological classification of existence will tell us whether presentness is a property, as Smith claims. So what is it for a thing to exist; more specifically, is existence a property?

III

Readers who are philosophers of religion will immediately realize that this question has played an important part in the centuries-old debate over the ontological argument, which in some versions treats existence as a great-making property which God, as the greatest conceivable being, must possess.[17] Kant's refutation centered on the denial that existence is a real predicate, or, more perspicuously, a property. In a familiar passage[18] Kant offers both a positive characterization of existence in terms other than its being a property, as well as a negative critique of the position that existence is a property.

In a detailed and careful analysis of the negative critique, Plantinga has argued persuasively that Kant fails to specify a sense of "predicate" such that, in that sense, it is clear both that existence is not a predicate and that Anselm's argument requires it to be one.[19] In point of fact, Plantinga is able to specify a sense in which existence is not a real predicate or property, though that sense turns out to be irrelevant to the cogency of the ontological argument (which is a matter of indifference for our present discussion). Plantinga interprets Kant's dictum that the content of an object and the content of the concept of that object must be the same along the following lines: call the concept whose content includes all (and only) the properties an object has the whole concept of that object. Call the concept whose content contains just those properties which are members of the content of the whole concept of an object *O* and neither entail nor are entailed by a property P the whole concept of *O* diminished with respect to *P*.[20] (Roughly, a whole concept diminished with respect to *P* is that concept minus *P*.) Now let C_1 be the whole concept of some object *O*, and let C_2 be the whole concept of *O* diminished with respect to some property *P*, say, pinkness, and let C_3 be the whole concept of *O* diminished with respect to existence (on the assumption that existence is a property). Kant's point is that existence is then unlike other properties like pinkness in that there are no possible worlds in which C_3, but not C_1, is exemplified, whereas there are worlds in which C_2, but not C_1, is exemplified. Since C_1 and C_3 are mutually entailing, they are equivalent concepts; in adding *existence* to C_3 one does not come up with a new concept. So if a predicate P represents a property only if it is not the case that a whole concept diminished with respect to P is equivalent to the corresponding whole concept, then existence, concludes Plantinga, is not a property. Thus, the Kantian critique, in Plantinga's assessment, makes a valuable, positive contribution to our understanding of existence and of certain fallacious versions of the ontological argument.

Refusing to allow property status to existence also affords us the benefit of an explanation of why existence, when construed as a property, is anomalous in comparison with other properties. These anomalies were noted in particular by C. D. Broad and G. E. Moore. Broad observed that negative existential propositions like *Dragons do not exist* cannot be what he called "characterizing propositions," that is, propositions which ascribe some property to an object.[21] For, in the example, there are no such things as dragons to possess the property of nonexistence. Moreover, if one tries to analyze negative existential propositions in familiar ways proposed for analyzing negative characterizing propositions, self-contradiction results. For example, on the conditional interpretation, the proposition at issue might be expressed by the sentence "If there were any dragons, none of them would exist" and on the instantial interpretation by the sentence "There are dragons and none of them exist," both of which are self-contradictory. Broad proceeded to argue that affirmative existential propositions like *Cats exist* are also not characterizing propositions. For on the conditional interpretation this proposition would be expressed by the sentence "If there were any cats, none of them would fail to exist" and on the instantial interpretation by the sentence "There are cats and none of them fail to exist," both of which are "mere platitudes,"[22] which *Cats exist* is not.

In his *God and Other Minds*, Plantinga protested Broad's assertion that there are no subject-predicate propositions about dragons as "plainly outrageous," furnishing the counter-example *Dragons are beasts of fable.*[23] But this seems to confuse sentences with propositions; Broad does not deny that there are sentences whose subject is dragons, but he denies that there are propositions ascribing properties to dragons. In his *Nature of Necessity*, Plantinga in effect comes to endorse Broad's position when he maintains that any world in which a predicative singular proposition is true is one in which the subject of that proposition has existence.[24] Negative existential propositions are not to be construed as ascribing the property of nonexistence to some (merely possible) object; rather they are to be interpreted as asserting that it is false that some object has the property of existing. Plantinga rejects "descriptivism" with regard to creatures of fiction; the descriptivist (falsely) assumes that in fiction we are talking about an object, say, Hamlet, and predicating a property of it. But Plantinga maintains that creatures of fiction are not subjects having properties in the actual world, nor do fictional names denote anything, being merely stylistic variants of variables. Fictional stories "are about nothing at all and the names they contain denote neither actual nor possible subjects."[25] Thus, Plantinga comes to affirm Broad's position, which he had earlier rejected, and his arguments that there are no merely possible objects go to confirm Broad's view. Still, this shows only that Broad was correct that negative existential propositions do not ascribe a property of nonexistence. One could agree with this and yet construe affirmative existential propositions (about real objects) as ascribing the property of existence.

In response to Broad's argument that affirmative existential propositions do not ascribe the property of existence, Plantinga points out that not all characterizing propositions, whether general or singular, are susceptible to either the conditional or instantial interpretation. For example, neither the general proposition *Buffalo once abounded on the Great Plains* nor the singular proposition *Cerberus is my favorite beast of fable* can be analyzed along these lines.[26] Broad, however, could respond to Plantinga that none of Plantinga's examples concern propositions which ascribe properties to individual objects either.[27] General propositions which fail to ascribe properties to every individual in a class or singular statements about creatures of fiction do not ascribe properties to objects and so provide no basis for thinking that affirmative existential propositions, in being insusceptible to the conditional and instantial analyses, nevertheless do ascribe the property of existence to objects.

Moore pointed out further peculiarities of affirmative existential propositions if they are construed as ascribing a property.[28] Unlike the statement "Tame tigers growl," the statement "Tame tigers exist" cannot be meaningfully interpreted as "All tame tigers exist" or "Most tame tigers exist;" it can only mean "Some tame tigers exist." Moreover, with other property ascriptions, as in "Some tame tigers growl," we may insert "do not" before the predicate and have a meaningful statement, but this cannot be done with the predicate "exist." If we do so with the statement "Some tame tigers exist," we have "Some tame tigers do not exist"—which appears to mean that there are some tame tigers which do not exist. Moreover, when one says, "Some tame tigers growl," this asserts that the propositional function *x is a tame tiger and growls* has a value which is a true proposition. By contrast, when one says, "Some tame tigers exist," this asserts not that the propositional function *x is a tame tiger and exists* has a value which is a true proposition, but simply that x is a tame tiger has a value which is a true proposition. These anomalies go to suggest that existence should not be taken to be a property.

Alston complains that Moore's anomalies do not show that existence is not a property, but only that it is a special property.[29] But this response is somewhat too facile. If existence is a property, then *why* is it so special? The only reason I can think of is that it would be an essential property of everything that exists. But that provides no explanation for the quantification anomalies noted by Moore. Moreover, other trivially essential properties, like *being colored if orange*, when ascribed to tame tigers, do not yield the same anomalies. Most importantly, the anomalies noted by Broad and Moore become perspicuous on the Kantian analysis that existence is not a property, whereas they remain simply mysterious on the "special property" view, which fact constitutes confirmatory evidence for the Kantian position.

A third consideration in favor of the view that existence is not a property is that things are explanatorily or logically prior to the properties that inhere in them or that are possessed by them, in which case a thing would have to exist logically prior to its having the property of existence, which is incoherent. Alston explains,

> Before we can attach any predicate to anything ("round," "heavy," "in my pocket," "belongs to Jones," "difficult to understand"), we must presuppose that it exists. If we were not making that assumption we could not even raise the question whether a given predicate attaches to it. . . .
>
> On this basis it is easy to show that "exists" cannot be a predicate. If the existence of the subject must be presupposed before we can set about attaching (withholding, wondering whether to attach) any predicate to (from) it, we will also be too late either to apply or to withhold a predicate of existence. . . . And the denial of such a predicate would contradict the essential conditions of any predication.[30]

This is a very powerful argument rooted in the nature of what it is to be a property of something. Alston considers the argument as it stands to be indecisive because things may have different modes of existence, such as existence in fiction or in the imagination, which do not presuppose the real existence of the subject logically prior to the inherence of properties in it. He then adjusts the argument by contending that real existence, at least, cannot be attributed to something possessing any non-real mode of existence. He writes, "no statement which attributes something to a non-real being can have the logical status (implications and so forth) of a statement of real existence."[31] This analysis would still preclude real existence's being a property, since real existence could not inhere in anything but a real existent, in which case the original problem resurfaces. But it is clear that we need not even concede so much as Alston does. Plantinga's analysis of merely possible objects and of creatures of fiction in particular shows convincingly that to speak, as Alston does, of "nonreal modes of existence"[32] is simply self-contradictory.[33] The only things that have existence are, obviously, things that exist in the actual world. But the very nature of what it is to be a property precludes existence's being a property of such things.

We have good reasons, then, for thinking that existence is not a property. But then what is it? Here we may recur to Kant's positive characterization of existence. According to Kant, existence is the positing of the subject with all its predicates as an object. Similarly, nonexistence is the rejection or absence of a subject's being posited with all its predicates.[34] How is Kant's brief characterization of existence and nonexistence to be understood? Since Kant repeatedly uses expressions like "we posit the thing as given (as existing)," "we are . . . required to posit the existence of its object," "if we reject subject and predicate alike," "we reject the thing itself with all its predicates," and so forth,[35] one might be led to the conclusion that existence is a positing in reality which we carry out. But Kant is here speaking from the first-person standpoint because he is exploring whether we can infer the existence of a being from the concept we form of it. It would be obviously absurd, however, to say that the absolutely necessary being, if it exists, depends for its existence upon our positing it in reality; nor can it coherently be said that an absolutely necessary being fails to exist for the anthropocentric reason that we reject it with its predicates. Rather Kant's positive characterization of existence may be understood as the equivalence of a thing's existing with the instantiating of the whole concept of a thing. On this view, for a thing to exist is for its whole concept (or, alternatively, its individual essence) to be instantiated. In either case, to exist is to exemplify properties.

Among philosophers of religion Thomas Aquinas stands out as the foremost advocate of the conception of being, not as a property, but as an act whereby a thing's essence is instantiated. Thomas's Islamic predecessors made a real distinction between a thing's essence and its existence, but they seemed to construe existence as an accident, that is, a nonessential property, which accrues to an essence, resulting in a substance's reality.[36] Ibn Sina, however, seems to have rejected the conception of existence as an accidental property, holding rather that existence is the instantiation of an essence in reality; it is, as he put it, a "happening," the actualizing of the potentiality of essence, resulting in a concrete thing.[37] Similarly, according to Thomas, "*Being* . . . is the name of an act, for a thing is not said to be because it is in potency but because it is in act."[38] He uses the verb *esse* ("to be") to indicate that being is not a static property or concept, but a dynamic act: "*To be* . . . mean[s] the act of essence.[39] Accordingly, existence is neither an essential nor an accidental property.[40] It is the actualizing or instantiating of a thing's essence.[41]

Broad and Moore defended somewhat similar views of existence. Although they did not speak in terms of the essences of things, like Ibn Sina or Aquinas, still they thought of existence, not as itself a property, but as the exemplification of properties. For example, according to Broad, existential propositions are not about things, but about properties, or characteristics:

> What they assert is that these characteristics do apply to something or that they do not apply to anything, as the case may be. "Cats exist" is equivalent to "The defining characteristics of the word 'cat' apply to something." Again "Dragons do not exist" is equivalent to "The defining characteristics of the word 'dragon' do not apply to anything." Suppose, e.g., that a "dragon" is defined as a reptile which flies and breathes fire. Then the statement that dragons do not exist is equivalent to the statement that nothing combines the three properties of being a reptile, of flying, and of breathing fire.[42]

Here Broad seems to treat existence as the exemplifying of the defining properties of a thing.

Moore also thought of existence in terms of the exemplification of properties:

> What "Lions are real" means is that some particular property or other—I will say, for the sake of brevity, the property of being a lion . . . does in fact *belong* to something—that there are things which have it, or, to put it another way, that the conception of being a lion is a conception which does apply to some things—that there are things which *fall under* it. And similarly what "Unicorns are *unreal*" means is that the property of being a unicorn belongs to *nothing*. Now, if this is so, then it seems to me, in a very important sense, "real" and "unreal" do not in this usage stand for any conceptions at all. . . .
>
> To say that it did would be to imply that it stood for some property of which we are asserting that everything which has the property of "being a lion" *also* has this other property. But we are not, in fact, asserting any such thing. We are not assert-

> ing of any property called "reality" that it belongs to lions, as in the assertion "lions are mammalian" we are asserting of the property of "being a mammal" that it belongs to lions. The two propositions "Lions are real" and "Lions are mammalian," though grammatically similar, are in reality of wholly different forms; and one difference between them may be expressed by saying that whereas "mammalian" does stand for a property or conception, the very point of this usage of "real" is that it does not.[43]

Moore's view is remarkably reminiscent of Thomas, in that he affirms that there is a property such as *being a lion* and denies that we even have, properly speaking, any conception of existence, since it is not a property.[44]

Such an understanding of existence is utterly unlike Russell's bizarre view that existence is a property of propositional functions rather than individuals,[45] an interpretation entertained by Smith in his effort to understand the "no property" view of presentness.[46] Smith argues that if presentness is a so-called second-order property of properties or propositional truth values, then it is also a first-order property of things and events. But not only is his argument loaded with gratuitous metaphysical freight,[47] it erroneously assumes that presentness is any sort of property at all. If we assimilate presentness to existence and construe a thing's existence as the act of exemplification of its properties, then presentness is neither a first nor second order property. If, like Smith, we wish to admit events into our ontology, then we should hold that what is existence for things is occurrence for events. Just as a thing exists by either its individual essence or whole concept's being instantiated, so an event occurs by the state of affairs describing it being instantiated. Thus, existence and occurrence are not properly construed as properties; as Thomas saw, they are acts.

IV

As already suggested, there is on a presentist ontology an extremely tight link between existence and presentness. If a thing's being present is the same as that thing's existing, then presentness is no more a property than existence. Indeed, some of the very arguments against existence's being a property recur in a parallel form with respect to presentness. For example, in order to possess the property of presentness, a thing or event must already be present, which is self-contradictory. On a presentist ontology past or future events cannot possess the property of presentness; the property of presentness can only inhere in events/things/times that are present. One might attempt to elude this problem along Smith's line by maintaining that presentness also inheres in the inherence of presentness in events, and also in the inherence of presentness's inherence in presentness's inherence in events, *ad infinitum.*[48] But it is clear that, in the case at hand, this is not a benign, but a vicious, infinite regress, since at every level, like the first, the contradiction exists that logically prior to the inherence of presentness in an event, the event must already be present in order to possess any properties.

Sosa raises a related problem when he asks about the nature of *having* the property of presentness.[49] Suppose we interpret the sentence "The flash is present" to ascribe to the flash the property of presentness. This property cannot be possessed tenselessly; otherwise either the present would be fixed at the time of the flash or the flash at one time would be present at all times. But if we say that the having of presentness is itself tensed, then the property of presentness seems superfluous. There would be no content in saying, "The flash now-has the property of being present" over and above saying "The flash now-is." Thus, the attempt to construe the propositional content of a present-tense sentence as a property of presentness is, in Sosa's words, self-defeating.[50]

A further difficulty is that a property like presentness would be a timeless universal, since it is possessed by many entities. But then it seems unintelligible how this timeless abstract object can be related to concrete particulars so as to locate them temporally. The same problem arises for the construal of presentness as any other kind of abstract object, say, a relation between a thing and its properties.[51] How can a timeless abstract object like a relation temporally connect other abstract objects like properties to a concrete object? Perhaps one could avoid the difficulty by holding abstract objects to be temporal, not timeless, or by distinguishing between abstract properties and temporal property instances; but it is not evident that these complications afford a more plausible alternative to presentness's not being a property. So, as I say, the grounds for regarding existence as not being a property echo on a presentist ontology for presentness as well. *To be present* seems to be identical with *to exist*.

Nevertheless, I think we would be hasty to embrace the strict identity of presentness and existence. For in the absence of some argument, we have no reason to regard timeless existence as impossible. In particular, we have no good reason to think that God cannot be timeless sans creation, and, of course, many philosophers think that a host of abstract objects exist timelessly. We cannot therefore rule timeless existence out of court by a definition. Rather presentness is properly construed as a mode of existence.[52] There are two modes of existing, timeless existence and temporal existence, and presentness is identical with this second mode. There are two ways of exemplifying properties: timelessly or presently. On a presentist ontology, to exist temporally is to be present. Since presentness is identical with temporal existence (or occurrence) and existence is not a property, neither is presentness a property. Presentness is the act of temporal being.[53]

Notes

1. John McTaggart Ellis McTaggart, *The Nature of Existence*, 2 vols., ed. C. D. Broad (Cambridge: Cambridge University Press, 1927; rep. ed.: 1968), 2: 10. See chapter 2 in this volume.

2. Ibid., 2: 11.

3. Ibid., 2: 10-11.

4. C. D. Broad, *An Examination of McTaggart's Philosophy*, 2 vols. (Cambridge: Cambridge University Press, 1938; rep. ed.: New York: Octagon Books, 1976), 2: 227.

5. Ibid., 2: 316.

6. C. D. Broad, *Religion, Philosophy and Psychical Research* (London: Routledge & Kegan Paul, 1953), p. 182; cf. Alvin Plantinga, *The Nature of Necessity*, Clarendon Library of Logic and Philosophy (Oxford: Clarendon Press, 1974), pp. 209-10.

7. Broad, *Examination*, 2: 302.

8. They thereby land themselves in a peculiar temporal version of the Problem of Temporary Intrinsics, and it is dubious whether they can escape. See my "McTaggart's Paradox and the Problem of Temporary Intrinsics," *Analysis* 59 (1999): 314-18.

9. *Contra* the presupposition of Eddy M. Zemach, "Time and Self," *Analysis* 39 (1979): 143-47; David Zeilicovici, "A (Dis)solution of McTaggart's Paradox," *Ratio* 28 (1986): 187; Quentin Smith, "The Infinite Regress of Temporal Attributions," *Southern Journal of Philosophy* 24 (1986): 383-96.

10. See E. J. Lowe, "The Indexical Fallacy in McTaggart's Proof of the Unreality of Time," *Mind* 96 (1987): 64-66, who seems, however, unduly diffident about tensed ascriptions of presentness to events, e.g., "E will be present," which is unobjectionable so long as "present" is not an indexical synonymous with "now," as Lowe mistakenly believes.

11. A. N. Prior, "Precursors of Tense Logic," in *Past, Present and Future* (Oxford: Clarendon Press, 1967), p. 18.

12. Ferrel Christensen, "McTaggart's Paradox and the Nature of Time," *Philosophical Quarterly* 24 (1974): 297.

13. Quentin Smith, *Language and Time* (New York: Oxford University Press, 1993), p. 166.

14. A. N. Prior, "The Notion of the Present," *Studium Generale* 23 (1970): 245. Similarly, Christensen has recently asserted, "To be present is simply to be, to exist, and to be present at a given time is just to exist at that time—no less and no more" (F. M. Christensen, *Space-like Time* [Toronto: University of Toronto Press, 1993], p. 168; cf. p. 226).

15. Prior, "Notion of the Present," pp. 246-47.

16. See the incisive critique by Alvin Plantinga, "On Existentialism," *Philosophical Studies* 44 (1983): 1-20.

17. For helpful anthologies and bibliographies, see Alvin Plantinga, ed., *The Ontological Argument*, with an introduction by Richard Taylor (Garden City, N.Y: Doubleday, 1965), and John H. Hick and Arthur C. McGill, eds., *The Many-faced Argument* (New York: Macmillan Co., 1967).

18. Limitations of space preclude its reproduction here. See Immanuel Kant, *Kant's "Critique of Pure Reason,"* trans. Norman Kemp Smith (London: Macmillan, 1929), pp. 504-5 (A598/B626-A599/B627); cf. p. 502 (A595/B623).

19. Alvin Plantinga, *God and Other Minds* (Ithaca, N.Y.: Cornell University Press, 1967), pp. 29-38.

20. Ibid., p. 35. Plantinga adds a further qualification which I omit for simplicity's sake.

21. C.D. Broad, "Arguments for the Existence of God," [1925], in *Religion, Philosophy and Psychical Research*, International Library of Psychology, Philosophy, and Scientific Method (New York: Humanities Press, 1969), pp. 182-83.

22. Plantinga objects to Broad's characterization of "There are cats and none of them fails to exist" as a platitude, since this sentence entails that cats do exist and so is not a mere tautology (Plantinga, *God and Other Minds*, p. 41). But perhaps we need not construe "mere platitudes" as tautologies; we could take Broad to be saying that the complex sentences analyzing existential propositions become redundant (of course, if there are cats, they exist!), while the sentences analyzing genuinely characterizing propositions do not.

23. Plantinga, *God and Other Minds*, p. 40.

24. Plantinga, *Nature of Necessity*, p. 150.

25. Ibid., p. 163; cf. 158-60.

26. Plantinga, *God and Other Minds*, p. 41.

27. See Jerome Schaffer, "Existence, Predication, and the Ontological Argument," in *Many-faced Argument*, p. 234. Schaffer's difficulties with Kant's argument that existence is not a property are implicitly met by Plantinga's analysis exposited above.

28. G. E. Moore, "Is Existence a Predicate?" [1936], in *Ontological Argument*, pp. 73-80.

29. William P. Alston, "The Ontological Argument Revisited," in *Ontological Argument*, p. 88.

30. Ibid., p. 89.

31. Ibid., p. 106.

32. Ibid., p. 103.

33. Plantinga, *Nature of Necessity*, chapters 7, 8. Plantinga notes the *prima facie* perplexity of the claim that there are objects that do not exist. Any non-actual possible world does not obtain, but it still exists as an abstract object, a maximal state of affairs. To say that an object which does not in fact exist exists in a possible world W is not to say that there is a nonexistent object, but that that object would exist if W were actual. The classical argument that there are nonexistent objects is based on the intuition that we can only talk about things that in some sense have existence. Because we can talk about them, there must be merely possible objects which have the property of nonexistence. Plantinga eludes the classical argument by distinguishing between predicative and impredicative singular propositions, affirming that the subject of the former, if true, has existence (*e.g., Socrates was snubnosed*), whereas the subject of the latter, even if true, need not exist (*e.g., Socrates was not snubnosed*, which is to be understood as *It is false that Socrates was snubnosed*). Thus, *Socrates has nonexistence* should be understood as *It is false that Socrates has existence*. Thus, Plantinga affirms that any world in which a singular predicative proposition is true is one in which its subject exists. But in a world in which an object does not exist, it has no properties at all, not even that of nonexistence. As for creatures of fiction, Plantinga rejects, as we have seen, the descriptivist position which holds that in speaking of fictions we are predicating properties of an object.

34. Kant, *Critique*, p. 502 (A594-595/B622-623).

35. Ibid.

36. For discussion, see my *The Cosmological Argument from Plato to Leibniz*, Library of Philosophy and Religion (London: Macmillan, 1980), pp. 120-21,195-96.

37. See helpful discussions by Fazlur Rahman, "Essence and Existence in Avicenna," *Mediaeval and Renaissance Studies* 4 (1958): 1-17; Parviz Morewedge, "Philosophical Analysis and Ibn Sina's Essence-existence Distinction," *Journal of the American Oriental Society* 92 (1972): 429.

38. Thomas Aquinas, *Summa contra gentiles* 1. 22. 7.

39. Thomas Aquinas, *Summa theologiae* la. 3. 5. ad 2. Of course, Thomas uses "essence" in the Aristotelian sense, not in the sense of an individual essence; but the salient point is that existence is an instantial act. See also his brief tractate *De ente et essentia*.

40. Thomas Aquinas, *Commentary on the "Metaphysics" of Aristotle* 1. 2. 556, 558; see also Frederick C. Copleston, *Aquinas* (Harmondsworth, England: Penguin Books, 1955), p. 107.

41. Because existence is a happening or act, it would be misconceived to say that an essence's being instantiated exists or a thing's exemplifying its properties exists, for this would be to treat *esse* as a substance which itself exists. To say that existence or being is an act is also not to say that it is an act performed by a thing, for then the thing must already exist explanatorily prior to its act. Rather it is the actualizing of essence, which constitutes a thing. Ultimately both Ibn Sina and Aquinas see the act of being as caused by God; it is He who instantiates essences and so causes things to exist, but who Himself exists necessarily.

42. Broad, "Existence of God," p. 183.

43. G. E. Moore, "The Conception of Reality," *Proceedings of the Aristotelian Society* 18 (1917-18): 114-16.

44. Cf. Thomas's denial that we have any conception of God's essence, since His essence is existence (*esse*) and so is incapable of being conceived (Thomas Aquinas, *Summa theologiae* la. 3. 4. ad 2).

45. Russell's confused statement of this view may be found in Bertrand Russell, "The Philosophy of Logical Atomism," [1918], in *Logic and Knowledge*, ed. Robert Charles Marsh (London: George Allen & Unwin, 1956), pp. 232-41. Russell says that if you take any propositional function like *x is a man* and assert that it is sometimes true, that gives you the fundamental meaning of "existence." When one says, "Men exist," what one means is that there is at least one value for x such that x is a man is true. So existence is a predicate or property of a propositional function. "It is of propositional functions that you can assert or deny existence" (ibid., p. 232). On Russell's view, then, individual objects do not exist. This seems crazy until we appreciate the fact that Russell has completely redefined "existence." When he says that it is a property of propositional functions, he is not using the word in the sense expressed when we say that there are propositional functions. Rather he redefines the word to indicate the property *being sometimes true*. We might be inclined to say that he is at liberty to define words as he pleases, but insist that there is some correlative concept to existence that would apply to individual objects in the world. But, incredibly, Russell denies this: "It is a sheer mistake to say that there is anything analogous to existence that you can say about them" (ibid., p. 241). Thus, Russell's view really is crazy. In any case, it is problematic because propositional functions are neither true nor false; propositions are. So existence must be the property *being of the form* x is a ——— (where "———" is filled by a common noun denoting a thing) and *having a value for x such that the resulting proposition is sometimes true*. Unfortunately, Russell identifies the notion of *sometimes* with the notion of the *possible*, so that even if the proposition is only possibly true, the sentence expressing it, e.g., "Unicorns exist" is true.

If we interpret Russell to mean that to say something exists is to say that a propositional function has the property *being satisfied*, then we have conflated semantics and ontology, just as Smith complained concerning Prior. The above only gives truth conditions for saying that something exists, but does not tell us what existence is.

46. Smith, *Language and Time*, pp. 168-69. This is unfortunate, since Russell's view is not representative of philosophers who hold that existence is not a property. See, e.g., Moore's rejection of Russell's characterization of existence as a property of propositional functions (Moore, "Is Existence a Predicate?" in *The Ontological Argument*, p. 81).

47. For example, Smith takes it for granted that "There is an explosion" ascribes the property *being exemplified* to the property of *exploding*. But there is no reason to think that the act of exemplification is a property (not to mention that no reason has been given to think that *exploding* is ascribed by this sentence either). Smith also assumes that the state of affairs described by the proposition *John is running* has the concrete individual John as a constituent. But since states of affairs are typically construed as abstract entities existing necessarily, John, as a concrete contingent, cannot be part of one.

48. Smith, *Language and Time*, p. 172.

49. Ernest Sosa, "The Status of Becoming: What Is Happening Now?" *The Journal of Philosophy* 76 (1979): 28.

50. Sosa's point also precludes construing existence or presentness as a second order property, viz., the property *being exemplified* or *being exemplified at t* ascribed to some essence, for such an analysis yields at best timeless or tenseless existence, not tensed existence.

51. Cf. Reinhardt Grossmann, *The Existence of the World* (London: Routledge, 1992), p. 20; cf. p. 102.

52. Interestingly, Christensen interprets Prior's view as holding that the notions of having occurred, occurring presently, and being going to occur are "modes of being, not relations (or properties)" and endorses this view himself (Christensen, *Space-like Time*, p. 122; cf. pp. 61, 123, 174). This interpretation is not quite right, however, since Prior held the past and the future to be unreal, so that these tense determinations cannot, like presentness, be construed as modes of being. Notice, too, that if we take necessity and contingency to be also modes of existence (ibid., p. 61), then things have complex, or mixed, modes of being: a thing may exist (i) atemporally and necessarily, (ii) atemporally and contingently, (iii) temporally and necessarily, or (iv) temporally and contingently. Whether all these mixed modes are broadly logically possible is a matter of debate.

53. My thanks to an anonymous referee for his comments on the first draft.

References

Alston, W. P. "The Ontological Argument Revisited." Pp. 86-110 in *The Ontological Argument: From St. Anselm to Contemporary Philosophers*, edited by A. Plantinga. Garden City, N.Y.: Doubleday, 1965.

Aquinas, T. *Commentary on the "Metaphysics" of Aristotle* 1. 2.

Broad, C. D. *An Examination of McTaggart's Philosophy*, 2 vols. Cambridge: Cambridge University Press, 1938; rep. ed.: New York: Octagon Books, 1976.

———. *Religion, Philosophy and Psychical Research*. London: Routledge & Kegan Paul, 1953.

———. "Arguments for the Existence of God." Pp. 175-201 in *Religion, Philosophy and Psychical Research*, International Library of Psychology, Philosophy, and Scientific Method. New York: Humanities Press, 1969.

Christensen, F. "McTaggart's Paradox and the Nature of Time," *Philosophical Quarterly* 24 (1974): 289-99.

———. *Space-like Time*. Toronto: University of Toronto Press, 1993.

Copleston, F. C. *Aquinas*. Harmondsworth, England: Penguin Books, 1955.

Craig, W. L. *The Cosmological Argument from Plato to Leibniz*, Library of Philosophy and Religion. London: Macmillan, 1980.

———. "McTaggart's Paradox and the Problem of Temporary," Intrinsics," *Analysis* 59 (1999): 314-18.

Grossmann, R. *The Existence of the World: An Introduction to Ontology*. London: Routledge, 1992.

Hick, J. H., and A. C. McGill, eds. *The Many-faced Argument: Studies on the Ontological Argument for the Existence of God*. New York: Macmillan Co., 1967.

Kant, I. *Kant's "Critique of Pure Reason,"* translated by N. K. Smith. London: Macmillan, 1929.

Lowe, E. J. "The Indexical Fallacy in McTaggart's Proof of the Unreality of Time," *Mind* 96 (1987): 62-70.

McTaggart, J. M. E. *The Nature of Existence*, 2 vols., edited by C. D. Broad. Cambridge: Cambridge University Press, 1927; rep. ed.: 1968. See chapter 2 in this volume.

Moore, G. E. "The Conception of Reality," *Proceedings of the Aristotelian Society* 18 (1917-18): 101-20.

———. "Is Existence a Predicate?" Pp. 71-85 in *The Ontological Argument: From St. Anselm to Contemporary Philosophers*, edited by A. Plantinga. Garden City, N. Y.: Doubleday, 1965.

Morewedge, P. "Philosophical Analysis and Ibn Sina's Essence-existence Distinction," *Journal of the American Oriental Society* 92 (1972): 425-35.

Plantinga, A., ed., *The Ontological Argument*, with an introduction by Richard Taylor. Garden City, N.Y: Doubleday, 1965.

———. *God and Other Minds*. Ithaca, N.Y.: Cornell University Press, 1967.

———. *The Nature of Necessity*. Oxford: Clarendon Press, 1974.

———. "On Existentialism," *Philosophical Studies* 44 (1983): 1-20.

Prior, A. N. "Precursors of Tense Logic." Pp. 1-19 in *Past, Present and Future*, by A. N. Prior. Oxford: Clarendon Press, 1967.

———. "The Notion of the Present," *Studium Generale* 23 (1970): 245-48.

Rahman, F. "Essence and Existence in Avicenna," *Mediaeval and Renaissance Studies* 4 (1958): 1-17.

Russell, B."The Philosophy of Logical Atomism." Pp. 177-281 in *Logic and Knowledge*, edited by R. C. Marsh. London: George Allen & Unwin, 1956.

Schaffer, J. "Existence, Predication, and the Ontological Argument." Pp. 226-245 in *The Many-faced Argument: Recent Studies on the Ontological Argument for the Existence of God*, edited by J. Hick and A. C. McGill. London: Collier Macmillan, 1968.

Smith, Q. "The Infinite Regress of Temporal Attributions," *Southern Journal of Philosophy* 24 (1986): 383-96.

———. *Language and Time*. New York: Oxford University Press, 1993.

Sosa, E.. "The Status of Becoming: What Is Happening Now?" *The Journal of Philosophy* 76 (1979): 26-42.

Zeilicovici, D. "A (Dis)solution of McTaggart's Paradox," *Ratio* 28 (1986): 175-95.

Zemach, E. M. "Time and Self," *Analysis* 39 (1979): 143-47.

5
Presentism: A Critique

L. Nathan Oaklander

The problems of time and change are inextricably connected, for change involves time and, Shoemaker (1969) notwithstanding, time involves change, or so McTaggart (1934) has argued. That they are related is not in doubt; how they are related is. For McTaggart they are related in such a way that if there is to be time and change, then there must be an A-series, and temporal becoming, but what is the A-series? And what is temporal becoming? These are not easy questions to answer, because there are many different versions of A-time and temporal becoming, and I do not intend to discuss them all. Rather, my aim will be to focus on one version of A-time, the presentist version, and argue that, contrary to its recent proponents, it does succumb to McTaggart's paradox.[1] Even within the limited scope of this chapter, the task of refuting presentism is complicated by there being several different versions of it. One would not think that this is so because all presentists maintain that only the present exists, whereas the past and the future do not exist. Nevertheless, there are different presentist versions of the A-theory, and, although I believe that in one way or another they are *all* susceptible to McTaggart's paradox, there is only one version that I shall endeavour to refute, namely that propounded by William Lane Craig in his recent trilogy on time: *The Tensed Theory of Time: A Critical Examination* (2000a), *The Tenseless Theory of Time: A Critical Examination* (2000b), and *Time and the Metaphysics of Relativity* (2001).

I chose Craig's defense of presentism for two reasons. First, A-theorists who follow Prior in adopting a presentist "metaphysic" are often criticized for lacking an ontology (see, for example, Oaklander 1984: 90-92; Smith 1993: 158-69; 1994a; 1999: 24-29; 2003; Tooley 1997: 165-70, 232-38). To say that the tenses do not refer to B-relations and do not ascribe A-properties is one thing; to say what then are the ontological correlates of the tenses is quite another. It is the latter task that Prior and his followers are commonly accused of shirking. Craig is an exception. He is sensitive to the "lack of ontology" criticism of Prior-based theories (see Craig 2000a: 192-94), and attempts to "found" or provide an ontological ground for both B-relations and A-determinations in the A-series, "tensed facts," and temporal becoming. For that reason, he provides his readers with a metaphysical theory to be evaluated.

I have a second reason for choosing to discuss Craig's version of presentism. Presentists typically explain, promote, and defend their view as being the temporal analogue of the serious actualist position with respect to possible words according to which only the actual world is real. By examining Craig's presentist metaphysic we can evaluate just how successful the marriage between presentism and actualism is. Since my ultimate goal is to argue that presentism, or at least Craig's version of it, is not immune from McTaggart's conundrum, I shall begin my discussion of Craig with an examination of McTaggart.

According to McTaggart, we ordinarily (or commonsensically) conceive of time as involving the notions of past, present and future (A-determinations) and earlier than/later than and simultaneous with (B-relations). Although McTaggart claims that the A-series (defined in terms of A-determinations) and the B-series (defined in terms of B-relations) are both essential to our *ordinary* conception of time, he believes that A-determinations and the A-series are more fundamental, more ultimate, and more essential to the *ontological* nature of time than B-relations and the B-series. In fact, his view is that the series is dependent on the A-series, not only because there would be no B-relations unless there were A-determinations, but more fundamentally because the B-series is ontologically reducible to the A-series and the non-temporal C-series. Thus, McTaggart claims that while the A-series and the C-series are each ultimate:

> [t]he B series, on the other hand, is not ultimate. For given a C series of permanent relations of terms, which is not in itself temporal and therefore is not a B series, and given the further fact that the terms of this C series also form an A series . . . it results that the terms of the C series become a B series, those which are placed first, in the direction from past to future, being earlier than those whose places are farther in the direction of the future (McTaggart 1934: 118).

I think that this passage makes it clear that, for McTaggart, there are no ontologically primitive or simple temporal relations. Metaphysically, time is entirely constituted by the A-series, and it together with the non-temporal but ordered C-series grounds the commonsense view of time as involving *both* A-determinations and B-relations.

Given his positive ontology of time, McTaggart's negative thesis can be recast by saying that, while the A-series and the C-series are necessary and sufficient for the existence of B-time, they are not sufficient for A-time or B-time, which is a contradiction. For time requires *change* and the A- and C-series cannot account for change without introducing some metaphysical correlate of temporal becoming. However, there is no consistent, non-circular way to metaphysically interpret temporal becoming so that change is not contradictory.[2] Since, for the A-theorist, B-time requires temporal becoming, and temporal becoming is contradictory or viciously circular, it follows that there is no B-time, and without B-time there is no time at all.

With this background we are ready to turn to Craig's discussion of McTaggart's paradox and his exposition of the metaphysics of presentism. Craig has basically two responses to McTaggart. The first is to claim that McTaggart mistakenly treats temporal becoming "as a sort of qualitative change insofar as he attempts to combine a B-theoretic ontology with A-theoretic becoming" (Craig 2000a: 179). On the pure A-theory that Craig adopts, "past and future events/things/times are not real or existent and, hence, do not exemplify properties like pastness or futurity. Rather, entities come to be and pass away absolutely, so that the only temporal entities that there are are the present ones" (Craig 2000a: 179). Craig's response gives rise to three central questions:

(1) If temporal becoming is not to be understood as a species of qualitative change, then how is it to be understood?
(2) If McTaggart mistakenly combines a B-theoretic ontology with the A-theory, how then does Craig attempt to analyze temporal relations between and among items existing in the B-series?
(3) If the past and the future do not exist, then what are the truthmakers of past- and future-tense statements?

This last question becomes particularly important given his second response to McTaggart's paradox.

Craig's second objection is that McTaggart's model of treating temporal becoming as the donning and doffing of the non-relational temporal properties of *pastness and futurity* is erroneous, because there are no such properties. To say that an event "is past" or "is future" is not to attribute a property to the event. "Rather such ascriptions should be parsed as asserting that the entity in question did or will exist. . . ." (Craig 2000a: 190).[3] Of course, to parse attributions of pastness and futurity in terms of statements about what *did* or *will* exist, or in terms of what *was* or *will be* the case, does not answer the question of how we are to understand grammatical ascriptions of pastness and futurity, but just raises it once again. For that reason, the account of the truthmakers of past- and future-tense statements in terms of "what is non-relationally present" is an ever-pressing concern.

It should be clear, therefore, that Craig's critique of Mellor's (1998: 70-78) version of McTaggart's problem is ineffectual. Craig repeats the familiar point against McTaggart, Mellor, and others, that no event has all three A-determinations timelessly or simultaneously but successively, and he reflects this by saying that no matter what level we start at we get a consistent set of propositions.[4] Suppose we start with

3. Fp*e* & N*e* & PF*e* (Craig 2000a: 203)

This is read as "*e* will be past" and "*e* is present" and "*e* was future." Craig claims there is no contradiction in 3. Perhaps not, but we are still left with the question: What are the truthmakers for the first and last conjunct? More specifically, what is the ontological difference between FP*e* and PF*e*, given that neither "F" nor "P" is a predicate that ascribes properties to *e?* Unless we are told, we cannot tell.[5] Without such an account, however, the appeal to grammatically consistent tensed statements is a vacuous response to McTaggart's paradox or Mellor's formulation of it.

The need for an account of the passage of time or temporal becoming is also urgent, and for basically the same reason. To see why consider the following passage:

> In his "McTaggart's Paradox Revisited," *Mind* 101 (1992): 323-326, Lowe synthesizes the A-theorist's position by saying that every event is such that it is or was or will be truly describable as past, and is or was or will be truly describable as present, and is or was or will be truly describable as future, which he symbolizes as
>
> 6**. (NT "Np" v PT "Np" v FT "Np") & (NT "Pp" v FT "Pp" v FT "Pp") & (NT "Fp" v PT "Fp" v FT "Fp").
>
> . . . surely (6**) does represent the passage of time, since the same tense operator in each conjunct cannot operate on the true disjunct, on pain of contradiction, so that differently tensed statements will be true in each conjunct. *This difference in tense does represent the flow of time.* (Craig 2000a: 205; emphasis added)

The appeal to truth predicates does not avoid the need to specify the grounds of truth. The fact that differently tensed statements will be true in each conjunct cannot adequately reflect the passage of time unless we have some account of the direction of becoming. More specifically, if NT "Np" & PT "Fp" & FT "Pp" then we want to know, given that the past and the future do not exist, what is the difference between PT "Fp" & FT "Pp"? What is the basis, in the metaphysics of presentism, for *p* being first future and then present and then past rather than the other way around? To answer that question we need some model upon which to understand temporal becoming.

Craig's explication of temporal becoming begins with an appeal to the serious actualist's conception of possible worlds as states of affairs that exist as abstract objects but are not instantiated. He then claims that "tensed possible worlds which *did*, *do*, or *will* obtain are tensed actual worlds" (Craig 2000a: 209; emphasis added). Of course, the appeal to tensed possible worlds which *did, do,* or *will* obtain can hardly provide a metaphysical explanation of what the tenses stand for in propositions reflecting temporal becoming. Leaving that difficulty aside for the moment, Craig (2000a: 209) continues by saying that "the tensed actual world at *t,* is the world which obtains when *t's being present* obtains, or more simply, when *t* is present," but *when* does *t's* being present obtain? Judging from his comments it appears that *t's* being present obtains *before t*'s* being present obtains (for any later *t*),* since Craig maintains that "[t]ensed actual worlds constitute the tensed history of the actual world a, for they are respectively comprised of all states of affairs entailed by a and each *successive t's* being present" (Craig 2000a: 209; emphasis added). Thus Craig's view is that there are possible worlds that exist whether they are instantiated or not, and as time flows possible worlds obtain or become actual by being *successively* instantiated. That the appeal to succession is integral to Craig's account of becoming is evident from other passages as well.

For Craig, temporal becoming is modeled on the different members of the A-series coming into existence successively, as successive times become present. He says, "the doctrine of objective becoming . . . could be graphically displayed as the *successive actualization of the history of the actual world.* It is this model of a *successively instantiated,* rather than tenselessly existing, actual world that precludes the existence of a "totality of facts" (Craig 2000a: 207; emphasis added). The appeal to succession implies the existence of temporal relations, and the appeal to possible worlds that *did* or *will obtain* implies the existence of past- and future-tense facts. Craig's prima facie commitment to B-relations and primitive past- and future-tense facts renders his version of "presentism" subject to McTaggart's paradox unless he can provide an ontological reduction of temporal relations and past- and future-tense facts to what is presently real. Thus, we are led once again to the question: What then, on a presentist metaphysics, are temporal relations, and what are the past- and future-tense facts that are the truthmakers of past- and future-tense statements?

Craig does attempt to answer these questions, and in so doing he diverges in many ways from temporal solipsism, "an idiosyncratic doctrine associated with the views of A. N. Prior and not logically connected with the A-Theory of time" (Craig 2000a: 214). One of the main ways in which Craig deviates from Prior's version of presentism is in his holding that there are past- and future-tense facts that are the truthmakers for past- and future-tense statements. I will let Craig speak for himself:

> On the presentist semantics given here, a future-tense statement is true iff there exists some tensed actual world at *t* in which the present-tense version of the statement is true, where *t has not elapsed by the present moment.* A past-tense statement is true iff there exists some tensed actual world at *t* in which the present-tense version of the statement is true, where *t has elapsed by the present moment.* Those are the truth-conditions of past- and future-tense statements; but they are not what make the statements true. Ultimately what makes the statements true is that reality *was* or *will be* as the statements describe; when the time comes, for example, a sea battle is going on, and therefore the statement made the day before, "There will be a sea battle tomorrow," was true. There are tensed facts corresponding to what tensed statements assert, but past- and future-tense facts exist because of the present-tense facts that *did* or *will* exist. (Craig 2000a: 213-14; emphasis added)

For Craig there are past- and future-tense facts, but they exist because purely present-tense facts, for example *a battle is being fought at Waterloo, did,* or *will* obtain. Alternatively, a fact is a future-tense fact if the *time t* at which it is present *has not elapsed by the present moment* (that is, *t is later than* the present moment), and a fact is a past-tense fact if the *time t* at which it is present *has elapsed by the present moment* (that is, *t is earlier than* the present moment). Thus, Craig's account either presupposes the existence of irreducibly past- and future-tense facts, or it assumes the existence of B-relations, or it leaves the tenses unanalyzed and so is guilty of the "lack of ontology" objection he and others have raised against Prior and his followers.

Look at it this way. On the one hand, Craig wants there to presently exist truthmakers for past- and future-tense statements. If a statement is *true now* then it must be true in virtue of some fact that *exists now.* On the other hand, he does not want to countenance past and future existents. He attempts to avoid the contradiction that a conjunction of those two views entails by claiming that past- and future-tense facts exist at present, but they are not ultimate. However, his attempt to show that past- and future-tense facts are not ultimate is either unsuccessful or it succeeds only at a cost of reintroducing a B-theoretic ontology that he sought to avoid, thus undermining presentism and making his A-theory susceptible to McTaggart's paradox. We can begin to see why this is so by noting that Craig claims that, if a past-tense statement is now true, then there is a present-tense fact that *did* obtain or there is a present-tense fact that exists at a time *t* that has elapsed by the present time. What, then, is involved in *t's* having elapsed by the present moment, or a present tense fact having obtained? I can think of several possibilities:

(1) The present moment is moving across the A-series of presently existing things/events/moments, and a present-tense fact *did* exist when the moving NOW has passed it by. So what exists now is the fact that the NOW (as a relation to a term outside the series or as a monadic property) has already passed (or has already been exemplified) by a given instantiated state of affairs, and that fact is the ground of the past-tense fact that X *was* F.

(2) To say that a present-tense fact did or will obtain at a time that has or has not elapsed by the present moment is to countenance the existence of presently obtaining primitive past- and future-tensed facts, X *was* F, and X *will be* F.

(3) If a past-tense statement is true, then there presently obtains the fact that a present-tense state of affairs exists at a time *t earlier than* the present moment *t**. Similarly, if a future-tense statement is true, then there presently obtains the fact that a present-tense state of affairs exists at a time *t later than* the present moment *t**.

(4) Finally, one can eschew ontology altogether and claim that the tenses are logical operators, or that the tenses and temporal becoming are conceptually primitive, and have no ontological significance whatsoever.

Clearly, the first two alternatives are unacceptable. The first involves a view of temporal becoming that McTaggart and many others, including Craig, have found reasons to reject.[6] The second is inconsistent with Craig's presentism, since if there are ultimate past- and future-tense facts then temporal objects must exemplify the properties of *pastness and futurity* and therefore must, in some sense, exist. The last alternative (4) is also explicitly rejected by Craig, who construes his version of presentism as providing an "ontological foundation" for temporal relations and the direction of time.

There remains the third interpretation, although it too raises questions. If only the present exists, then how can there presently obtain a temporal *earlier than* or *later than* relation between two temporal objects, at least one of which does not exist? Nevertheless, there is reason to believe that Craig adopts the alternative (3), which analyzes past- and future-tense facts in terms of what is *earlier* or *later than* the present moment, since he expresses sympathy with such a view about the ontological status of the past and future put forth by Alfred Freddoso (1983). Freddoso maintains that "the proposition "Socrates drank hemlock" is now temporally necessary, since "Socrates drinks hemlock" is a member of a past submoment which obtains *prior* to the present in any world sharing the same history *prior* to the present with our world" (Craig 1991: 180; emphasis added; see also Craig 2000a: 214, fn. 140). And referring to future-tense propositions Freddoso says, "a proposition *p* is necessary *per accidens* at *t* in world *w* just in case *p* is true at *t* and at every moment *after t* in every possible world which shares the same history . . . with *w* at *t*" (Freddoso 1983: 266; emphasis added, quoted in Craig 1991: 180). The appeal to "prior" times implies a temporal relation between a past event or time and the present, and the statement "every moment *after t*" implies a temporal relation between a later event or time and the present moment. If, however, Craig appeals to *unanalyzable* temporal relations to account for the truthmaker of past- and future-tense facts, then Craig contradicts himself, since he claims that a B-theoretic ontology coupled with A-theoretical becoming renders McTaggart's paradox inescapable. It is not surprising, then, that he attempts to provide an ontological reduction of B-relations in terms of A-determinations, the A-series and temporal becoming (Craig 2000b: 149-58). In the final part of this chapter I shall critically examine Craig's attempt.

In *The Tenseless Theory of Time: A Critical Examination,* Craig (2000b) agrees with McTaggart's positive view of time that "on the A-Theory of time, the obtaining of the temporal relations *earlier than/later than* among temporal particulars can be derived from the objectivity of A-determinations and the A-series" (Craig 2000b: 150). Paradoxically, Craig interprets Mellor as also maintaining that "the very temporal relations which lie at the heart of the B-Theory are derivable from the A-series [and A-determinations]" (Craig 2000b: 151). It is true that Mellor offers various possible reductions of the B-series to the A-series, but there are two important facts to note about his "definitions." First, they presuppose the existence of McTaggart's A-series and A-determinations. More specifically, on Mellor's interpretation of the A-theory, and temporal becoming, "Futurity, temporal presence, and pastness are all supposed to be real non-relational properties that everything in time successively possesses, changing objectively as it exchanges each of properties for the next" (Mellor 1981a: 89-90). Craig explicitly rejects this interpretation of the A-theory, arguing, as does Mellor, that it leads inevitably to McTaggart's paradox.

Second, Mellor has argued that McTaggart has shown that A-change is contradictory and thus the A-theory of temporal becoming is absurd. As he puts it,

"What disproves all A-theories is a contradiction inherent in their concept of change" (Mellor 1998: 70). Thus, although Mellor would agree that *if* the A-theory is true then B-relations could be defined in terms of the A-series, the point is moot since the A-theory is false. Clearly, Mellor does not believe that B-relations can be defined in terms of A-determinations. Craig, on the other hand, claims to be defending an ontological reduction of temporal relations that "goes all the way back to McTaggart" (Craig 2000b: 150), but the analyses that McTaggart and Mellor propose imply that A-determinations are either properties of events/moments/things or relations to some term outside the temporal series. Craig (2000a) denies the existence of A-determinations as characteristics of events/things/moments, whereas in Craig (2000b) the definitions of B-relations he offers require those properties. Thus, his appeal to Mellor's definitions to support an ontological reduction of B-relations to A-determinations is inconsistent with his presentist metaphysic according to which there are no such properties. Furthermore, I shall argue that, in adopting the A-account of B-relations endorsed by McTaggart and spelled out by Mellor, Craig's analysis of temporal relations does not avoid the difficulties McTaggart raises since he is committed to a theory that is contradictory, circular, or vacuous.

Craig's first ontological reduction of *earlier than/later than* relations is as follows:

D_1': *e is* earlier than e* ≡ *e* is more past or less future than *e**.

e is later than *e** ≡ *e* is more future or less past than *e**.
(Craig 2000b: 153)

According to Craig, *more past/future than* are A-relations and not monadic properties. They are relations that presently obtain between terms that occupy different positions in the A-series:

> Thus, for example, if *e is* earlier than *e** and it happens that *e* is present, then *e* is less future than *e**. Similarly, in the case that one of the events is past and the other future, we should think of each one as having none of the A-determinations of its *relatum*. Thus, for example, if *e* is past and *e** is future, then *e is* earlier than *e** just because *e* is more past than *e**. (Craig 2000b: 153)

Craig (2000b: 154) maintains that *more past* and *more future* are primitive concepts. What, then, are its relata? And if the *relata* of an A-relation are *e* is past and *e* is future, then what is the ontological status of those relata? Clearly, if *being future* and *being past* are non-relational properties of past and future events, then his view is inconsistent with presentism and, by his own lights, susceptible to McTaggart's paradox (see Craig 1998). On the other hand, if past and future events have no ontological status, so that neither *e's being past* nor *e's being future* exists, then we have an A-relation without relata, which is absurd. Finally, if Craig attempts to analyze *e* is past and *e* is future in accordance with the possible worlds analysis he offered previously (Craig 2000a), then the truthmaker for, say, "It was raining" is that the present-tense fact It is raining obtains at a moment of time *t* that has elapsed by the present moment. In that case, however, his ontological reduction of so-called B-relations is obviously circular, since, as we have seen, there is no acceptable account of "time *t* has elapsed by the present moment," other than that time *t is earlier than* the present moment.

Furthermore, his account of relations, sketchy as it is, raises serious problems concerning his notion of A-relations. He says:

> relations are abstract objects which plausibly do not exist in time at all. Non-contemporaries stand in a relation at their respective times and the timelessly existing relation reaches across time to relate the two individuals. As for the individuals themselves, we could ascribe to them relational properties: Socrates, at the time he existed, had the property of *going to be referred to by William Craig* or the property of *being referred to by William Craig at t.* He no longer has that property, but I now have the property of *referring to Socrates.* The *relation between us can be analyzed in terms of such relational properties or said to exist timelessly in virtue of such properties.* (Craig 2000a: 212; last emphasis added)

The first problem with this account of relations is that it is incompatible with his account of A-relations, and his presentist ontology. Craig claims that "the A-theorist is at liberty to stipulate that the above concepts [*more past* and *less future*] are among his theoretical primitives" (Craig 2000a: 154). Perhaps so, but if A-relations are theoretical primitives, then they cannot be analyzed in terms of relational properties, or be said to exist in virtue of such properties. On the other hand, if A-relations are analyzable in terms of *relational* properties, then what else could they be if not parasitic on the B-relations that he is attempting to analyze? Finally, the very notion that we could treat *being past* and *being future* as *relational properties* of the terms of A-relations contradicts his previous claim that "The construal of pastness and futurity as relational predicates *should not be taken to mean that these are relational properties inhering in events*" (Craig 2000a: 190; emphasis added). Given the inconsistency between Craig's accounts of relations in general and temporal A-relations in particular, it is debatable whether or not there are any terms of A-relations or, indeed, whether there are any A-relations at all.

To see what is involved in this last point, note that Craig claims that "If the relations *earlier than/later than* can be truly and tenselessly ascribed, it is because and only because the A-relations *more past/future than* and *less past/future than* can be truly and presently ascribed" (Craig 2000a: 152). My question is this: What is the ground of the truth of statements that assert the existence of an A-relation between events/things/moments, and what are A-relations presently ascribed to? What presently exemplifies those relations? I do not think that Craig has a consistent set of answers to these questions. Since past and future things/events/moments do not exist, they cannot be the terms of A-relations nor can A-relations ascribe a tense to *them.* For the presentist, what does not presently exist cannot presently exemplify properties, including tensed properties. But then, since A-relations, like all relations, are timeless, there is nothing that presently exists that could provide an ontological foundation for affirming the existence of the (tenseless) temporal relations of *earlier than/later than*. To put the difficulty otherwise, Craig is faced with a dilemma. If past and future temporal objects do not exist, then there is nothing for A-relations to presently ascribe objective tense to. If past and future temporal objects do exist, then presentism is false. Thus, given Craig's version of the A-theory, whether there are past and future temporal objects or not, there are no A-relations, there are no B-relations, and there is no time. For these reasons, his first reductive analysis of B-relations is unsuccessful.

Craig's second definition or reduction of B-relations to A-determinations is also unsuccessful. His second definition is as follows:

> D_2": *e is* earlier than e* $=_{df.}$ There is some time *t* such that at *t* it is an objective fact that has presentness and e* is future. (Craig 2000b: 156)
>
> *e is* later than e* $=_{df.}$ There is some time *t* such that at *t* it is an objective fact that *e* has presentness and e* is past. (Craig 2000b: 156)

Richard Gale has claimed that to relativize A-determinations to times in this way is circular because "The predicates "___ *is* past at ___" and "___ *is* future at ___" . . . express a timelessly true or false statement about a B-relation between two events, i.e., they make B-statements" (Gale 1968: 90-91; quoted in Craig 2000b: 155). Craig dismisses Gale's claim on the grounds that (I) a tenselessly true statement such as the *definiens* of D_2" can refer to an A-determination and (2) being "at a time" does not "illicitly smuggle in the so-called B-relation of *simultaneous with* . . . [since] being "at a time" is foundational to the notion of simultaneity, rather than the other way around" (Craig 2000b: 156, fn. 22). Craig concludes that the "*definiens* thus should not be construed in terms of the ascription of any so-called B-relations" (Craig 2000b: 156-57). There are two problems with Craig's second definition and his response to the objections. First, Craig never specifies what are the ontological correlates of "*e** is past" and "*e* *is future" in the *two definientia* of D_2". He does say that "According to (D_2"), at *t e* has the premier A-determination of presentness, and the *definiens* in each case refers to an objective tensed fact . . . [and] therefore refers to an A-determination" (Craig 2000b: 156), but what are the objective tensed facts in this case? If "*e** is past" and "*e** is future" attribute A-determinations (properties) to *e**, then the past and future must exist in order to exemplify those properties, and that is incompatible with his professed presentism. On the other hand, if the past and the future do not exist so that past- and future-tense facts are not ultimate, but analyzable in terms of present-tense facts that have or have not yet elapsed, then his analysis is circular, since there is no analysis of "time *t* has elapsed by the present moment" that is both consistent and does not reintroduce B-relations, and this leads to another problem with his second reductive analysis of B-relations.

Craig misses the main point of Gale's charge of circularity. The introduction of time and, in Craig's case, absolute time or moments is crucial if we are to avoid a contradiction.[7] For if the *definientia* are, as Craig says, tenselessly true statements, then, in order to avoid a contradiction, time must be included in what the *definientia* of D_2" express. Otherwise we would get *e is* earlier than e* $=_{df.}$ It is an objective fact that *e* has presentness and *e** is future.

and

e is later than *e** $=_{df.}$ It is an objective fact that *e* has presentness and *e** is past.

Obviously, those two objective facts contradict each other and fail to account for whether the direction of time is from *e* to *e** or from *e** to *e*. The introduction of some time *t* and *t** at which the objective facts mentioned in the *definiens* are at provide such an account *if and only if t* and *t** are members of a *temporal* sequence, that is a sequence with an intrinsic direction. Thus, even if the predicates, "is past at," "is present at," and "is future at" do not presuppose the existence of B-relations, Craig's analysis is still circular because the existence of "at time *t*" in his analysis does presuppose the existence of B-relations.

Finally, let us turn to Craig's third attempt to ground the existence of B-relations on the reality of tensed facts.

D_3': *e is* earlier than *e** ≡ *e* becomes present first and *e** becomes present second.

e is later than *e** ≡ *e** becomes present first and *e* becomes present second.

Craig considers one objection to this account, and his reply is telling.

> Oaklander objects that the use of "first" and "second" conceal so-called B-relations (Oaklander 1996: 211); but a moment's reflection shows that this is not the case. There are ordinal numbers that are wholly atemporal and can characterize spatial or abstract objects as well as temporal particulars. Given the order of their temporal becoming, the temporal ordering of the two events in question necessarily follows. (Craig 2000b: 157)

Admittedly, *spatial* or *abstract objects* can be characterized as "first" and "second" without presupposing temporal relations, but it does not follow that "first" and "second" can characterize *temporal objects* without presupposing temporal relations. Indeed, temporal relations between and among particulars are intrinsically different from all other instances of one-dimensional order, such as that of points on a line and numbers in order of magnitude, in that only a temporal series has an intrinsic direction. The terms "first," "second," "third," and so on, can give a spatial series an *order,* but they cannot give spatial objects a *direction.* For that reason, to say that *e* becomes present *first* and *e** becomes present *second* is either irrelevant to determining their B-relation to one another or assumes that *e* and *e** become present in a given direction; it does not account for it. Thus, Craig's third account of B-relations is circular unless we eliminate the "first" and "second" from it. In that case, however, Craig's analysis is inadequate since from *e* becomes present and *e** becomes present we cannot infer that *e is* earlier than *e** or vice versa.

One last, related, criticism. Suppose that *e* becomes present first, *e** becomes present second, *e*** becomes present third, *e**** becomes present fourth, and so on. Since this conjunction is tenselessly true the *definiens* in D_3' leaves out the information about which event is present NOW. Indeed, what could be the ground of the *definiens* being true now, unless there is the further fact that *e* is present NOW.* However, since to become present is an act of a temporal being, it follows that all of the terms in the A-series obtain (present tense) at some time. But what accounts for their direction? Which events become present *before* the others? Unless the NOW moves *successively* along the series of events that obtain (present tense) at some time or other, there is no change, and without change there is no time. Unfortunately, the notion of the *successive* actualization of the terms of the A-series presupposes precisely what Craig is attempting to analyze, namely B-relations. For that reason, Craig's analysis is viciously circular and the circularity cannot be avoided by positing another A-series of events or times at which the terms of the first series undergo becoming on pain of a vicious infinite regress.

According to Craig, "the A-theorist can account for the existence of so-called B-relations by *founding them on the reality of tensed facts*; thus far, *McTaggart's argument seems to be vindicated*" (Craig 2000b: 157).The problem is twofold: first, McTaggart's view on B-relations implies the existence of the A-series either as a series of terms that have an A-relation to a term outside the series or as a series of terms that have the A-properties of *pastness, presentness*, and *futurity*. Thus, in so far as McTaggart's view is vindicated, Craig's presentist metaphysics is refuted, since the two are incompatible, and if his reductive analysis of temporal relations depends on McTaggart's positive view of time being vindicated, then his analysis is refuted once again. Second, if Craig rejects McTaggart's view of the A-series and temporal becoming, then it is unclear how he has accounted for the existence of B-relations by founding or ontologically grounding them on the reality of tensed facts, because it is not clear what tensed facts exist on a presentist metaphysics. If the only tensed facts there are are present-tense facts—those that exist NOW—then there are no present-tense facts that could ground the truth of statements about what is earlier or later than now or about what is past or future. Clearly, the appeal to present-tense facts that *did* obtain, or *will* obtain, or to present-tense facts that obtain at a time that *has or has not elapsed by the present moment* is either to eschew ontological commitment altogether or to appeal to precisely those past- and future-tense facts or B-relations that Craig sought to avoid. In any case, on Craig's presentist version of the A-theory, time is unreal. Craig claims to "have an ontological foundation in [his] metaphysic of time for affirming the existence of the (tenseless) temporal relations *earlier than/later than*" (Craig 2000b: 159). On the basis of my critique of Craig's metaphysics of presentism, it would appear that he has not provided an ontological foundation for temporal relations. I conclude that Craig's A-theoretical account of time, change, and becoming *is* subject to McTaggart's paradox and must therefore be rejected.

Acknowledgement

I wish to thank the Faculty Development Fund of the University of Michigan - Flint for its support of the research for this chapter.

Notes

1. For recent defenses of presentism see, for example, Prior (1968), Chisholm (1981), Bigelow (1991; 1996), Hinchliff (1996), Craig (1998), Zimmerman (1998), Ludlow (1999), and Percival (2002). For criticisms see Oaklander (1984; 1994a; 1999; 2002), Le Poidevin (1991; 1999), Smith (1993; 1999; 2003), Tooley (1997), and Mellor (1998; 2001), among others.

2. For interpretation and defense of McTaggart's negative argument against A-time see Le Poidevin and Mellor (1987), Mellor (1998), and Oaklander (1996).

3. Craig also argues that presentness is not a property: "Since presentness is identical with temporal existence (or occurrence) and existence is not a property, *neither is presentness a property.* Presentness is the act of temporal being" (Craig 2000a: 202; emphasis added; compare with Craig 2000a: 191-201). Note, however, that Craig is not consistent on this matter since he also claims "one need not use tensed statements alone to talk about tense; for example, "The A-determination *presentness* is an absolute property, not a mere relation" is tenselessly true (or false), but refers to an objective A-determination" (Craig 2000b: 156; compare with Craig 2000a: 222).

4. This common critique misses the mark because McTaggart does *not* start off by *assuming* that every event is past, present, and future. On the contrary, McTaggart begins by *insisting* that an event or moment in time can have *one and only one* A-determination. Thus, he says: "And we *must* say that a series is an *A* series when each of its terms has, to an entity *X* outside the series, *one and only one,* of three indefinable relations, pastness, presentness, and futurity. . . ." (McTaggart 1927: 2, 20; emphasis added). And, again in "The unreality of time"; "Past, present, and future are incompatible determinations. *Every event must be one or the other, but no event can be more than one.* . . . And, if it were not so, the A series would be insufficient to give us, in combination with the C series, the result of [B-] time" (McTaggart 1934: 123; emphasis added). The further claim that every event/thing/moment has all three A-determinations is not assumed but is implied by the view—endorsed by A-theorists—that change requires temporal becoming.

5. For the most carefully worked out A-theoretical account of the ontological significance of the tenses see Smith (1993; 1994a, b). For a critique of Smith see Oaklander (1996). Smith (2002) has recently modified his views.

6. For my criticism of this model of becoming see Oaklander (1984: chaps. 2 and 3) and essays in Oaklander and Smith (1994: Part II). For criticisms of other non-presentist accounts of becoming see Oaklander (1994b, c; 2001) and Mellor (1981a, b; 1998).

7. Craig (2001) argues for absolute time.

References

Bigelow, J. "Worlds Enough for Time," *Noûs* 45: (1991): 1-20.

———. "Presentism and Properties," *Philosophical Perspectives* 10 (1996): 35-52. Chapter 8 in this volume.

Chisholm, R. M. *The First Person*. Minneapolis: University of Minnesota Press, 1981.

Craig, W. L. *Divine Foreknowledge and Human Freedom: The Coherence of Theism: Omniscience*. New York: E. J. Brill, 1991.

———. "McTaggart's Paradox and the Problem of Temporary Intrinsics," *Analysis* 58 (1998): 122-27.

———. "Oaklander on McTaggart and Intrinsic Change," *Analysis* 59 (1999): 319-20.

———. *The Tensed Theory of Time A Critical Examination*. Dordrecht: Kluwer, 2000a.

———. *The Tenseless Theory of Time: A Critical Examination*. Dordrecht: Kluwer, 2000b.

———. *Time and the Metaphysics of Relativity*. Dordrecht: Kluwer, 2001.

Freddoso, A. J. "Accidental Necessity and Logical Determinism," *Journal of Philosophy* 80 (1983): 257-83.

Gale, R. M. *The Language of Time*. London: Routledge, Kegan & Paul, 1968.

Hinchliff, M. "The Puzzle of Change." Pp. 119-136 in *Philosophical Perspectives, 10, Metaphysics*, edited by J. Tomberlin. Cambridge, Mass.: Blackwell Publishers, 1996.

Le Poidevin, R. *Time, Cause and Contradiction*. Basingstoke: Macmillan, 1991.

———. "Egocentric and Objective Time," *Proceedings of the Aristotelian Society* 99 (1999): 19-36.

Le Poidevin, R. and D. H. Mellor. "Time, Change and the Indexical Fallacy," *Mind* 96 (1987): 534-38.

Ludlow, P. *Semantics, Time and Tense: An Essay in the Metaphysics of Natural Language*. Cambridge, Mass.: MIT Press, 1999.

McTaggart, J. E. M. *The Nature of Existence: Vol. 2*. Cambridge, U.K.: Cambridge University Press, 1927. See chapter 2 in this volume.

———. "The Unreality of Time," *Mind* 18 (1908): 457-74; reprinted in *Philosophical Studies*, edited by S. V. Keeling. London: Edward & Arnold & Co., 1934.

Mellor, D. H. *Real Time*. Cambridge, UK: Cambridge University Press, 1981a.

———. "McTaggart, Fixity and Coming True." Pp. 79-97 in *Reduction, Time and Reality*, edited by R. Healey. Cambridge, UK: Cambridge University Press, 1981b.

———. *Real Time II*. London: Routledge, 1998.

———. "Real Time II: Replies to Hinchliff, Paul and Perry." Pp. 95-104 in *The Importance of Time*, edited by L. N. Oaklander. Dordrecht: Kluwer, 2001.

Oaklander, L. N. *Temporal Relations and Temporal Becoming*. Lanham, Md: University Press of America, 1984.

———. "Bigelow, Possible Worlds and the Passage of Time," *Analysis* 54 (1994a): 244-48.

———. "McTaggart, Schlesinger, and the Two-Dimensional Time Hypothesis." Pp. 221-28 in *The New Theory of Time*, edited by L. N. Oaklander and Q. Smith. New Haven, Conn.: Yale University Press, 1994b.

———. "Zeilicovici on Temporal Becoming." Pp. 252-256 in *The New Theory of Time*, edited by L. N. Oaklander and Q. Smith. New Haven, Conn.: Yale University Press, 1994c.

———. "McTaggart's Paradox and Smith's Tensed Theory of Time," *Synthese* 107 (1996): 205-21.

———. "Craig on McTaggart's Paradox and the Problem of Temporary Intrinsics," *Analysis* 59 (1999): 314-18.

———. "Tooley on Time and Tense." Pp. 3-12 in *The Importance of Time*, edited by L. N. Oaklander. Dordrecht: Kluwer, 2001.

———. "Presentism, Ontology and Temporal Experience," Pp. 73-90 in *Time, Reality, and Experience*, edited by C. Callender. Cambridge, UK: Cambridge University Press, 2002.

Oaklander, L. N., and Q. Smith, eds. *The New Theory of Time*. New Haven, Conn: Yale University Press, 1994.

Percival, P. "Mellor on Time," Pp. 91-118 in *Time, Reality, and Experience*, edited by C. Callender. Cambridge, UK: Cambridge University Press, 2002.

Prior, A. N. *Papers on Time and Tense*. New York: Oxford University Press, 1968.

Shoemaker, S. "Time without Change," *Journal of Philosophy* 66 (1969): 363-81.

Smith, Q. *Language and Time*. New York: Oxford University Press, 1993.

———. "The Infinite Regress of Temporal Attributions," Pp.180-94 in *The New Theory of Time*, edited by L. N. Oaklander and Q. Smith. New Haven, Conn.: Yale University Press, 1994a.

———. "The Logical Structure of the Debate about McTaggart's Paradox." Pp. 202-10 in *The New Theory of Time*, edited by L. N. Oaklander and Q. Smith. New Haven, Conn.: Yale University Press, 1994b.

———. "The 'Sentence-Type' Version of the Tenseless Theory of Time," *Synthese* 119 (1999): 233-51.

———. "Time and Degrees of Existence: A Theory of 'Degree Presentism.'" Pp. 119-36 in Time, Reality, and Experience, edited by C. Callender. Cambridge, UK: Cambridge University Press, 2002.

———. "Reference to the Past and the Future." Pp. 357-90 in A. Jokic and Q. Smith (eds) *Time, Tense and Reference*, Cambridge, Mass.: MIT Press, 2003.

Tooley, M. *Time, Tense and Causation*. Oxford: Clarendon Press. 1997.

Zimmerman, D. "Temporary Intrinsics and Presentism." Pp. 206-19 in *Metaphysics: Big Questions*, edited by D. Zimmerman and P. v. Inwagen. Cambridge, Mass.: Basil Blackwell, 1998.

Part II

Defining Presentism: The Triviality Objection

Introduction: Defining Presentism: The Triviality Objection

Ernâni Magalhães

The triviality problem faced by presentism can be understood as a constructive dilemma with the conclusion that presentism is not worth discussing. Either presentism is true but trivial, or it is substantive but obviously false. If it is trivial and true then it is not worth discussing. If it is substantive and obviously false then it is not worth discussing. So presentism is either not worth discussing or it is not worth discussing. Therefore, presentism is not worth discussing.

The problem for presentism arises at the very outset of the discussion, when the presentist sets out her view. Presentism seems to be the view that

(1) Everything that exists is present.

But what does "exists" mean in (1)? It seems any plausible way of spelling it out leaves (1) either trivial or obviously false—either way, not worth discussing.

"Exists" may mean "exists now," in which case presentism is the view that

(2) Everything that exists now is present.

There seems to be no way something can exist now unless it is present. But (2) cannot be the interpretation of (1) that makes it worth discussing since no sensible non-presentist could possibly want to deny it. The non-presentist wants to affirm that some things are not present, not that some things that exist now are not present. Thus (2) seems not to be what the presentist can be after.

Another suggestion is that "exists" means "existed, exists, or will exist." On this interpretation, presentism is the view that

(3) Everything that exists, existed, or will exist is present.

Unlike (2), (3) seems not to be trivial. It can be easily imagined that a non-presentist would deny (3) because she might claim that some things that existed or will exist are not present. But (3) seems to fall prey to the other horn of the dilemma. It seems too easy for the non-presentist to deny (3), and thus hopeless for the presentist to affirm it. (3) is a universal generalization that says of any thing that exists, existed, or will exist that it is present. Given the triviality of (2) it will not be possible to find examples of things that exist—i.e., exist now—which are not present. But there seem to be obvious examples of things that existed or will exist that are not present. Julius Caesar is an example of something that existed but is not present, and my first grandson is an example of something that will exist but is not present. Because these examples are so obvious, if (3) expresses presentism then it is not worth discussing. Any discussion would seem likely to end with a quick agreement that presentism is false.

A final possibility is that "exists" in (1) means "exists *simpliciter*" or "just plain exists." The thought is that the sense of "exist" at issue in the presentist's thesis is so basic as to resist analysis into either the translations proposed in (2), (3), or any other. So understood, presentism is the view that

(4) Everything that exists *simpliciter* is present.

In rejecting this proposal, Ulrich Meyer (chapter 6, p. 101) argues that if something exists *simpliciter* then it either exists temporally or outside of time. Since presentism is not a doctrine about atemporal things, (4) must imply that everything that exists temporally is present. But something exists temporally, according to Meyer, only if it either exists, existed, or will exist. Thus (4) implies (3); i.e., (4) is true only if (3) is true. But since (3) has been rejected as substantive but obviously false, so (4) should be rejected as substantive but obviously false.

A number of responses have been offered on behalf of presentists. The more popular response is to endorse one of the interpretations and deny that it is either trivially true or obviously false. A less popular alternative identifies the core thesis of presentism in something other than (1). This response then claims that this core thesis is worth discussing. A more indirect response claims that there must be some error in the triviality argument since it implies that any ontological debate is merely verbal, which seems wrong.

According to the alternative thesis view, the triviality objection fails because it identifies the thesis of presentism with (1), the claim that everything that exists is present. According to this view, presentism is only secondarily a view about what exists. Primarily, presentism is a view about existence or "exists." It says that existence is present existence; or that "exists" means "exists now." Trenton Merricks is an advocate of this view. Merricks identifies "*existing at the present time*" with "*existing*" (2007: 125). This thesis, it is claimed, is substantial and worth discussing.

Two versions of the first strategy have been defended. Theodore Sider argues in a number of places that presentism is expressed by "exists *simpliciter*" as in (4) (2001, 2006). He denies, however, that (4) is an obvious falsehood.

Thomas Crisp (chapter 7) accepts interpretation (3) but denies that it is obviously false. (3), recall, claims that

(3) Everything that exists, existed, or will exist is present.

Crisp expresses (3) with

(Pr_b'') For every *x*, *x* is a present thing (4),

where the quantifier—"every *x*"—ranges over anything that exists in time; that is, anything that exists, existed, or will exist. Crisp maintains that presentism should be understood as the view that everything in this very inclusive domain is a present thing. Thus Crisp's (Pr_b'') entails (3). Crisp denies that Julius Caesar or the Roman Empire are obvious counterexamples to (Pr_b''). The claim that the Roman Empire existed but is not present—which he calls "(RE)"—can be interpreted in two ways:

(RE_1) *WAS*(for some *x*, *x* is the Roman Empire and *x* will not exist in t_α)

or

(RE_2) For some *x*, *x* was the Roman Empire and x is no longer present.

(RE_1) uses the past-tense operator—"WAS"[1]—to assert that it was the case that something, *x*, is the Roman Empire and will not exist in t_α, i.e., the present time. (RE_2) says there is something that was the Roman Empire and that thing is not (any longer) present. While (RE_1) is an obvious truth—it only requires that the sentence "something is the Roman Empire and will not exist in 2010" was true at some time—Crisp maintains that (RE_2) is far from obvious and well worth discussing. (RE_2) is true only if the most inclusive temporal quantifier—the one that includes all temporal things—includes something that used to be the Roman Empire. Since it is not obvious that it does, it is not obvious that the past existence of the Roman Empire is a counterexample to (Pr_b''), which expresses presentism. The triviality objection, Crisp concludes, fails to show that presentism is either a trivial truth or obvious falsehood.

According to Crisp, in the sense that (RE) is obvious it is not inconsistent with presentism. This means that, for Crisp, in its obvious sense, "the Roman Empire existed without being present" does not imply that there is something which was the Roman Empire. This, Meyer argues, is to hold that existential generalization—inferring that something is F from the fact that *x* is F—is only valid where a name presently has a referent. But in that case, Meyer goes on, "the notion of existence employed is really *exists now* (and not *exists temporally*)" (p. 101), which means the presentist is back to affirming the trivial (2)—everything that exists now is present. If Meyer is right then the only way for Crisp to block the inference from the past existence of the Roman Empire to its inclusion among existing things is to restrict the most inclusive temporal domain to things that exist now. Crisp, on the other hand, believes it is not obviously mistaken to have a domain which does not preclude non-present things but which includes none of them simply because there are none. To get a sense for Crisp's view, compare the domain of things which are either supernatural or round with the domain of things that are round. According to many people, the class of things that are either supernatural or round includes many round things and no supernatural things. According to others, it includes both. But the domain itself is "open" to both. The domain of things that exist temporally is like this, according to Crisp's understanding of presentism. The domain is defined in terms of things that are either (wholly) past, present, or (wholly) future. But for the presentist the only things included are those that are present.

Another response to the triviality objection is to claim that it generalizes in an undesirable way. If presentism is either trivially true or obviously false, then non-presentism either denies a triviality or affirms an obvious truth. (The most popular version of non-presentism—eternalism—holds that there are sundry past and future objects in addition to the present ones.) In either case, non-presentism would appear to suffer from the same defect as presentism. The triviality objection therefore suggests that the dispute between presentism and its opponent is merely verbal (see Dorato 2006b). Each understands (1) in a different way. The presentist thinks every thing that exists now is present, which the non-presentist does not deny. And the non-presentist thinks some things that existed, exist now, or will exist are not present, which the presentist will not deny. Generalized, this suggestion explains why presentists and non-presentists agree on so many things, such as the past existence of Caesar and the future existence of 2022. But if the triviality objection succeeds in showing that the presentism/non-presentism dispute is merely verbal, then, according to this response, it also implies that many other apparently meaningful ontological disputes are merely verbal. Consider the dispute about whether any merely possible things exist. (A merely possible thing is something that could exist but actually does not. A talking donkey is an example.) The possibilist affirms, while the actualist denies, that merely possible things exist. But what does "exist" mean? Does it mean "actually exist"? If so, then the possibilist seems to be obviously wrong: it is clearly not true that merely possible (= possible but not actual) things actually exist. If "exist" means "possibly exist," then possibilism seems to emerge as the obvious truth, since some things possibly exist which are not actual.

Meyer denies that holding the presentism/non-presentism dispute to be merely verbal implies that the actualism/possibilism dispute is only verbal (pp. 104-5). The core modal dispute, he thinks, is between ersatz theorists, who construct possibility out of abstract objects like propositions or states of affairs, and modal realists, who build up possibility and "possible worlds"[2] out of real concrete individuals. Ersatzism and modal realism are competing interpretations of modal discourse. Because neither view is either trivially true or obviously false, Meyer claims, the dispute is substantive. However, there seems to remain an ontological dispute between ersatz and modal realism on Meyer's conception. The ersatzer and modal realist are said to differ in their "ontological commitments" (p. 104). The modal realist believes, while the ersatzer denies, that there are concrete *possibilia*.[3] This formulation of the dispute, however, is susceptible to the same dilemma as the original formulation. Is the dispute about whether there *actually* are concrete possibilia or whether there *possibly* are concrete possibilia? The former interpretation seems to yield an easy victory for the ersatzer; the latter an easy victory for the modal realist.

Notes

1. Operators affect the meaning of whole sentences. The operator "it is not the case that" takes a sentence and affirms its negation. The tense operators, "WAS," "NOW," and "WILL," take sentences and apply a tense to them. Arthur Prior is credited with developing tense logic. See Prior 1968.

2. A possible world is a complete actual or non-actual situation.

3. Possibilia are possible but not actual entities. Barack Obama's (possible) son is an example.

References and Further Readings

Callender, C. "Shedding Light on Time," *Philosophy of Science* 67 (2000): S587-S599.

Chisholm, R. M., and D. Zimmerman. "Theology and Tense," *Noûs* 31 (1997): 262-65.

Crisp, T. "Presentism." Pp. 211-45 in *Oxford Handbook of Metaphysics*, edited by D. Zimmerman and M. Loux. Oxford: Oxford University Press, 2003.

———. "On Presentism and Triviality." Pp. 15-20 in *Oxford Studies in Metaphysics*, edited by D. Zimmerman. Oxford: Oxford University Press, 2004. Chapter 7 in this volume.

———. "Reply to Ludlow." Pp. 37-46 in *Oxford Studies in Metaphysics*, edited by D. Zimmerman. Oxford: Oxford University Press, 2004.

Dolev, Y. *Time and Realism*. Cambridge, Mass: MIT, 2007.

Dorato, M. "The Irrelevance of the Presentism/Eternalism Debate for the Ontology of Minkowski Spacetime." Pp. 93-110 in *The Ontology of Spacetime*, edited by D. Dieks. Amsterdam: Elsevier, 2006a.

Dorato, M. "Absolute Becoming, Relational Becoming and the Arrow of Time: Some Non-conventional Remarks on the Relationship between Physics and Metaphysics," *Studies in the History and Philosophy of Modern Physics* 37 (2006b): 559-76.

Lombard, L. "On the Alleged Incompatibility of Presentism and Temporal Parts," *Philosophia* 27 (1999): 253-60.

———. "Time for a Change: A Polemic against the Presentism/Eternalism Debate." In *Time and Identity: Topics in Contemporary Philosophy, Vol. 6*, edited by J. K.

Campbell, M. O'Rourke, and H. Silverstein. Cambridge, Mass.: MIT, forthcoming.

Ludlow, P. "Presentism, Triviality and the Varieties of Tensism." Pp. 21-36 in *Oxford Studies in Metaphysics*, edited by D. Zimmerman. Oxford: Oxford University Press, 2004.

Mason, F. "What Is Presentism?" *Southern Journal of Philosophy* 44 (2006): 107-28.

Merricks, T. *Truth and Ontology*. Oxford: Oxford University Press, 2007.

Meyer, U. "The Presentist's Dilemma," *Philosophical Studies* 122 (2005): 213-25. Chapter 8 in this volume.

Prior, A. *Papers on Time and Tense*. Oxford, UK: Clarendon, 1968.

Savitt, S. "Presentism and Eternalism in Perspective." Pp. 111-27 in *The Ontology of Spacetime*, edited by D. Dieks. Utrecht, the Netherlands: Elsevier, 2006.

Sider, T. *Four-Dimensionalism: An Ontology of Persistence and Time*. Oxford: Oxford University Press, 2001.

———. "Quantifiers and Temporal Ontology," *Mind* 115 (2006): 75-97.

Williams, C. "A Bergsonian Approach to A- and B-Time," *Philosophy* 73 (1998): 379-93.

Zimmerman, D. "Temporary Intrinsics and Presentism." Pp. 206-219 in *Metaphysics: The Big Questions*, edited by P. v. Inwagen and D. Zimmerman. Malden, Mass.: Blackwell Publishers, 1998. Chapter 17 in this volume.

6
The Presentist's Dilemma

Ulrich Meyer

Abstract

This paper defends three theses: (i) that presentism is either trivial or untenable; (ii) that the debate between tensed and tenseless theories of time is not about the status of presentism; and (iii) that there is no temporal analogue of the modal thesis of actualism.

Presentism, we are told by its advocates, is the following thesis about the relation between time and existence:

> P: Nothing *exists* that is not present.

In the recent literature, Bigelow (1996) [Chapter 8 in this volume—*Editors*] and Markosian (2004) defend presentism as the commonsensical view of time. Sider (1999) is impressed by the ontological benefits of presentism, but ultimately rejects it in his (2001, ch. 2). However, a majority of authors is opposed to the view: Putnam (1967), Sklar (1981), Savitt (2000), and Saunders (2002) all argue that presentism is incompatible with the theory of relativity, and thus false a posteriori.

There is little common ground amongst these competing views, but the one thing that all sides of the debate seem to take for granted is that the status of presentism raises an important metaphysical question. It is this shared assumption that I want to challenge. In this paper, I argue that presentism is either trivially true or obviously false.

1. The Dilemma

The first point I want to make is that the presentist's thesis is ambiguous between different interpretations of the "exists" that occurs in it. On one reading it is trivial and on the other obviously false, but there is no reading on which it expresses a substantial metaphysical truth. Similar problems were raised by Zimmerman (1998, p. 209) and Callender (2000), but the objection has not yet received the attention that I think it deserves, and Zimmerman only presents it in order to reject it immediately. I want to give what I hope to be a sharper and more persuasive formulation of the argument. (I will briefly discuss Zimmerman's defense of the presentist in Section 3.)

As I want to present the problem, there are two main ways of understanding the presentist's thesis. The first possibility is to regard the "exists" in P as the ordinary present tense of the verb to exist:[1]

P1: Nothing *exists now* that is not present.

On this reading, presentism is true, but utterly trivial. To exist now and to be present are the very same thing. Since the presentist clearly meant to advance a substantial thesis, it thus seems safe to conclude that P1 is not what he had in mind. But what could he have meant? The apparent problem with P1 is that it interprets the "exists" in too narrow a sense. To get a non-trivial account of the thesis, we would need a more inclusive notion of existence.

Say that an object *exists temporally* just in case it exists *at* some time or other. By regarding the "exists" in P as expressing temporal existence, we then get a second reading of the presentist's thesis:

P2: Nothing *exists temporally* that is not present.

Since it does not follow from an object's existing temporally that it exists now, this thesis is indeed non-trivial. But consider:

JC: Caesar crossed the Rubicon.

Caesar only could have crossed the Rubicon if he did exist. Yet if he did exist then he does exist at some past time, and therefore does exist temporally. But Caesar does not exist now; he died on the Ides of March in 44 BC. Hence if Caesar crossed the Rubicon (and he did) then there is an object (namely Caesar) that exists temporally, but is not present. If JC is true (and it is) then P2 is false.

Both P1 and P2 are readings on which the "exists" in P encodes some temporal information. The presentist might object that by "exists" he just means *exists simpliciter*, and that neither P1 nor P2 are acceptable interpretations of his thesis. But suppose an object *a* exists simpliciter (whatever that amounts to). Then either *a* exists in the actual world or it exists in some other possible world (if there are such worlds). And if *a* exists in the actual world then it either exists temporally or it exists "outside" time altogether (*at* no time, if that's possible). So whatever existence simpliciter amounts to, the following conditional is necessarily true, because its consequent is already necessary:

$$\begin{aligned} a \text{ exists simpliciter} \supset (&a \text{ exists temporally} \\ &\vee a \text{ exists outside time} \\ &\vee a \text{ exists in some other possible world}) \end{aligned}$$

Hence the only way in which a temporally "unindexed" notion of existence could differ from temporal existence is that objects outside time, or those in other possible worlds, might exist simpliciter, but not temporally. However, neither of the following claims is a recognizably presentist thesis:

> Nothing *exists outside time* that is not present.
> Nothing *exists in other possible worlds* that is not present.

At best, these are roundabout ways of denying that there are objects outside time, or objects in other possible worlds. The only respects in which existence simpliciter could differ from temporal existence are thus irrelevant for current purposes. For assessing presentism, temporal existence is the most general notion of existence that needs to be considered.

The presentist might grant this point and still object to my discussion of P2. What is at issue, he might claim, is whether names like "Caesar" have a referent. I seem to assume that the question has already been settled in his opponent's favor by using the existential generalization from JC to "There is something that exists temporally." But suppose we follow this suggestion and permit existential generalizations only for names that presently have a referent. Then the notion of existence employed is really *exists now* (and not *exists temporally*), and we're back to the trivial P1. If we don't impose such a present referent condition then we do get P2, but we also get the above problems with JC. So the problem for the presentist arises independently of our views about reference.

Perhaps the presentist has a more radical view in mind, and is prepared to reject claims like JC in order to save P2. But if that were the proposal, nothing more would need to be said in response than to repeat that it *does* deny the true claim JC. Whatever we say about the semantic value of proper names used at times at which their ostensible referents do not exist, any account that ends up rejecting all historical claims like JC is *eo ipso* unacceptable as a theory of names.

We have seen that the presentist's thesis becomes untenable when spelled out in terms of temporal existence, as we did in P2 above. While P1 interprets the "exists" too narrowly, P2 interprets it too liberally. There remains the question of whether an intermediate notion of existence could improve matters.

We can define such a notion (call it *existence**) by allowing the existential generalization

$$\text{EG: } \frac{a \text{ is } F \text{ at time } t}{\therefore \quad a \text{ exists*}}$$

only for certain choices of t, and not for all times, as we would in the case of temporal existence. For example, we might only permit the inference for times later than three minutes ago. In this case, Caesar's existence* would not follow from JC. This is a positive feature, but the revised presentist thesis

P3: Nothing *exists** that is not present,

avoids the problems of P2 only to the extent to which it is trivial. Unless we permit the inference EG only for the present time (in which case P3 would just be a notational variant of the trivial thesis P1), we would still run into problems with objects that exist*, but do not exist now.

There is thus an entire spectrum of interpretations of the presentist's thesis, with the trivial P1 and the clearly false P2 at either extreme. In between, there are views like P3, which combine the disadvantages of the two main views. Since these readings exhaust all possible alternatives, presentism is therefore either trivial, untenable, or a balanced mix of the two.[2]

2. Tensed Theories of Time

Some readers will object that the discussion in the previous section misses the point at issue. In their view, it is only when we try to spell out the, so far unanalyzed, notion of *existence at a time* that the real issue emerges. There are two ways of accounting for such time-relative existence claims (and time-relative property attributions more generally): the *tensed* and the *tenseless* view of time. The choice between these two views, they would argue, that's what the presentism debate is really all about.

I disagree. While I am happy to concede that the choice between the two accounts of time raises substantial philosophical questions, the status of presentism is not amongst them. The widely held view that tensed accounts of time endorse presentism while tenseless theories reject it is simply mistaken.

Tensed views of time, advocated most prominently by Prior (1968), propose to account for time-relative property attributions in terms of a family of primitive sentential tense operators such as:

P It was the case that. . . .	*M* At some time, ...
F It will be the case that. . . .	*L* At all times, ...

Our claim JC would be formalized as "*PRc*," where "*c*" is a singular term that denotes Julius Caesar, and "*R*" stands for the predicate ". . . crosses the Rubicon." Sentences without any tense operators are reserved for making claims about the present moment. The untensed "*Rc*" thus claims (falsely) that Caesar is crossing the Rubicon now. There is also a monadic existence predicate "*E*" that is used to account for time-relative existence claims. So "Caesar did exist" would get formalized as "*PEc*."

Tenseless theories, on the other hand, trade tense operators for quantification over substantival time points.[3] They analyze "Caesar did exist" as "$\exists t\ (t<t^* \ \&\ Ect)$," where t is a variable ranging over times, t* the present time (time of utterance), $<$ the earlier-than relation, and E a temporal "location" relation that objects bear to times.

The tensed and the tenseless view offer rival accounts of what it is for an object to exist at a time, and provide different means of defining present and temporal existence in terms of this basic notion. They can be used to spell out the theses P1 and P2 in two different ways:[4]

	P1	P2
tensed theory	$\forall x(Ex \supset Ex)$	$\forall x(MEx \supset Ex)$
tenseless theory	$\forall x(Ext^* \supset Ext^*)$	$\forall x(\exists t\, Ext \supset Ext^*)$

However, this merely explains the alternatives of the presentist's dilemma; it does nothing to dissolve it. Whichever view of time we adopt, the options for the presentist remain exactly the same. There is the trivial view P1 that nothing exists now that is not present, and the false view P2 that only present objects exist temporally. Explaining what it is for an object to exist at a time has no impact on the dilemma.

The tensed theory of time does accord a privileged position to the present moment in a way that the tenseless theory does not. But that does not show that it endorses a non-trivial version of presentism. To use a different example, suppose there were a special feature that all unmarried people have in common. To find out about this property would be a fascinating discovery, but it would not make "All bachelors are unmarried" less trivial, or "All Italians are unmarried" less false. The same holds in our case. Singling out the present as "metaphysically special" does not change the triviality of P1, nor does it alter the falsity of P2.

There are other tensed and tenseless theories of time, which provide further ways of spelling out the alternatives of the presentist's dilemma. Some tenseless views postulate temporal parts instead of time points, and some tensed accounts use different primitive tense operators than the ones listed above. There are also tensed accounts that employ a Kripke-style semantics on which the existential quantifier ranges not over all temporally existing objects, but only over those that exist at the time of evaluation. Such accounts can do without an existence predicate. But while the choice amongst these competing views of time raises significant philosophical problems (they all differ with regard to their ontological and ideological costs), the status of presentism is not amongst them. Tensed theories neither support nor require a non-trivial form of presentism, and even the proponent of the tenseless theory of time has to accept the trivial presentist thesis P1.

I noted above that opponents of presentism often argue that the view conflicts with the theory of relativity. But if there is no non-trivial version of presentism, as I have argued, then the question of its compatibility with relativity should never really arise. Empirical considerations of this type do have a role to play in the choice between tensed and tenseless views of time, where some candidate solutions seem to contradict the theory of relativity. But since the choice amongst these various accounts has no bearing on the presentist's dilemma, neither does relativity.

3. Actualism

I think the argument in Section 1 can stand as it is, but there is one further objection that is worth discussing. Some authors argue that presentism is the temporal analogue of actualism, the modal thesis that nothing exists that is not actual. Since actualism is a non-trivial thesis, they claim, the same must hold for presentism. Any argument to the contrary is bound to contain a mistake.[5]

The modal analogue of the tensed theory of time is modal primitivism, which accounts for modal claims in terms of two primitive modal operators "□" ("It is necessary that. . . .") and "◇" ("It is possible that. . . ."). Its main rival is David Lewis's modal realism (1986), which advocates a strategy similar to that of the tenseless theory of time by eliminating modal operators in favor of quantification over concrete possible worlds.

If these were the only ways of accounting for modality then we would get an actualist's dilemma to mirror the presentist's dilemma discussed earlier. If we interpret the "exists" in the actualist's thesis as actually exists then the thesis is trivial. But if we use the more liberal possibly exists (which is the analogue of temporal existence in the modal case) then actualism is clearly false. That elephants might have been pink is true, but it can only be true if pink elephants possibly exist, and pink elephants are not actual.

However, modality admits a third kind of view, which is what Lewis calls "ersatz views" of modality. Like the modal realist, the ersatz theorist eliminates modal operators in favor of quantification over possible worlds. The difference is that the ersatzer does not claim that possible worlds are aggregates of concrete individuals. On his view, possible worlds are "ways the world might have been," abstract objects that exist in the actual world.[6] Let π be the non-modal claim that elephants are pink. According to the ersatz theorist, π is possibly true, $\Diamond\pi$, just in case there is one of his abstract possible worlds such that π is true according to it. If we abbreviate "π is true according to w" as "$A(w, \pi)$" then the ersatzer offers "$\exists wA(w, \pi)$" as an analysis of the claim that elephants might have been pink—without committing himself to the existence of any non-actual objects. Whether this view of modality ought to be preferred over its primitivist and realist rivals is a different question, but ersatz actualism is at least a contender.[7]

Since the ersatzer is still taking the notion of *ways the world might have been* as primitive, he does not eliminate all modal concepts. But that wasn't his project, anyway: his aim was to make a point about the ontological commitments of modal discourse, not to reduce it to non-modal talk. The ersatz theorist successfully avoids commitment to concrete *possibilia*, but he does not avoid commitment to concrete *actualia*. Amongst his possible worlds, there is one that is special. It describes not how the world might have been, but the way the world actually *is*, and which possible world this is depends on the contingent features of the *actualia*. A possible world is actual just in case any statement ϕ that is true according to it is true simpliciter:[8]

ACTUAL: w^* is *actual* if and only if $\forall\varphi\ (A(w^*, \varphi) \supset \varphi)$.

The ersatzer cannot eliminate all unmodalized claims in favor of something like "$A(w^*, \varphi)$" ("According to the actual world, φ") because he needs to appeal to such claims in order to specify which possible world *is* the actual one.[9]

This point becomes important when we try to adapt the actualist's construction to the temporal case. Impressed by the success of the ersatz theorist, it is tempting to propose a similar strategy to the presentist. The idea would be to interpret times as "ways the present might have been"; as abstract representations of "possible presents." But this project quickly runs into trouble. While every way the world might have been qualifies as a possible world, not every way the present might have been is a time. At best, times are ways the present *was*, *is*, or *will be*, and not every way the present might have been is of this kind: some possible presents never happen. To pick out past, present, and future times amongst all possible presents, we need something like ACTUAL, only more complicated:

$$\text{TIMES: } p \text{ is } \left\{ \begin{array}{c} \textit{past} \\ \textit{present} \\ \textit{future} \end{array} \right\} \text{ if and only if } \left\{ \begin{array}{c} \forall\varphi(A(p,\varphi) \supset P\varphi) \\ \forall\varphi(A(p,\varphi) \supset \varphi) \\ \forall\varphi(A(p,\varphi) \supset F\varphi) \end{array} \right\}.$$

To illustrate this, consider our sample sentence JC, and let "γ" denote the untensed claim that Caesar crosses the Rubicon. We cannot say that JC is true just in case there is *some* possible present according to which γ is true. At best, JC asserts that there is a possible present p such that γ is true according to p *and* p is now past. The problem lies with the second conjunct, which reintroduces the temporal modifier that we are trying to get rid of.

Similar remarks apply to Dean Zimmerman's proposal for eliminating tense operators in favor of individual essences (1998, p. 211). Zimmerman suggests that problematic claims like "There is something that has existed but does not presently exist" can be analyzed as "There is an individual essence that is not now exemplified but was once exemplified." In specifying which individual essences were once exemplified, we again need to use the past tense operator, and we're back to where we started.

If we accept TIMES then tense operators cannot be eliminated in favor of quantification over abstract possible presents. It is a contingent matter which possible presents are times, and to pick out past, present, and future times amongst all possible presents, we would need to employ the very apparatus of tenses that we are trying to eliminate. So there is no temporal analogue of ersatz actualism. Presentism is either trivial or obviously false, and actualism is neither.

4. Primitive Times

The discussion in the previous section naturally leads to the question whether the presentist could improve his lot by rejecting the problematic principle TIMES. Could he not claim that it is a primitive fact which possible presents are past, present, and future times?

Someone who advocates a view like this (even though he does not put it in these terms) is Bigelow (1996). He would disagree with what I have said in Section 1, and argue that the non-trivial thesis P2 is compatible with JC. Let n be the now-stuff, the mereological sum of everything that exists at the present moment. Then what makes JC true, according to Bigelow, is that the now-stuff has the property C of being-such-that-Caesar-crossed-the-Rubicon. This is an unusual property (to say the least), but "Cn" has the advantage of being only committed to the temporal existence of the now-stuff, which is present. Moreover, Bigelow can even accept that JC entails that Caesar did exist, for that would just mean that n also has the property of being-such-that-Caesar-did-exist. While his opponents postulate "the existence of past, present and future individuals," Bigelow "trades off this ontological expenditure on individuals against a lavish outlay on properties of present individuals" (1996, p. 47) [p. 136].

In Bigelow's proposal, properties like C play a similar role as possible presents did in the previous section. The difference is that there is no principle TIMES to spoil the presentist's game. Which of these properties is a "time" only depends on the primitive fact of whether or not the now-stuff *n* possesses it. Bigelow thus successfully avoids asserting the (temporal) existence of past or future objects. But the problem is that this proposal leaves it completely unclear why we should accept "Cn" as an *analysis* of JC. On the face of it, the former has nothing to do with the latter. All we're told is that *n* has some properties that stand in certain inferential relations to one another, and that Bigelow likes to *call* them "being-such-that-Caesar-crossed-the-Rubicon" and "being-such-that-Caesar-did-exist," respectively. What we haven't been given is an explanation of what *n*'s possessing these properties has to do with Caesar and his crossing of the Rubicon.

The underlying point is that the presentist cannot relinquish TIMES any more than the ersatzer could do without ACTUAL. It is only thanks to these principles that possible presents and possible worlds are even relevant to the task of analyzing tensed and modal claims. By giving up TIMES, the presentist would abandon the project of analyzing tensed talk in terms of possible presents. Yet unless "Cn" is a plausible analysis of JC, the presentist cannot avoid rejecting the latter by asserting the former.

Faced with similar problems, Sider (1999) and Markosian (2004) suggest that the appropriate response by the presentist is to deny claims like JC. Bigelow's "Cn" should not be regarded as a proposed analysis of JC, but as a substitute. Such an exchange of the problematic JC for a presentist surrogate would have the benefit of reducing ontological commitment to philosophically suspect entities (past and future objects). However, while most philosophers would agree that we ought to simplify our ontology whenever we can do so, what we are being asked to do here is to indulge our preference for a frugal ontology at the expense of truth. Even if the presentist can offer us "Cn" as a substitute, he is still rejecting JC, and that's still bad. To paraphrase Lewis (1991, p. 59), how would you like the job of having to tell historians that they need to rewrite their history books now that philosophy has discovered that Caesar didn't cross the Rubicon because Caesar didn't exist? Caesar did cross the Rubicon; that's what started the civil war. Any philosophical view that forces us to deny claims like JC is for that reason unacceptable.[10]

Notes

1. In P1, the "nothing" is to be understood as negation, not as an existential quantifier in addition to the "exists now." Put semi-formally, P1 claims that: ~ (*exists now x*) ~ *x* is present.

2. Here, as elsewhere in this paper, I am ignoring presentists who are Dummettian antirealists about the past by identifying truth with warranted assertability. Michael Dummett's discussion of this view raises important issues, but it would go beyond the scope of this paper to try to address them here. See his (1978, 2003) for more details.

3. Versions of these views are defended by Smart (1955), Quine (1960, §36), and Mellor (1998).

4. Some readers might ask what the initial quantifiers range over. For the two readings of P1, this question does not matter, for they are in any case trivial. But for the P2 readings to be any different from the trivial P1 accounts, we have to be permitted to instantiate them with names that lack a present referent. (See also the discussion on page 3 above [p. 101].)

5. This argument is suggested in Sider (1999, sec. 1) and Zimmerman (1998, p. 211) [chapter 17].

6. Also the modal primitivist might talk about abstract "possible worlds," but he would do so in the context of providing an abstract model theory for his modal logic, and not as part of a proposed elimination of modal operators. Unlike what the realist and the ersatzer propose, this use of possible worlds is ontologically innocuous.

7. Ersatz theorists disagree amongst themselves about the precise nature of possible worlds and the "according to" relation *A*, but these differences are irrelevant here. For further details, see Stalnaker (1976) and Lewis (1986).

8. Here I am tacitly assuming that possible worlds satisfy a suitable maximality constraint that ensures that there is exactly one possible world that is actual. The details of this constraint do not matter for current purposes.

9. Note that, on the ersatz view, the actual world is a proper part of the aggregate of everything that exists actually. It is that abstract possible world that accurately describes the actually obtaining states of affairs.

10. Earlier versions of this paper were read at Wake Forest University, Lafayette College, Colgate University, and at the Joint Meeting of the Aristotelian Society and the Mind Association in Belfast. I would like to thank my audience at these places for their very

helpful comments. Special thanks are due to Adrian Bardon, David Barnett, George Graham, Ralph Kennedy, and an anonymous referee for *Philosophical Studies*.

References

Bigelow, J. "Presentism and Properties," *Philosophical Perspectives* 10 (1996): 35-52. Chapter 8 in this volume.

Callender, C. "Shedding Light on Time." Pp. S587–S599 in *PSA98,Part II: Symposia Papers*, edited by D. Howard. East Lansing: Philosophy of Science Association, 2000.

Dummett, M. "The Reality of the Past." Pp. 358–74 in *Truth and Other Enigmas*, by M. Dummett. Cambridge, Mass.: Harvard University Press, 1978.

———. "Truth and the Past: The Dewey Lectures 2002," *Journal of Philosophy* 100 (2003): 5–53.

Lewis, D. *On the Plurality of Worlds*. Oxford: Blackwell, 1986.

———. *Parts of Classes*. Oxford: Blackwell, 1991.

Markosian, N. "A Defense of Presentism." Pp. 47-82 in *Oxford Studies in Metaphysics*, edited by D. Zimmerman. Oxford: University Press, 2004.

Mellor, D. H. *Real Time II*. London: Routledge, 1998.

Prior, A. *Papers on Time and Tense*. Oxford: Clarendon Press, 1968.

Putnam, H. "Time and Physical Geometry," *Journal of Philosophy* 64 (1967): 240–47. Chapter 13 in this volume.

Quine, W. V. *Word and Object*. Cambridge, Mass.: The MIT Press, 1960.

Saunders, S. "How Relativity Contradicts Presentism." Pp. 277–92 in *Time, Reality & Experience*, Vol. 50 of *Royal Institute of Philosophy Supplement*, edited by C. Callender. Cambridge: Cambridge University Press, 2002.

Savitt, S. "There's No Time Like the Present (in Minkowski Spacetime)." Pp. S563–S574 in *PSA98, Part II: Symposia Papers*, edited by D. Howard. East Lansing: Philosophy of Science Association, 2000.

Sider, T. "Presentism and Ontological Commitment," *Journal of Philosophy* 96 (1999): 325–47.

———. *Four-Dimensionalism: An Ontology of Persistence and Time*. Oxford: Clarendon Press, 2001.

Sklar, L. "Time, Reality, and Relativity." Pp. 129–42 in *Reduction, Time, and Reality*, edited by R. Healy. Cambridge: Cambridge University Press, 1981.

Smart, J. J. C. "Spatialising Time," *Mind* 64 (1955): 239–41.

Stalnaker, R. "Possible Worlds," *Noûs* 10 (1976): 65–75.

Zimmerman, D. "Temporary Intrinsics and Presentism." Pp. 206–19 in *Metaphysics: The Big Questions*, edited by P. van Inwagen and D. Zimmerman. Oxford: Blackwell, 1998. Chapter 17 in this volume.

7
On Presentism and Triviality

Thomas Crisp

One sometimes hears the following argument put against presentism. Presentists say: "Only present things exist." But consider the verb "exist" in the quoted sentence. Is it tensed or untensed? Suppose tensed: then the presentist's thesis amounts to the claim that only present things exist now—a trivial truism. Suppose untensed: then her thesis is equivalent to the claim that only present things existed, exist, or will exist—a manifest falsehood: plenty of things (e.g., the Roman Empire) existed, or will exist which aren't present. The upshot: presentism is a trivial truism or an obvious falsehood. Either way, it's not a very interesting piece of metaphysics.[1]

Let us call this the "triviality argument" against presentism. I shall argue for two claims in this paper. First, the triviality argument is unsound. And second, its conclusion is false: presentism is neither banal nor manifestly false. I begin by spelling out the triviality argument more carefully.

1. The Triviality Argument Explained

Presentists say that only present things exist.[2] But the sentence

(Pr) Only present things exist

is ambiguous. Its verb "exist" plausibly admits of a tenseless and two tensed readings. Each reading of the verb yields a slightly different thesis. Presentism, let us assume, is one of these three theses.

We begin with the tensed readings of (Pr). There are two plausible options. We could take its verb as present tensed, yielding:

(Pr_a) Only present things exist now (i.e., at present),

But this, presumably, is not what the presentist has in mind since (Pr_a) is a trivial truism: for what is a "present thing" but a thing which exists at present?

Perhaps the presentist means (Pr)'s "exist" is to be read *disjunctively*, yielding:

(Pr_b) Only present things existed, exist, or will exist.

But (Pr_b) is manifestly false: the Roman Empire existed but isn't a present thing; my great grandchildren will exist but aren't present things. So it's false that only present things existed, exist, or will exist.

If "exist" isn't plausibly regarded as tensed, maybe it's a tenseless verb, one which indicates nothing about pastness, presentness, or futurity. Some who "take tense seriously" will object that there *are* no tenseless verbs.[3] But let us suppose they're wrong and that presentism is the claim that

(Pr_c) Only present things (tenselessly) exist.

A brief comment about (Pr_c). Some think that numbers, properties, and the like exist *timelessly* or *atemporally*. (Pr_c) should not be read as the claim that only present things exist in this way. To be a present thing is to be a temporal thing, a thing which exists *in time*. The presentist may be benighted, but he doesn't mean to claim that only present things exist timelessly or atemporally.

How then are we to understand (Pr_c)? The idea here is that we can speak of a thing as "existing" without thereby saying anything about whether it exists now, in the past, or in the future. To say of x that it exists in this sense is to claim merely that our most inclusive quantifiers range over x. x may be past, present, or future; we say nothing about *which* when we say of x that our widest quantifiers "pick it up." So, to say that only present things (tenselessly) exist is to say something like: for every x (using our most unrestricted quantifier), x is a present thing.

Could the presentist have this reading of (Pr) in mind? Once again, presumably not since it's a simple matter of logic that if, for every x, x is a present thing, then for every x, x existed, exists, or will exist only if x is a present thing. But notice: to claim that, for every x, x is a G only if x is an F is just to claim that only Fs are Gs. So to say that, for every x, x existed, exists, or will exist only if x is a present thing is to say that only present things existed, exist, or will exist. And the latter claim, recall, is just (Pr_b), a trivial falsehood. (Pr_c) implies a trivial falsehood. Since it's obvious that this is so, it's obvious that (Pr_c) is false.

We can now put the triviality argument precisely. First, we suppose that presentism is one of the theses yielded by disambiguation of (Pr). There are three plausible disambiguations, (Pr_a), (Pr_b), and (Pr_c); so

(P1) Presentism is either (Pr_a), (Pr_b), or (Pr_c).

But

(P2) (Pr_a) is trivially true, and

(P3) (Pr_b) and (Pr_c) are trivially false.

Accordingly

(C) Presentism is either a trivial truism or a manifest falsehood.

Fortunately for the presentist, the foregoing argument is unsound.

2. The Triviality Argument Examined

(P3) says that it is manifestly false that only present things existed, exist, or will exist. Says the objector: this is manifestly false because there were and will be things which aren't now present—e.g., the Roman Empire, my great grandchildren. Since it's manifestly true that there were and will be things which aren't now present, it's manifestly false that only present things existed, exist, or will exist.

But here there is confusion. Again, to say that only Fs are Gs is to say that, for every x, if x is a G, then x is an F. Thus to say that only present things existed, exist, or will exist is to say that

(Pr_b') For every x, if x existed, exists, or will exist, then x is a present thing.

(Pr_b') invokes an *unrestricted* quantifier, one that ranges over *everything*. The presentist who construes her thesis thus proposes of each thing x in our most inclusive domain of quantification that if x existed, exists, or will exist then x is a present thing. We can state the same thing differently by shifting to a restricted quantifier, one whose domain is restricted to the class of all things in time, the class of all things which existed, exist now, or will exist. (Of course, for those of us who think *all* things are temporal things, this amounts to no restriction at all.) Then we get:

(Pr_b'') For every x, x is a present thing,

where here we quantify restrictedly over the domain of all things in time—henceforth, D_t.

The triviality objector proposes that (Pr_b'') is trivially false and points to the Roman Empire as an obvious counterexample. The idea, presumably, is that D_t includes something—viz., the Roman Empire—which existed but is not present. But is it an obvious truth that D_t includes something identical with the Roman Empire which existed and is no longer present? Clearly not. Here we need to distinguish between the *de dicto* assertion that

(RE_1) *WAS* (for some x, x is the Roman Empire and x will not exist in t_α),

where "t_α" names the present moment, and the *de re* assertion that

(RE2) For some x, x was the Roman Empire and x is no longer present.

RE_1 is a *de dicto* claim predicating past truth of the proposition [for some x, x is the Roman Empire and x will not exist in t_α]. Most everyone—presentist or no—will grant that RE_1 is an obvious truth.[4] RE_2 is a *de re* claim to the effect that the open sentence "x was the Roman Empire and x is no longer present" is satisfied by some *res* in D_t. This claim isn't nearly as obvious. It's certainly no Moorean fact[5] that the domain of temporal things is still populated with something non-present and identical with the Roman Empire. Were it a Moorean fact that *eternalism*—the view that our most inclusive domain of quantification includes past, present, and future entities—is true, I suppose it would be a Moorean fact that D_t includes the Roman Empire. But it's not just *obvious* that eternalism is true: it's not just obvious that our widest domain of quantification still includes wholly past objects like the Roman Empire. If eternalism is true, we need serious argument to see that it is.

If there is an obvious truth in the neighborhood, then, it's the truth expressed by RE_1. But this truth is consistent with the claim that, quantifying over all things in time, for every x, x is present. It thus provides no counterexample to (Pr_b''). Similar comments apply to the suggestion that (Pr_b'') is false because there will be things—e.g., my great grandchildren—which aren't yet.

The triviality objector's argument for the claim that (Pr_b'') is trivially false is less than impressive. Moreover, (Pr_b'') is *not* trivially false. No doubt some philosophers take themselves to have good reason for thinking eternalism true and (Pr_b'') false. Perhaps they do have good reason. Again though, it's not just *obvious* that the domain of temporal things includes non-present objects. Eternalism might *be* true, but seeing this requires substantive metaphysical argument.

I conclude that my opponent is mistaken in thinking that (Pr_b'') is manifestly false. Thus I reject premise (P3) of her argument. The triviality argument is unsound.

3. Neither Trivially False nor Trivially True

Not only is the triviality argument unsound, its conclusion is false. Let us think of presentism as the thesis expressed by (Pr_b'').[6] According to (C), this thesis is either a trivial truism or a manifest falsehood. I've already argued that it's not manifestly false. It's obvious, I think, that it's also not trivially true. Suppose that eternalism is true. Then our most inclusive domain of quantification includes wholly past and wholly future objects, where these are objects that occupy past times or future times, but not the present time. If so, then it's false that, for every x—quantifying over all temporal entities—x is a present thing. The Roman Empire existed, for instance, and it isn't a present thing. Since, according to the eternalist, the domain of temporal things includes the Roman Empire, if eternalism is true, (Pr_b'') isn't. But eternalism isn't manifestly false: I, at any rate, can't just *see* that the domain of temporal things includes no wholly past or future entities. Since eternalism isn't manifestly false, (Pr_b'') isn't trivially true.

I conclude that the triviality argument against presentism fails. Its failure is two-fold: it is unsound and its conclusion is false.

Notes

1. Lawrence Lombard raises a version of this objection in his 1999, 254-55, as does Craig Callender in his 2000, S588-S589. For discussion, see Sider 1999, 325-27; Zimmerman 1998, 209-10 [chapter 17 in this volume—*Editors*]; Merricks 1995, 523 and 1999, 421-22; Hinchliff 2000, S576-S577 [chapter 14]; and Rea 2003.

2. This way of putting the thesis, or something close to it, is fairly common in the literature. Bigelow characterizes presentism as the thesis that "nothing exists which is not present" (1996, 35) [chapter 8, p. 125]; Markosian thinks of it as the view that "necessarily, it is always true that only present objects exist" (2004, n. 1). Merricks suggests that presentism is the thesis that "all that exists, exists at the present time" (1995, 523). Zimmerman characterizes it as the thesis that "the only things that exist are those that exist at present" (1998, 209) [p. 308].

3. See for instance Smith 1993, chapter 6; Tichy 1980, 177-79.

4. Let us set aside those of us in the grip of philosophy who doubt there are such things as empires.

5. To borrow from Armstrong: "one of the many facts which even philosophers should not deny" (1978, 440-41).

6. This way of putting presentism isn't quite right. To put the thesis most accurately, we need the "ALWAYS" operator: presentism is the claim that it's always the case that, for every x, x existed, exists, or will exist only if x is a present thing. Else, for a brief moment, presentism is true in a Broad/Tooley-style growing block universe with a first moment (see Broad 1923: 53-84; Tooley 1997). Some think that a further emendation is needed, the addition of a "□". I won't take a stand on that here.

References

Armstrong, D. M. *Universals and Scientific Realism*. Cambridge, New York: Cambridge University Press, 1978.

Bigelow, J. "Presentism and Properties," *Philosophical Perspectives* 10 (1996): 35-52. Chapter 8 in this volume.

Broad, C. D. *Scientific Thought*. London: Kegan Paul, 1923.

Callender, C. "Shedding Light on Time," *Philosophy of Science* 67 (Proceedings) (2000): S587-S599.

Hinchliff, M. "A Defense of Presentism in a Relativistic Setting," *Philosophy of Science* 67 (Proceedings) (2000): S575-S586. Chapter 14 in this volume.

Lombard, L. "On the Alleged Incompatibility of Presentism and Temporal Parts," *Philosophia* 27 (1999): 253-60.

Markosian, N. "A Defense of Presentism." Pp. 45-82 in *Oxford Studies in Metaphysics*, edited by D. Zimmerman. Oxford: Oxford University Press, 2004.

Merricks, T. "On the Incompatibility of Enduring and Perduring Entities," *Mind* 104 (1995): 523-31. Chapter 18 in this volume.

Merricks, T. "Persistence, Parts, and Presentism," *Noûs* 33 (1999): 421-38.

Rea, M. "Four Dimensionalism." Pp. 246-80 in *The Oxford Handbook for Metaphysics*, edited by M. Loux and D. Zimmerman. Oxford: Oxford University Press, 2003.

Sider, T. "Presentism and Ontological Commitment," *The Journal of Philosophy* 96 (1999): 325-47.

Smith, Q. *Language and Time*. Oxford: Oxford University Press, 1993.

Tichy, P. "The Transiency of Truth," *Theoria* 46 (1980): 165-82.

Tooley, M. *Time, Tense and Causation*. Oxford: Oxford University Press, 1997.

Zimmerman, D. "Temporary Intrinsics and Presentism." Pp. 206-19 in *Metaphysics: The Big Questions*, edited by P. van Inwagen and D. Zimmerman. Malden, Mass.: Blackwell Publishers, 1998. Chapter 17 in this volume.

Part III

Overviews of Presentism

Introduction: Overviews of Presentism

Ernâni Magalhães

Although Arthur Prior is historically the most influential presentist, the numerous serious obstacles facing the view only began to be confronted in a systematic way with the second wave of presentists in the last thirty years or so. The selections in part III represent three of the most comprehensive of these efforts. Together they indicate some of the possibilities for developing presentism, as well as some of the more significant hurdles facing the view. Although they disagree in many ways, these selections agree that presentists must explain the ontological basis for various aspects of temporal reality—truths about the past, especially—that seem to go beyond the present. In this they disagree with another group of presentists, who believe the view is better served by either providing no account at all of such facts or as minimal an account as possible. See the references to Jonathan Tallant, Brad Monton, Brian Kierland, Trenton Merricks, and Ben Caplan and David Sanson for representative works in this spirit.

Bigelow

John Bigelow's development of presentism in "Presentism and Properties" is organized around his consideration of the objection from relations, an objection that is considered in more detail by the selections in section 1 of part V. The objection notes that present things appear to be related to non-present things. Present things are caused by non-present (e.g., past) things; they resemble non-present things; and they occur after non-present things. Assuming the plausible principle that X cannot be related to Y unless Y exists, it would seem to follow that the non-present things which cause, resemble, and precede present things must exist. So, contrary to presentism, non-present things must exist.

Bigelow's response reflects the views of the Stoics and Lucretius. There is such an object as the world and there are properties which involve tense. The world is the concrete object which is the sum of all the other concrete objects. A tensed property may be either present, past, or future. Something can be presently eating, but it can also exemplify the past- or future-tense versions of that property. It can exemplify *having eaten* or *going to eat*. When they are exemplified by the world such tensed properties can characterize an entire situation. The world exemplifies the property *being such that I am writing*. The same is true for past- and future-tense properties. The world exemplifies the property *being such that Helen was abducted* and *being such that a certain man will die*. Exemplifying these properties is a way for the world to be "burdened with a certain sort of past, or ... pregnant with a certain sort of future" (chapter 8, p. 136).

Bigelow's solution to the problem of causal relations—the relation he focuses on—allows that causation is a genuine relation but denies that its relata ever exist at different times. On Bigelow's view, the world's property *being such that this man was wounded in the heart* causes the world's property *being such that the man dies*. In general, causation relates simultaneously exemplified world properties.

Bigelow's appeal to tensed properties is the basis for a general account of presentism. As Bigelow sees it, there is a symmetry between the main competing theories of time: "[Eternalism] postulates the existence of past, present and future individuals: presentism trades off this ontological expenditure on individuals against a lavish outlay on properties of present individuals" (p. 136). Bigelow's presentist ontology treats the past and the future as on a par. Past states are manifested as present past-tensed world properties. Future states are manifested as present future-tensed world properties. (Craig Bourne and M. Oreste Fiocco (chapters 9 and 10), on the other hand, maintain that a presentist should treat the future as having less of a basis in reality than the past (if it has any at all).) Bigelow's ontology is also dynamic. This dynamic character is understood not in terms of the coming into and going out of existence of concrete objects, but in terms of the changing exemplification of tensed properties (p. 136). The tensed properties that are exemplified now are different from those that were exemplified when I began writing this sentence.

Bigelow's conception of presentism allows that truths about the past and the future have some basis in reality. The world itself is different in virtue of Socrates' having been a philosopher or my grandson's future activities. Supposing there are such things as world properties, it would seem the world has the property *being such that I am typing* because I am typing. That is, the world has this property because there are real objects which are undertaking real activities. The same was true of the world and Helen's abduction. At one time, the world exemplified the property *being such that Helen is being abducted*. Later it came to exemplify the property *being such that Helen was abducted*. What is a legitimate question in the case of my exemplification of the property *writing* seems also to be a legitimate question with regard to the property about Helen. What is it in virtue of which the world exemplifies this property? It cannot be true in virtue of Helen or any of her activities, since neither are real. Bigelow can insist that there is a causal explanation for this fact, but there seems to be no ontological explanation for it. From the metaphysical perspective, it seems to be a brute fact about the world that it exemplifies the property *being such that Helen was abducted*. Of course, every theory must have its primitives, but one might complain that making such properties primitive leaves too much unexplained. For one thing, as Craig Bourne contends in chapter 9, a plausible theory of presentism—any theory of time for that matter—should provide an account of the temporal links between the exemplification of properties. (Bourne puts the point in terms of links between the truth-value of propositions.) There is an intimate connection between the world's property *being such that I am writing* and the world's property *being such that it will be the case that it was the case that I am writing*. What, if anything, about these properties guarantees that the exemplification of the former entails the exemplification of the latter?

The tensed properties Bigelow appeals to can be construed as either simple or complex. If the properties are complex then they must have components. What could these possibly be in the case of *being such that Helen was abducted*? Because there is (now) no such abduction, the complex property better not involve any such event. Since Helen is no longer, the property seemingly cannot include her as a constituent, either.[1] On the other hand, if Bigelow's tensed properties are simple, then connections which seem to arise from the nature of the properties are all left primitive. But a plausible theory about temporal relations should not allow that it is a primitive fact that the exemplification of *being such that Helen was abducted* entails the exemplification of *being such that someone was abducted*. Indeed, eternalists appear to be better off than Bigelow in answering this question. Since eternalists appeal to real individuals as constituents of events, they can make use of the connection between the properties *being Helen* and *being someone* to explain the relevant entailment.

Similarly, one wants there to be some account of the way in which *being such that Helen was abducted* fits with *being such that Brian was abducted* in a way that it does not with *being such that Brian will be abducted*. The first pair has something in common that the second does not, while the second has something in common that the first does not. How can this be accounted for on the view that the ontological basis for truths about the past and future are simple properties?

Bourne

Helpfully, Bourne describes three desiderata of any acceptable version of presentism: "(a) It must preserve our views about which statements are true and which false; (b) It must be transparent what the truthmakers are for those statements; (c) It must explain the truth-value links between various times" (p. 141). (a) reflects the desire that the theory not yield counterintuitive results about what statements are true or false. Presentism should not overrule the evident truth that Socrates was a philosopher. Since Bourne wants to capture truths about the past, (b) implies that the theory should ground these truths in reality in a perspicuous way. (c) is the requirement, mentioned above, that an acceptable theory explain why there are certain logical entailments among tensed truths. The truth of "Mary is singing" entails, for example, the truth of "It will be the case that it was the case that Mary is singing." In particular, the connection should be revealed by the nature of the facts that make these sentences true.

Bourne claims that primitive or Priorian presentism isn't up to the task. In responding to the request for an ontology, the Priorian Presentist holds that "there *is* nothing more to say about the nature of time than is said by a perspicuous tense logic" (p. 142). Appealing to a primitive fact that Socrates taught Plato won't do the trick because we're not told anything about the "constituents" or "structure" of these facts (p. 143). Reductive presentism, which identifies the truthmakers for truths about the past with present evidence and the laws of nature, is also rejected. On reductive presentism, past-tense statements are made true by the evidence of past events (p. 145). "Mary ate a sandwich" is made true by the presence of the detritus of the meal in her stomach, by the uneaten portions on the plate, and so on. One major difficulty for the view is finding a basis in present reality for truths about what laws of nature there are. These truths concern not only the present but the past and future. But there is little hope for supposing that present evidence entails that the future and past are as the laws of nature require them to be.

After rejecting these alternatives, Bourne puts forward his own proposal, which is similar in many ways to Bigelow's. (Although Bourne appeals to propositions rather than tensed properties, Bigelow in fact says that a world property just is a proposition.) Where the eternalist posits the existence of numerous concrete past and future times, the ersatzer presentist maintains there are abstract objects which serve the function of the eternalist times—ersatz times. By analogy with ersatz worlds, which stand in for real but non-actual possible worlds, ersatz times are presentist-friendly abstract objects. An ersatz time, on Bourne's conception, is a pair set made up of a maximally consistent proposition and a date, where the set stands in the ersatz *earlier than* relation to some other ersatz times (pp. 148-49). Ersatz precedence is understood to be the abstract analogue of precedence. Intuitively, the maximal propositions that constitute ersatz times describe the concrete events that go on at a certain concrete time. Where the eternalist bases truths about other times in concrete non-present times, the ersatzer bases such truths in abstract descriptions of non-present times. Past-tense sentences are made true in virtue of pairs involving propositions whose date index is smaller than the date index of the time that obtains, i.e., is present. Bourne's ontology for the future is more complicated. Whereas the E-relation is one-one in the past direction, it is one-many in the future direction. The ersatz times that are E-earlier than a given time form a line. The ersatz times that are later compose branches. Bourne accepts a version of the open future account of time on which contingent future-tense statements are not (determinately) true or false at the time of their utterance.

Bourne claims as an advantage of his theory that it can accommodate the truth-value links between truths at different times. This is accomplished by the structure of the E-relation. But how does ersatzer presentism fare in accounting for plausible logical and ontological connection between facts about the past? That Socrates was a philosopher guarantees that Socrates was conscious. What in Bourne's ontology can account for this relationship? There is a proposition, *Socrates is a philosopher*, and there is a proposition, *Socrates is conscious*. Why does the presence of the former proposition in an ersatz time E-earlier than the present guarantee the presence of the latter proposition at the same time? If propositions involved as constituents the items and properties they are about, then the connections between them would be intelligible. The exemplification of *being a philosopher* entails the exemplification of *being conscious*. If both properties are present in the relevant propositions, and they have a subject in common, then it would make sense that the proposition about Socrates' being a philosopher would entail the one about his being conscious. But if these propositions do not involve Socrates or his properties as constituents—as they seemingly cannot, given presentism—then it would seem the entailment between them is a brute fact.

Fiocco

Fiocco thinks of himself as advocating a non-orthodox version of presentism, one which is not consistent with the traditional claim that everything is present. Although he does accept the existence of non-present things, in literal conflict with the thesis, in fact, presentists typically allow that timeless things are not to be counted among the range of "everything."[2] The non-present things Fiocco believes are real are in fact not temporal at all. Nonetheless, Fiocco's is an unorthodox variety of presentism. This is because the timeless entities he believes are real are past. Because past entities are traditionally held to be temporal, conceding their existence would typically be felt to violate not only the letter but the spirit of presentism. For Fiocco, the present moment is the only real moment. Nothing exists at it but present things. As present things cease to exist they leave a mark on the world. This mark takes the form of facts: The fact that Caesar crossed the Rubicon, the fact that Roosevelt was inaugurated, and so on. These are real past facts. Although they are timeless, they are not always equally existent. Because of the unreality of the future, prior to his crossing it, there was no such fact as Caesar crossing the Rubicon. The fact only came to be thereafter. But a past fact does not come to be at any given moment since it does not exist at any moment. So temporal reality is exhausted by present items; but reality as a whole includes many non-present, past things.

According to Fiocco, one of the strengths of the view is its ability to explain the "structure of temporal reality" (p. 163). Why are some things true of time and others not? A plausible theory of time should explain why it is true that

"Socrates was a philosopher" but not true that "Socrates was a fisherman." This explanation, moreover, should be grounded in reality. The truth of the first statement, Fiocco says, is accounted for by the existence of the past fact. "Socrates was a fisherman" is not true because there is no such fact. Other truths about the past as well as relations between the present and the past are grounded in the qualities and relations of these past facts.

How does Fiocco's theory fare in explaining why Socrates' having been a philosopher entails that he was conscious and other logical or ontological connections between truths about the past? Fiocco's past facts are simple (p. 164). Although they have properties and stand in relations, they have no constituents. So in particular neither Socrates nor the property of being a philosopher is a constituent of the fact that Socrates was a philosopher. For Fiocco, both the fact that Socrates was a philosopher and that he was conscious are real. Why should the obtaining of the one entail the obtaining of the other? Since these facts have no structure there can be nothing in the structure of either fact that guarantees the connection between them. Fiocco does postulate that facts have properties and stand in relations. The fact that Socrates was a philosopher concerns philosophy. But unless facts also have intrinsic properties, all these relations must be external—they must arise independently of the nature of the facts related. But if any relation depends on the nature of the related items it is surely entailment. Because they lack any structure, all the properties and relations of facts must be brute. Whereas it seems many properties and relations of facts are amenable to further explanation.

Notes

1 . Taking the property to involve Helen's individual essence—the unique property that can only be exemplified by her—seems similarly unpromising. Since it was Helen and not her essence that was abducted, it seems irrelevant to the real character of the past that the world should exemplify a property connecting her essence with abduction.

2. See Crisp (2004), Mason (2006), and Craig (2003).

References and Further Readings

Caplan, B., and D. Sanson. "The Way Things Were." *Philosophy and Phenomenological Research* (forthcoming).

Craig, W. L. *The Tensed Theory of Time*. Dordrecht: Kluwer Academic Publisher, 2000.

———. "In Defense of Presentism." Pp. 390-408 in *Time, Tense and Reference*, edited by Q. Smith and A. Jokic. Cambridge, Mass.: MIT Press, 2003.

Crisp, T. "Presentism and the Triviality Objection." Pp. 15-20 in *Oxford Studies in Metaphysics*, volume 1, edited by D. Zimmerman. Oxford: Oxford University Press, 2004. Chapter 7 in this volume.

Le Poidevin, R. *Change, Cause and Contradiction*. Basingstoke: Macmillan, 1991.

Ludlow, P. *Semantics, Tense, and Time*. Cambridge, Mass.: MIT Press, 1999.

Markosian, N. "A Defense of Presentism." Pp. 47-82 in *Oxford Studies in Metaphysics*, edited by D. Zimmerman. Oxford: Oxford University Press, 2004.

Mason, F. "What Is Presentism?" *Southern Journal of Philosophy* 44 (2006): 107-28.

Merricks, T. *Truth and Ontology*. Oxford: Oxford University Press, 2007.

Monton, B., and B. Kierland. "Presentism and the Objection from Being-Supervenience," *Australasian Journal of Philosophy* 85 (2007): 485-97.

Mozersky J. "Presentism." In *The Oxford Handbook on Time*, edited by C. Callender. Oxford: Oxford University Press, forthcoming.

Prior, A. "Changes in Events and Changes in Things." Pp. 1-14 in *Papers on Time and Tense*, by A. Prior. Oxford: Clarendon Press, 1968. Chapter 3 in this volume.

———. "The Notion of the Present," *Studium Generale* 23 (1970): 245-48.

Smith, Q. "Time and Degrees of Existence: A Theory of 'Degree Presentism.'" Pp. 119-36 in *Time, Reality & Experience*, edited by C. Callender. Cambridge: Cambridge University Press, 2002.

Tallant, J. "Ontological Cheats Might Just Prosper," *Analysis* 69 (2009): 422-30.

Zimmerman, D. "The Priviledged Present: Defending an 'A-Theory' of Time." Pp. 211-25 in *Contemporary Debates in Metaphysics*, edited by D. Zimmerman, T. Sider, and J. Hawthorne. Malden, Mass.: Blackwell Press, 2008.

8
Presentism and Properties

John Bigelow

I am a presentist: nothing exists which is not present. I say that this was believed by everyone, both the philosophers and the folk, until at least the nineteenth century; it is written into the grammar of every natural language; and it is still assumed in everyday life even by philosophers who officially deny it. The content of this doctrine is not easy to grasp, is best explained by comparing presentism with its rivals, and will I hope become somewhat clearer during the course of this paper.

Presentism was assumed by everyone everywhere, until a new conception of time began to trickle out of the high Newtonianism of the nineteenth century. The Christians' Holy Bible says that there is no new thing under the sun but this is not true, at least not in the sense which first comes to mind out of context. The so-called fourdimensionalist theory of time was something genuinely new, when it gradually came into being last century. Anti-presentist theories began to be articulated by philosophers late in the nineteenth century and have been expounded through the twentieth century. Bergson denied presentism and influenced literary figures like Proust and Eliot. Russell, Reichenbach, Smart, Quine, Lewis, and many other anglophonic, analytic philosophers also denied presentism though in a different way. (I make no pretense to understand where Heidegger's *Being and Time* stands with respect to presentism and fourdimensionalism.) The defenders of fourdimensionalism are very impressive. I have, however, been encouraged in my hankering for presentism by papers and conversations with Phil Hanson, Steve Savitt, Mark Hinchliff, Andrew Holster, Caroline Lierse, David Oderberg, George Schlesinger, and Quentin Smith and others.

Furthermore, the exponential explosion of time travel stories in the popular media, beginning late in the nineteenth century, is an indication that a very new conception of time is brewing in the Zeitgeist. The utter absence of any time travel stories whatsoever prior to the nineteenth century is a profoundly puzzling fact. See references for Ash, Bailey, Davenport, Philips, and Scholes and Rabkin for material on the emergence of time travel stories. I suggest that this is at least partly explained by the utter universality of presentism prior to the nineteenth century and by the utter absence of any rivals to presentism.

Presentism is a metaphysical doctrine, and it is hard if not impossible to prove. I do not propose to try. When properly understood, presentism is also hard to disprove. The emergence of a rival to presentism is not to be explained by the appearance of any cogent refutation, but by factors more diffuse than explicit argumentation. In any case, how would you go about refuting a doctrine like presentism? Two strategies of refutation suggest themselves: you could try to refute presentism by scientific observation and experiment; or you could try to refute it by sheer logic. If presentism could be proved false by a scientific experiment then, it would seem, presentism would not be a "metaphysical" matter after all but would be a merely empirical hypothesis. Initially, that seems improbable, though we should return to this option. So should we try to refute presentism a priori? One way to do that is by reductio ad absurdum, showing that presentism entails a contradiction. Another way is by showing that presentism is meaningless. Yet, I say, presentism is neither meaningless nor self-contradictory. This leaves us in a typically metaphysical predicament. It is hard to see how to settle the issue: presentism is hard if not impossible to prove, yet it is also hard if not impossible to disprove. Other metaphysical predicaments are very similar in this respect; I think particularly of issues surrounding what is known as modal realism, the doctrine impressively defended by David Lewis. (See references for Adams, Lewis, Loux, and Plantinga for more on actualism and modal realism or possibilism.)

Yet when not properly understood, it may appear that presentism is not so hard to disprove. There is a very impressive argument which may be mounted against presentism: I will call it the argument from relations. This argument is too good to be true: presentism is not the sort of doctrine which could be refuted by such a simple argument.

I used to think I knew a way of answering the doubts raised by the argument from relations. Then I read some ancient Greek texts and found that I was going up almost the same blind alley that others had gone up before. Past presentists had worked out the raw materials for a much better answer to the argument from relations. I will draw on ideas from Stoics and Epicureans of ancient Greece and Rome.

In defending presentism against the argument from relations, an opportunity arises for me to give reasons for two propositions which some contemporary philosophers have been inclined to doubt. In some moods, I feel embarrassed to find myself arguing for these propositions: they seem so obvious that it strikes me as frivolous to waste time defending them. They seem obvious, for instance, when I imagine myself talking to scholars of ancient thought. Yet in other moods, for instance when I imagine myself talking to contemporary analytic philosophers, it feels as though doubts about these propositions are very profound. The two propositions I want to defend in passing, while discussing the argument from relations, are the following. First, I think the discussion will show that there is such a doctrine as presentism, and that it is more than verbally distinct from a rival doctrine which is very influential, nay orthodox, in certain circles. And secondly, I will be showing that this doctrine has actually been believed, and that indeed presentism was accepted unquestioningly by the people I will be discussing. It is instructive, to analytic philosophers at least, to see not only that at least some of the ancient thinkers were presentists, but also that even when the opportunity to doubt it is handed to them on a platter, that option is not taken seriously.

Here is how the argument from relations runs:

> Take as a first supposition that, in order for a relation to hold between two things, both of those two things will have to exist. Call this the principle that all relations are existence entailing. Add as a further premise the supposition that relations sometimes hold between a present thing and something else which is not present. The conclusion follows ineluctably, that some things exist which are not present.

The first premise in this reasoning is one which initially strikes me as implausible. I will first give reasons why it is probably best to accept this premise, despite some initial intuitive objections which can be mustered. Then I will explain how the argument from relations could be rewritten, if you are unconvinced by the reasons for thinking that all relations have to be existence entailing.

Consider first the principle that all relations are existence entailing, that if a relation holds between two things then both of those things must exist. A two-place relation may be said to hold between a relator and a relatee, and the role of the relator may be different from that of the relatee. When Othello loves Desdemona, for instance, there is a directedness to the relation, and the role of the lover is different from that of the beloved. When wondering if all relations must be existence entailing, there are two questions we should ask. There is a question about the relator, and a question about the relatee. As we might put it: Is every relation existence entailing on the left? and is it existence entailing on the right?

To get a grip on these questions, apply them to Othello's relation to Desdemona. Is it true that, in order for Othello to love Desdemona, Othello must exist? That seems problematic. As a matter of fact, Othello does not exist: he is just a fictional character from a play by Shakespeare. Yet that does not stop it from being true to say that Othello loves Desdemona. It is not obvious that his nonexistence prevents it from being a truth that one can assert about him, that he loves Desdemona. So it is not obvious that the relation of loving is existence entailing on the left. Now ask whether it is existence entailing on the right. Does the fact that Othello loves Desdemona entail that Desdemona exists? Surely not, not even if Othello had been a real person and not just a character in a play. Love is one of the network of notions in folk psychology, and a great many of these notions have a characteristic called intentionality. That is, many mental states are directed upon an object, and the relation between a mental state and its object is a queer one, very unlike ordinary physical relations among objects. The term "intentionality" comes to us through Brentano and Meinong, and an important feature of intentionality that they highlighted is that intentional relations are not existence entailing on the right. You can think about, pity, or love something which does not exist. The magic of consciousness, especially of the imagination, is that it can conjure up things which do not exist: things which never did, do not, and never will exist, and even things which could not possibly exist. See references for Routley and Parsons for more on nonexistent objects. Thus, it would seem, relations do not have to be existence entailing on either the left or the right, but it is especially striking that intentional relations do not have to be existence entailing on the right. So the first premise of the argument from relations cannot stand up to scrutiny, and this attempted refutation of presentism fails.

Sound-minded philosophers should not be persuaded by this defense of presentism. There is something fishy about it. The principle that relations are existence entailing can be defended against the counterexamples which are raised against it. The defenses are, however, very tricky. The level of confidence that there is something wrong with the counterexamples is very high, among many sensible philosophers; yet the ability to articulate what is wrong with the counterexamples is very low. It is not the case that people notice an obvious flaw in the counterexamples, and then conclude that the argument resting on them does not work. On the contrary, people seem to know a priori that there must be something wrong with the counterexamples, because they know a priori that all relations must be existence entailing.

What reasons could people have for the high degree of confidence in the principle that all relations must be existence entailing? Here is a train of thought that may lie behind the principle. There is a deep assumption behind much of our thought, that every truth (or at least, every simple truth about how one thing is related to another) requires a truthmaker: whenever something is true (or at least, every simple truth about how one thing is related to another), there must exist some thing or things in the world in virtue of which this is true. As I prefer to put it in general: truth supervenes on being—there could not be a difference in what is true unless there were a difference in what exists.

This truthmaker principle has a strong grip on us. Its grip is strong enough that it can be used to guide us in diagnosing when we are really faced by a truth about how things are related to one another, and when we are faced by a truth of some other kind. Consider for example the truth that Othello loves Desdemona. If we think that this is true even though Desdemona does not exist, does this show that the relation of love is not existence entailing on the right? Or does it show that the truth in question is really just a truth about Othello, not about Desdemona or about how Othello is related to Desdemona? The latter answer is the better one. If you say that Othello loves Desdemona, in a sense which does not require the existence of Desdemona, then what is said about Othello is really just that he has a certain property, not that he stands in a certain relation to Desdemona. And if we think it to be a truth that Othello loves Desdemona even though Othello does not exist, does this show that the relation of love is not existence entailing on the left? Or does it show that the truth in question is not really a truth about Othello at all? The latter answer is the better one. If you say that Othello loves Desdemona, in a sense which does not require the existence of Othello, then what is said is really not something about Othello at all, but about something else, perhaps we are just saying something about Shakespeare's play (whatever that is) or about Shakespeare and what he said.

It is, I maintain, an a priori truth that a two-place relation can only be manifested when it holds between two things, and in order for this to be so there must be two things which stand in the relation. And in saying "there must be" two things which stand in the relation, one really is asserting that "there must exist" two things—one is committed to the existence of those things. The principle of the existence entailment of relations is an a priori truth. Hence the first premise in the argument from relations, against presentism, should be granted. Or so I say.

Yet suppose that I am wrong about the existence principle. Suppose that some relations can hold between two things even if one or both of those things are nonexistent. Even if some relations are not existence entailing, there may yet be other relations which are existence entailing. Consider causation. Perhaps relations involving intentionality do not require the existence of both related things, but maybe the causal relation is different. In order for the causal relation to hold between cause and effect, it might be maintained, both cause and effect have to exist.

Consider an objection: it is true that Iago deceived Othello, and this is a causal claim, yet neither Iago nor Othello ever did or will exist. So, it seems, not even the causal relation is existence entailing. Even if we grant this, there is another, more indirect existence entailment waiting in the wings. Suppose we were to grant that the causal relation might hold either between two existents or between two nonexistents. What of the other two possible permutations: could the causal relation hold between an existent and a nonexistent, or between a nonexistent and an existent? Surely not. If the causal relation holds between two things and the cause exists, then the effect must exist as well. And if the causal relation holds between two things and the effect exists, then the cause must exist as well. This at least seems pretty secure. The causal relation is, we might say, existence symmetric.

The existence symmetry of causation is all that we need in order to mount an effective objection to presentism. The argument from relations begins with an extremely strong premise: that no relation can hold between any two things unless both of the related things exist. Replace that strong claim with the much weaker claim that the causal relation is existence symmetric. There will then be a very plausible replacement we can make for the second premise of the argument, and the argument will validly entail the falsity of presentism. The argument will run like this:

> Causation is existence symmetric: if an event exists and it is a cause of some other event, then that other event exists; and if an event exists and it is caused by some other event, then that other event exists. Some present events are caused by events which are not present. And some present events are the causes of other events which are not present. Therefore things exist which are not present.

This is a forceful argument, I admit. Call it the argument from causation.

Michael Tooley has recently championed a theory which is tailored to preserve as much as possible of presentism, while giving a huge concession to the argument from causation. Tooley grants that some present events exist which were caused by other events, and that therefore past events exist. This is an abandonment of strict presentism. Tooley's world contains all past events together with the present. Yet Tooley denies the existence of any future events. Tooley is not alone in holding a lopsided ontology with a full past and an empty future: Bergson was one of the earliest and most influential advocates of such a theory. Broad was another advocate of such a theory.

Tooley blocks the argument for the existence of future events by denying that it is ever true that a present thing is the cause of some future thing. It may turn out that a present thing brings about a future thing, and when the future thing happens then you will be able to say that it is the effect of the earlier event. But when the earlier event was happening, it was not the cause of the later event simply because at that time the later event did not yet exist. According to Tooley, it is true of a present event that it is the effect of some earlier event, but it is not true of any present event that it is the cause of a future event. By contradicting the third of the premises in the argument from causation, Tooley manages to block inference to the existence of future events. Yet, he argues, it would be unreasonable to deny the second premise as well: to deny that present events are the effects of prior causes. That would be unreasonable because it would banish causation altogether from the world. But you can deny the existence of the future, without giving up on causation.

Ancient philosophers of the Hellenistic period were not tempted by the Bergson, Broad, and Tooley theory. Yet they held onto presentism, even though they were well aware of the argument from causation. How did they do it? Did they abandon the principle of existence entailment for relations in general, or of the existence symmetry of causation? No, surprisingly, they managed to retain both the principle of existence entailment and presentism. Here is how the trick can be carried off, as explained by Sextus Empiricus in an exposition of the doctrines of the Stoics. To begin with, he says:

> In addition, they [the Stoics] say that the sign must be a present sign of a present thing.

This is a striking sentence. I was struck by the use of the term "sign" here. Are the Stoics denying that a symbol, a representation, a word, a concept, an idea, a thought can refer to something which is not present? If so, then they are concerned to defend the most general principle of existence entailment, that all relations, even ones which are alleged to involve intentionality, must be existence entailing. As the passage continues, it becomes clear that the word "sign" is being understood as a something like a causal relation. This does not mean that the concerns expressed do not touch on the intentionality of "signs" in the sense of representations. I have long been sympathetic to a causal theory of reference. Language, however, is not in fact the immediate focus of the doctrine under scrutiny. The "signs" that the Stoics are concerned with are natural signs, as when we say that a footprint is a sign of the animal which made it. Here is how the passage continues:

> For some people make a mistake and want a present thing to be also a sign of a past thing, as for instance, "If this man has a scar, this man has had a wound." For "He has a scar" is something present, since it is evident, but his having had a wound is past, since there is no longer a wound. They also want a present thing to be a sign of a future thing, as is encompassed in a conditional like, "If this man has been wounded in the heart, this man will die." For they say that the wound in the heart is already present, but the death is in the future.

What we have here are two examples, in one an effect is taken as a sign of the cause, in the other, a cause is taken as a sign of an effect. The latter example weaves in a further complexity, by its reference to conditionals, and by following this lead we will probably find that one thing can be a "sign" of another even if it is neither cause nor effect of that other. As long as there is a law which correlates two things, one will be a "sign" of the other, even if the law is some sort of law of conservation, or of symmetry, or of say the harmony between the microcosm with the macrocosm, a law which says nothing about the causal mechanisms if any which link the correlated entities. Or one thing may be a sign of another when both have a common cause. However, let us stick to the simple cases where one thing is a sign of another because there is a causal relation between them.

The Stoics say that it is a mistake to think that the causal relation can hold between a present thing and some other thing which is not present. That is, they say that the causal relation is presence symmetric. Why do they think causation is presence symmetric? I am sure that this assumption rests on the deeper assumption that causation is existence symmetric. And the reason they assume that causation is existence symmetric is that they assume the fully general principle of existence entailment: that all relations are existence entailing on both the left and the right.

For present purposes, however, we have enough to worry about with causation alone. The Stoics were clearly aware of the argument from causation against presentism. Yet they did not revise their commitment to presentism. As the exposition given by Sextus Empiricus continues:

> Those who make such statements [that a scar in the present is a sign of a wound which is past, or that a wound in the present is a sign of a death which is future] do not realize in fact that, though the past and the future are different, the sign and its object, even in these cases, is a present thing of a present thing.

How did the Stoics get away with this stubborn adherence to presentism, in the face of the devastating argument from causation? Here is how:

> . . . the sign and its object, even in these cases, is a present thing of a present thing. . . . For in the earlier example . . . the wound has already happened and is past, but "This man has had a wound," which is a proposition, is present, though it is said about something which has happened. And in the case of, "If this man has been wounded in the heart, this man will die," the death is in the future, but the proposition, "This man will die," is present, though it is said about what is future, because it is true even now.

This is the Stoic defense against the argument from causation: the causal relation holds between true propositions, and a proposition exists provided that it is true. Note that the passage does not imply that false propositions exist. The proposition, "This man will die," is said to be present "because it is true even now." Thus, causation is construed, not as a relation between a present thing, a wound, and a future thing, a death, but as a relation between two presently true propositions. The causal relation holds between the true proposition that this man is wounded in the heart, and the true proposition that this man will die. Later, when the man is dying, the causal relation holds between two different true propositions: that this man was wounded in the heart, and that this man is now dying. As time passes, some things pass out of existence and others come into being, and among the things which pass away or come into being are true propositions. A proposition which was once true, that the man will die, ceases to be and a new true proposition comes into being, that this man is dying. The Stoic ontology includes no things which are not present, but it does include present true propositions, and it requires that which of these true propositions exist is something which changes as time passes. The Latin word here translated as "true proposition" is *lekton* (plural, *lekta*).

In recent times Arthur Prior has articulated a theory which is in some respects reminiscent of the Stoic theory, and this is probably no accident: Prior was well versed in classical sources of logic. Prior, however, did not think of propositions as coming into or going out of existence as time passes. As a first step in grasping Prior's version of presentism, think of propositions as existing eternally, but as changing in truth value as time passes. Prior would agree with the Stoic that the proposition "This man will die" is present—but not that it is present "because it is true even now." Even after the man dies, the proposition "This man will die" is present. But before the man died this proposition was true—after the man died it ceased to be true. For Prior, true propositions do not go out of existence—but sometimes they do cease to be true. Prior's theory is a little different from the Stoic theory reported by Sextus. Yet the difference should not be exaggerated. Both preserve presentism, and both do so by appeal to the present existence of propositions. Cause is not, they say, prior to effect: both cause and effect are present propositions.

Thus, the Stoics faced the argument from causation squarely, but they did not renounce presentism. They did not, for instance, consider adopting the Bergson, Broad, and Tooley picture, in which past things exist as well as the present, though the future does not. Nor did they consider the Russell and Quine picture in which a linear sequence of past, present, and future things all exist. Nor did they consider a picture in which what exists includes a linear past and a branching future of possibilities. The idea of branching futures is in Aristotle, and is something which would have been taken seriously by the Stoics, but only if it is understood in such a way as to be consistent with presentism. Although the Stoics would have taken it as a serious question whether there are branching future possibilities, they would not have taken seriously the idea that any future things exist which are not present. Nor would Aristotle. None of the ancient Greek thinkers took seriously the idea that there might exist things which are not present.

Recently, Storrs McCall has articulated a contemporary version of the Aristotelian picture of branching future possibilities; but in the wake of Russell and Quine it is not easy for him to do so without leading us into thinking of a picture which is not presentist, and hence is not quite Aristotle's. On Aristotle's theory, there exist in the present many distinct potentialities; but that does not mean that there exist any things which are not present. The potentialities, you see, are present. The Stoics would understand this as meaning that there are present, true propositions about potentialities. Believing as they do in stern necessity, the Stoics would question whether there are indeed such present, true propositions about branching potentialities for the future, but they would understand what was being claimed. They were not, however, taking seriously the idea that it is true in the present that future things exist even though they are not yet present.

The Stoics preserve presentism, and they keep all non-present things out of their ontology, but only at the cost of positing the existence of true propositions. Some people might not be happy with that ontological commitment. The Epicurean philosopher Lucretius, for instance, was very unhappy about allowing the existence of such things. His ontology was very sparse indeed. Lucretius says that:

> nothing exists that is distinct both from body and from vacuity and could be ranked with the others as a third substance.

All that exist are just atoms in the void, and properties of these:

> You will find that anything that can be named is either a property or an accident of these two. A property is something that cannot be detached or separated from a thing without destroying it, as weight is a property of rocks, heat of fire, fluidity of water, tangibility of all bodies, intangibility of vacuum. On the other hand, servitude and liberty, poverty and riches, war and peace, and all other things whose advent or departure leaves the essence of a thing intact, all these it is our practice to call by their appropriate name, accidents.

The ontology of Lucretius is very sparse. If he is to maintain presentism, he must deny the existence of things which are not present—but he cannot trade off this ontological economy against an added ontological commitment to extra things, like the true propositions posited by the Stoics, not at any rate if true propositions are construed as things other than atoms, vacuity, or properties or accidents of these. The ontology of Lucretius does not include the true propositions posited by the Stoics. And so, it might seem, Lucretius is in a very tight bind indeed if he wants to hang onto presentism. How can Lucretius consistently maintain presentism in the face of the arguments from relations, and from causation? Here is how Lucretius responds to one interesting, and presumably Stoic argument for the existence of something more than bodies, vacuity, or properties and accidents of these:

> Again, when men say it is a fact that Helen was ravished or the Trojans were conquered, do not let anyone drive you to the admission that any such event is

> independently of any object, on the ground that the generations of men of whom these events were accidents have been swept away by the irrevocable lapse of time. For we could put it that whatever has taken place is an accident of a particular tract of earth or of the space it occupied.

It is interesting that he adds "or of the space it occupied." Without this addition, a plausible counterexample would have come to mind: What if the earth had been removed? For Lucretius the earth could not have ceased to exist, but could only have been moved to another location, or scattered. Perhaps being scattered would not debar it from still bearing the accident which he attributes to it: perhaps it would still be true of this scattered earth that it is the earth on which Helen was ravished or the Trojans were conquered. But even if you would rather not try to plant such an accident on such scattered earth, we can in any case rely on the continuing existence of the region of space which that earth once occupied. It is true of this region that it is the region in which Helen was ravished or the Trojans were conquered.

Lucretius goes on to argue for his proposal that "whatever has taken place" should be taken to be an "accident" of matter or of space. He does so by arguing that it has only a dependent existence in the same way that servitude and liberty, poverty and riches, war and peace have only a dependent existence: none of these things can exist unless there are individuals who "have" them. Likewise for past events:

> If there had been no matter and no space or place in which things could happen, no spark of love kindled by the beauty of Tyndareus' daughter would ever have stolen into the breast of Phrygian Paris to light that dazzling blaze of pitiless war; no Wooden Horse, unmarked by the sons of Troy, would have set the towers of Ilium aflame through the midnight issue of Greeks from its womb. So you may see that events cannot be said to be by themselves like matter or in the same sense as space. Rather, you should describe them as accidents of matter, or of the place in which things happen.

Lucretius is translated here as speaking of facts or events rather than true propositions, but the issue is really the same as that faced by the Stoics. He reasons here, that past events cannot be taken as existing in the same way as matter and space, and what this means is that they cannot be taken to be "substances" in the way that matter and space can. Substances are things whose existence does not depend on the existence of anything else. The facts, or true propositions, or events concerning Helen and so forth are not substances, Lucretius argues, because they have only a dependent existence. They could not have existed if space and matter had not existed. Hence they cannot be substances, but must be properties or accidents, and in the present case it is clear that they must be accidents rather than properties in Lucretius's sense, that is, essential properties.

Thus, Lucretius displaces the Stoic postulation of true propositions by a postulation of accidents of matter or of space. The rival postulation, however, does one of the same jobs that the Stoic postulation was designed to do: it enables

Lucretius to preserve presentism in the face of the argument from relations. What exists in the present is a tract of land which has the accident of being the earth on which a spark of love kindled a dazzling blaze of pitiless war. Or, what exists in the present is a region of space which has the accident of being the space within which a Wooden Horse set the towers of Ilium aflame through the midnight issue of Greeks from its womb. All that is present are matter, vacuity, and properties and accidents of these; and among these properties and accidents are some which are expressed in language by using the past or the future tense. We do not need to suppose the existence of any past or future things, however, only the possession by present things of properties and accidents expressed using the past or future tenses. And furthermore, time itself, as well as past and future things, has to be construed as nothing more than properties or accidents of present things. As Lucretius puts it:

> . . . time by itself does not exist; but from things themselves there results a sense of what has already taken place, what is now going on and what is to ensue. It must not be claimed that anyone can sense time by itself apart from the movement of things or their restful immobility.

This is reminiscent of Aristotle's dictum that "time is the measure of motion with respect to before and after": present things have present properties, like motion or rest, and these are the ontological ground of the past, the future, and the passage of time. I suggest a modification to the Lucretian doctrine. One of the things that exists is the whole world, the totality of things that exist. The world can have properties and accidents, just as its parts may have. It is a present property of the world, that it is a world in which Helen was abducted and the Trojans were conquered. A Stoic's true proposition can be identified with a property of the world as a whole. This identification of propositions with world properties has a lot to be said for it, and it meshes very neatly with work in semantics. Since Frege, we have been accustomed to let sets play as surrogates for properties (or accidents). In place of a property of things, we talk of the set of things which have that property, the *extension* of the property. (See especially "Logic and the reification of universals" in Quine (1953) for the shift from properties to sets.) The formal semantical stand-in for a property of a whole world, therefore, will be a set of worlds. And in formal semantics, a set of possible worlds is called a *proposition*. (See references for Stalnaker, Powers, and Forrest for more on propositions and world properties.) I think it is a mistake to conflate a property with its extension, and I think it is a mistake to conflate a world property with a set of possible worlds. But it is not a mistake to identify a proposition with a property of the world taken as a whole. If we construe propositions in that way, the difference between the Stoics and Lucretius evaporates.

What emerges from this merging of Stoics with Lucretius is an economical answer to the challenge presented by the argument from relations, and the argument from causation. The causal relation does not, in fact, ever hold between things that exist at different times. At any given time the causal relation holds between properties, perhaps between world properties, each of which is present and is presently instantiated. These properties may include things like the property of being burdened with a certain sort of past, or (as Leibniz put it) being pregnant with a certain sort of future. (In thinking about these properties, my mind is drawn to McTaggart, and to remarks on McTaggart in papers by Dummett and Shoemaker.) But the admission of such properties does not amount to the abandonment of presentism.

Presentism is not refuted by a simple *reductio ad absurdum*: it is a logically consistent theory. Its rival postulates the existence of past, present, and future individuals: presentism trades off this ontological expenditure on individuals against a lavish outlay on properties of present individuals. (The same thing arises, I urge, in the metaphysics of modality: actualism trades off an ontological expenditure on merely possible individuals against a lavish outlay on mere potentialities, properties of actual individuals.) What we have in metaphysical disputes like this are pairs of metaphysical world-views, each of which is self-consistent yet neither of which is compatible with its rival. How do we decide which theory to believe in such cases? Not, I urge, through a simple refutation of presentism by the argument from relations. Nor, I fear, by a simple refutation of its rival. (Likewise for the rivalry between actualism and its rival, possibilism.) Rational choice here is not easy; but it is not, I think, impossible.

The past no longer exists; yet there is a sense in which the past can never be lost: the world will always be one with the property of having once been thus and so. Likewise the future does not exist yet; yet there is a sense in which the future will be what it will be: the world has always been one with the property of being a world which is going to be thus and so. At any given time, you can grasp truths which transcend your present and describe the world *sub specie aeternitatis*, from the standpoint of eternity. This can be managed by a simple logical trick. At any time you can collect all the truths about what was, is, or will be. Truths about what was were once truths about what will be; and truths about what will be will eventually be truths about what was; and so truths about past, present, and future keep changing. But for every truth about what was, there is a truth about what was, is, or will be. This truth about what was, is, or will be was once true because it was future; then it was true because it was present; now it is true because it is past. The ground of its truth keeps changing. But it is always true. And of course this holds not only for what was but also for what is: for every truth about what is, there is a truth about what was, is, or will be. And likewise also of course for what will be: for every truth about what will be, there is a truth about what was, is, or will be. There are truths about what is past, and these truths were not always true. There are truths about what is present, and these truths are fleeting. There are truths about what is future, and these will not last. Yet there are also truths which we can grasp at any time at all, concerning what is past or present or future. And these truths never change.

The truths about what was change, the truths about what is change, the truths about what will be change: yet the truths about what was, is, or will be never change. From the standpoint of presentism the rival theory, fourdimensionalism, is just a partial picture of reality. It only acknowledges the eternal truths about what was, is, or will be. The things it acknowledges to be true are indeed true as far as they go, but they are all accommodated as an abstracted, logical consequence of what is real according to the presentist. Hegel did not quite get it right, in his *Philosophy of Nature*:

> only the Present is, while what is before and after is not. The concrete Present is, however, the result of the Past, and is pregnant with the Future. The true Present is therefore identical with eternity.

There is no identity between eternal truths about what was, is, or will be, taken together disjunctively, and truths severally about the past, or about the present, or about the future. So the true Present, the world, is not identical with eternity. The world is, rather, a changing ground for unchanging truths.

Plato is closer to articulating a good presentist theory of the eternal, in the Timaeus: in this dialogue, the character Timaeus says that this world is a moving image of eternity. The presentist is not compelled, however, to accept the causal order implicit in the use of the word "image" here. Plato sees the order of explanation as proceeding from the eternal to the changing image. A more this-worldly presentist might choose not to take eternal truths as the ground of changing truths about the transience of life in this temporal world. Rather, the presentist may take this world as a changing ground of eternal truths. Eternity is an unchanging image cast, through a little logical lens, by a changing world.

References

Adams, R. H. "Theories of Actuality," *Noûs* 8 (1974): 211-31.

Ash, B., ed. *The Visual Encyclopedia of Science Fiction*. Pan Books, London UK, 1977.

Bailey, J. O. *Pilgrims through Space and Time: Trends and Patterns in Scientific and Utopian Fiction*. New York, N. Y.: Argus, 1947.

Bergson, H. *Time and Free Will*, trans. by R. L. Pogson. London UK: George Allen and Unwin, 1910.

———. *Matter and Memory*, trans. by N. M. Paul and W. S. Palmer. London UK: George Allen and Unwin, 1912; and New York, N. Y.: Macmillan, 1912.

Bergson, J. *An Introduction to Metaphysics*, trans. by T. E. Hulme. New York, N. Y.: G. P. Putnam's Sons, 1912.

Bigelow, J. C. Review of T. Parsons, *Nonexistent Objects*. New Haven, Conn. and London: Yale University Press, 1980.

———. "Real Possibilities," *Philosophical Studies* 52 (1987): 85-112.

———. "The World Essence," *Dialogue* 29 (1990): 205-17.

———. "Worlds Enough for Time," *Noûs* 25 (1991): 1-19.

Bigelow, J. C., and R. J. Pargetter. "Beyond the Blank Stare," *Theoria* 53 (1987): 97-114.

Bigelow, J. C., and R. J. Pargetter. *Science and Necessity*. Cambridge, UK: Cambridge University Press, 1990.

Brentano, F. *The True and the Evident*, edited by O. Kraus. London: Routledge and Kegan Paul, 1966.

———. *The Theory of Categories*, trans. by R. M. Chisholm and N. Guterman. Boston, London, UK, and The Hague: Martin Nijhoff, 1981.

Broad, C. D. *Scientific Thought*. London, UK: Kegan Paul, 1923.

———. *An Examination of McTaggart's Philosophy*. Cambridge, UK: Cambridge University Press, 1938.

Bury, R. G. *Transl. of Sextus Empiricus, Against the Physicists*, Vol 3. Cambridge, Mass.: Harvard University Press, 1960.

Davenport, B. *Inquiry into Science Fiction*. New York, N. Y.: Longmans, Green, 1955.

Davenport, B., et al. *The Science Fiction Novel*, 2nd edition. Chicago, Ill.: Advent, 1964.

Dummett, M. "Can an Effect Precede its Cause?" *Proceedings of the Aristotelian Society*, Supplementary Volume 28 (1954): 27-44.

———. "A Defense of McTaggart's Proof of the Unreality of Time," *Philosophical Review* 69 (1960): 497-504.

———. "Bringing about the Past," *Philosophical Review* 75 (1964): 338-59.

———. "The Reality of the Past," *Proceedings of the Aristotelian Society* 13 (1969): 239-58.

———. *Truth and Other Enigmas*. Cambridge, Mass.: Harvard University Press, 1978.

Earman, J. *World Enough and Space-Time*. Cambridge, Mass.: MIT Press, 1989.

Eliot, T. S. *The Complete Poems and Plays*. London, UK: Flood, 1969.

Flood, R., and M. Lockwood, eds. *The Nature of Time*. Blackwell, Oxford UK, 1986.

Forbes, G. *The Metaphysics of Modality*. Oxford, UK: Clarendon, 1985.

Forrest, P. "Ways Worlds Could Be," *Australasian Journal of Philosophy* 64 (1986): 15-24.

Frege, G. *Die Grundlagen der Arithmetik*. Breslav, Germany: Wilhelm Koebner, 1884. (Translated by J. L. Austin as *The Foundations of Arithmetic*, 2nd rev. ed. Oxford: Blackwell, 1959.)

Gale, R. M., ed. *The Philosophy of Time: A Collection of Essays*. Garden City, N. Y.: Doubleday, 1967.

Hegel, G. *The Philosophy of Nature*. Oxford, UK: Clarendon, 1970.

Hinchliff, M. "The Puzzle of Change." Pp. 119-36 in *Philosophical Perspectives, 10, Metaphysics*, edited by J. Tomberlin. Cambridge, Mass.: Blackwell Publishers, 1996.

Hinkfuss, I. *The Existence of Space and Time*. Oxford, UK: Clarendon, 1975.

Holster, A. *Physical Time Flow*. PhD dissertation, Massey University, New Zealand, 1991.

Horwich, P. *Asymmetries in Time*. Cambridge, Mass.: MIT Press, 1987.

Kneale, W. "Time and Eternity in Theology," *Proceedings of the Aristotelian Society* 61 (1960-61): 87-108.

Latham, R. *Transl. of Lucretius, On the Nature of the Universe*. Baltimore Md.: Penguin, 1951.

Le Poindevin, R., and M. MacBeath, eds. *The Philosophy of Time*. Oxford, UK: Oxford University Press, 1993.

Leibniz, G. W. *Discourse on Metaphysics, Correspondence with Arnauld, and Monadology*, trans. by G. R. Montgomery. La Salle, Ill.: Open Court, 1962.

Lewis, D. K. "The Paradoxes of Time Travel," *American Philosophical Quarterly* 13 (1976): 145-52.

———. "Truth in Fiction," *American Philosophical Quarterly* 15 (1978): 37-46.

———. *Philosophical Papers*, 2 vols. Oxford, UK, and New York, N. Y.: Oxford University Press, 1983.

———. *On the Plurality of Worlds*. New York, N. Y.: Blackwell, 1986.

Long, A. A., and D. N. Sedley. *The Hellenistic Philosophers, Vol. 1, Translations of the Principal Sources with Philosophical Commentary*. Cambridge, UK: Cambridge University Press, 1987.

Loux, M. J., ed. *The Possible and the Actual*. Ithaca, N. Y., and London, UK: Cornell University Press, 1979.
Lovejoy, A. O. "The Problem of Time in Modern French Philosophy," *Philosophical Review* 21 (1912): 11-31, 322-43, and 527-45.
Lucretius, see Latham, and Munro.
McCall, S. "Temporal Flux," *American Philosophical Quarterly* 3 (1966): 270-81.
———. "Objective Time Flow," *Philosophy of Science* 43 (1976): 337-62.
———. "A Dynamic Model of Temporal Becoming," *Analysis* 44 (1984):173-76.
———. *A Model of the Universe*. Oxford, UK: Clarendon, 1994.
McTaggart, J. M. E. "The Unreality of Time," *Mind* 17 (1908): 457-74.
———. *The Nature of Existence*, vol. 2. Cambridge, UK: Cambridge University Press, 1927. See chapter 2 in this volume.
———. *Philosophical Studies*. London, UK: Edward Arnold, 1934.
Mellor, D. H. *Real Time*. Cambridge UK: Cambridge University Press, 1981.
Munro, H. A. J. *Transl. of Lucretius, On the Nature of Things*. New York, N. Y.: Washington Square Press, 1965.
Nerlich, G. *The Shape of Space*. Cambridge, UK: Cambridge University Press, 1976.
Newton-Smith, W. H. *The Structure of Time*. London, UK: Routledge and Kegan Paul, 1980.
Oderberg, D. S. *The Metaphysics of Identity over Time*. London, UK: Macmillan, 1993.
Parsons, T. *Nonexistent Objects*. New Haven, Conn., and London, UK: Yale University Press, 1980.
Philips, M., ed. *Philosophy and Science Fiction*. Buffalo, N. Y.: Prometheus, 1984.
Plantinga, A. "Actualism and Possible Worlds," *Theoria* 42 (1976): 139-60. (Reprinted in M. J. Loux, ed., *The Possible and the Actual*. Ithaca, N. Y.: Cornell University Press, 1979.)
———. "Two Concepts of Modality: Modal Realism and Modal Reductionism." Pp. 189-231 in *Philosophical Perspectives*, Vol. 1: Metaphysics, edited by J. Tomberlin. Atascadero, Calif.: Ridgeview, 1987.
Plato. *Timeaus*, trans. by J. Warrington. London, UK: Dent, 1965.
Powers, L. "Comments on Stalnaker's 'Propositions.'" Pp. 93-103 in *Issues in the Philosophy of Language*, edited by A. F. MacKay and D. D. Merrill. New Haven, Conn., and London, UK: Yale University Press, 1976.
Prior, A. N. *Time and Modality*. Oxford UK: Clarendon, 1957.
———. "Time after Time," *Mind* 67 (1958): 244-46.
———. "Thank Goodness that's Over," *Philosophy* 34 (1959): 12-17.
———. "Identifiable Individuals," *Review of Metaphysics* 13 (1960): 684-96.
———. *Past, Present and Future*. Oxford, UK: Clarendon, 1967.
———. *Papers on Time and Tense*. Oxford, UK: Clarendon, 1968.
———. "The Notion of the Present," *Studium Generale* 23 (1970): 245-48.
Prior, A. N., and K. Fine. *Worlds, Times, and Selves*. Amherst, Mass.: University of Massachusetts Press, 1977.
Proust, M. *Remembrance of Things Past*, 3 volumes, trans. by C. K. Scott Moncrieff. New York, N. Y.: Random House, 1934.
Quine, W. V. *From a Logical Point of View*. Cambridge, Mass.: Harvard University Press, 1953.
———. *Word and Object*. Cambridge, Mass.: MIT Press, 1960.
Reichenbach, H. *Elements of Symbolic Logic*. London, UK: Macmillan, 1947.
———. *The Direction of Time*, edited by M. Reichenbach. Berkeley, Calif.: University of California Press, 1956.

———. *The Philosophy of Space and Time*, trans. by M. Reichenbach and J. Freund. New York, N. Y.: Dover, 1958.

———. *Introduction to Symbolic Logic and Its Applications*. New York, N. Y.: Dover, 1958.

Routley, R. *Exploring Meinong's Jungle and Beyond: An Investigation of Noneism and the Theory of Items*, interim ed. Canberra, A.C.T.: Departmental Monograph, Australian National University, 1980.

Russell, B. "On the Experience of Time," *The Monist* 25 (1915): 212-33.

———. *Human Knowledge: Its Scope and Limits*. New York, N. Y.: Simon and Schuster, 1948.

———. *Mysticism and Logic*. London, UK: Unwin Books, 1963.

Schlesinger, G. N. *Aspects of Time*. Indianapolis, Ind.: Hackett, 1980.

Scholes, R., and E. S. Rabkin. *Science Fiction: History, Science, Vision*. Oxford, UK: Oxford University Press, 1977.

Sextus Empiricus, see Bury, and Long and Sedley.

Shoemaker, S. "Time without Change," *The Journal of Philosophy* 66 (1969): 363-81.

Smart, J. J. C. "The River of Time," *Mind* 58 (1949): 483-94.

———. "Spatialising Time," *Mind* 64 (1955): 239-41.

———. *Philosophy and Scientific Realism*. London, UK: Routledge and Kegan Paul, 1963.

———, ed. *Problems of Space and Time*. New York, N. Y.: Macmillan, 1964.

———. *Between Science and Philosophy*. New York, N. Y.: Random House, 1968.

Smith, Q. *Language and Time*. New York, N. Y., and Oxford UK: Oxford University Press, 1993.

Sorabji, R. *Time, Creation and the Continuum*. London, UK: Duckworth, 1983.

Stalnaker, R. "Possible Worlds," *Noûs* 10 (1976): 65-75. (Reprinted in M.J. Loux, ed., *The Possible and the Actual*. Ithaca, N. Y.: Cornell University Press, 1979.)

Swinburne, R. *Space and Time*. London, UK: Macmillan, 1968. (2nd ed. 1981)

———, ed. *Space, Time and Causality*. Dordrecht-Holland: Reidel, 1983.

Tooley, M. *Causation: A Realist Approach*. Oxford UK: Clarendon, 1987.

van Inwagen, P., ed. *Time and Cause*. Dordrecht-Holland: Reidel, 1980.

Williams, D. C. "The Myth of Passage," *The Journal of Philosophy* 48 (1951): 457-72.

———. *Principles of Empirical Realism*. Springfield, Ill.: Charles C. Thomas, 1966.

Zalta, E. *Abstract Objects: An Introduction to Axiomatic Metaphysics*. Dordrecht-Holland: Reidel, 1983.

———. *Intensional Logic and the Metaphysics of Intentionality*. Cambridge, Mass.: MIT Press, 1988.

9
A Theory of Presentism

Craig Bourne

I. The Parameters of the Problem

Most of us would want to say that it is true that Socrates taught Plato. According to realists about past facts,[1] this is made true by the fact that there is, located in the past, i.e., earlier than now, at least one real event that is the teaching of Plato by Socrates. Presentists, however, in denying that past events and facts exist[2] cannot appeal to such facts to make their past-tensed statements true. So what is a presentist to do?

There are at least three conditions that would ideally be met in a satisfactory solution to this problem:

a) It must preserve our views about which statements are true and which false
b) It must be transparent what the truthmakers are for those statements
c) It must explain the truth-value links between various times.

I shall survey two different families of proposals for the presentist's truthmakers, recent examples of each being advocated by Craig (2003) and Ludlow (1999), and show that they fail at least one of these three conditions. This is not entirely negative, for it shows us what an adequate solution to the problem would look like. I go on to show where presentists can find suitable objects that satisfy these conditions, and in this way give a clear statement of presentism, something that is lacking in the literature.[3]

II. Priorian Presentism

The truth of our statements depends on the way the world is: contingent truths require truthmakers. Everyone should agree with this. What people may disagree about is what truthmakers are. But so long as we all agree that truths depend in some way on the contents of the world (whether they are states of affairs, as in Armstrong's work (e.g., his (1997)) or particulars, as in Lewis's (2003)), then that is enough to generate the concerns of this paper.[4] Thus, given that there are true past-tensed statements, these too require truthmakers. But where in the presentist's world are these to be found? An obvious response is to say that, since only the present exists, these truthmakers are to be found in the present. But what sort of truthmakers could these be?

Suppose we take our cue from Arthur Prior, and use the tense operators, P, N and F (read "It was the case that," "It is (now) the case that," and "It will be the case that," respectively), to pick up the tense of an interior tensed proposition. However complex the tense of the initial proposition, it can be analyzed as a basic present-tensed interior proposition together with the appropriate iteration of the tense operators. Since the interior proposition is always present-tensed, there is a redundancy in the use of the operator N to signify the present tense. This leaves only the F and P operators as non-superfluous; we can always drop N when it appears outside the scope of P or F (see Kamp (1971); McArthur (1976)). Because of this, Prior held a "redundancy theory of the present" (Prior (1957, 9-10); (1967, 8-10, 14-15); (1968a, 17-23); (1968b, 101)): everything that is presently true, is true simpliciter; N is superfluous in the same way as "It is true that" is redundant according to the redundancy theory of truth. Indeed, he says "reality consists in the absence of a qualifying prefix" (1970).

However, this is not to say (pace Tooley (1997, 166)) that we should not think of it as being there for, according to Prior:

> "Socrates taught Plato" (**P***p*)

means

> "It is *now* the case that Socrates taught Plato" (**NP***p*),

where *p* is the proposition that Socrates is teaching Plato. This has to be correct because it is a *present* truth that Socrates taught Plato. Thus, the tacit presence of an **N** outside the scope of all other operators implies that *all* tensed propositions *are* present-tensed; that is, implies, e.g., that the proposition that **FP***p* *is* the proposition that **NFP***p*; and the first stage of the presentist agenda to reduce past- and future-tensed propositions to present-tensed propositions is easily and naturally achieved.

But the essential question is: what *makes* such truths true? What are the constituents of the facts that make them true? Prior denies that these questions need answering: for Prior, there *is* nothing more to say about the nature of time than is said by a perspicuous tense logic (e.g., Prior (1996, 45)). Nevertheless, I (with others, e.g., Smith (1987, 188-91); Tooley (1997, 166-67); Le Poidevin (1998, 38-39)) feel more needs to be said, especially since the issue of truthmaking is independent of the theory of truth: truthmaking arises just as much for deflationary theories of truth as it does for the correspondence theory (as Lewis (2001) and Smith (2003) show). Thus, even if Prior adopts a deflationary theory of *truth* (and the present) this will not help him avoid the issue of *truthmaking*; he still needs to say more.

But if we press for an ontology, Prior is ultimately left in an uncomfortable position on one of the prongs of a dilemma. For given his doctrine that propositions are themselves "logical constructions" out of the objects they are about (Prior (1971, chp. 1)), *how* is it possible for the proposition that Socrates taught Plato to be true? Which particulars can be invoked as the constituents of such a fact? Not Socrates or Plato—they don't exist. Nor can we invoke a present past-Socrates—what a mysterious object that would be! The alternative is to invoke the primitive present fact that Socrates taught Plato. But without being able to say how this fact is structured (for its constituents are certainly not Socrates or Plato), this move is far from satisfactory. Thus either option leaves us with an obscure ontology.

We might think that these are just problems peculiar to Prior's position and of which later presentists must surely have said something satisfactory. But it appears not. Craig (2003) carries on in the same vein. He claims that reference to past objects is possible, since "the proper name 'Socrates' expresses an individual essence of Socrates rather than denotes nonconnotatively the actual object Socrates and so does not require Socrates to exist in order for the name to refer" (395). This move, following Plantinga's work on proper names (e.g., Plantinga (1974, 71-81; 137-44; 149-63)), would allow propositions like *Socrates taught Plato* to be about Socrates and Plato because "Socrates" and "Plato" express the essences of Socrates and Plato, and thus refer to them. Now, this sounds incomprehensible to those who subscribe to the view that in order for genuine reference to an object to take place, the object must exist: for according to this view, if Socrates does not exist, "Socrates" does not refer. But Craig is at pains to draw a distinction between "reference" and "correspondence" (394). Correspondence is a relation between world and word and thus requires an object to fall under a name; but reference, we are told, doesn't, since it has "to do with how terms serve to pick out individuals" (395), something which, it is claimed, doesn't require the existence of the individual. But even if we concede this, it hardly addresses the question of which objects *are* required for *truthmaking*, a quite separate issue. If it isn't the concrete object that figures in truthmaking, is it the "individual essences" expressed by the proper names, essences of objects which exist even though the concrete objects themselves don't? No, for whatever these "individual essences" are supposed to be, Craig uses them only to argue for the possibility of "reference" to past objects, not in order to spell out the truthmakers for past-tensed propositions. So what are we left with? For Craig, what makes the proposition that *Socrates taught Plato* true is the "tensed fact" (400) that Socrates taught Plato. End of story. And quite literally so, because nothing more is said about this fact. It certainly cannot have Socrates and Plato as constituents (regardless of whether "reference" can be made to them), since it is a *present* fact that Socrates taught Plato and, presently, they don't exist. And it is no good complaining that they *did* exist, because we want to know what *does exist* to make the proposition true now. (Note that this is not the unjustifiable requirement that what makes propositions true at a particular time must also be *located* at that time; since what makes the proposition that *Socrates taught Plato* true now, according to the tenseless view, is something located in the past, and this seems an entirely reasonable view. The requirement is rather that there *be* something in existence to make the proposition true; so this cannot be something involving Socrates and Plato, for the presentist, regardless of whether they *did* exist.) We are simply not told what the constituents of these facts are or how they are structured. It seems that no real progress has been made on this issue with this version of presentism since Prior's work.

But the view that the present contains within it all of the facts required to make past-, present- and future-tensed propositions true creates another, less obvious, concern. We can represent times as sets of present-tensed propositions. Suppose *p* is a true present-tensed proposition. Now, merely from considering the truth-value links which must hold across times, the following set of true present-tensed propositions must also hold of that time: {*p*, **FP***p*, **PF***p*, ...}.[5] But what guarantees that a later time preserves these links? Given the present time represented by {*p*, **FP***p*, **PF***p*, ...}, there must be a later time represented by {**P***p*, **PFP***p*, **PPF***p*, ...}. But how can a Priorian Presentist even guarantee this, let alone explain it? Of course, **PFP***p* must hold in the future, if *p* holds presently: every adequate theory of time must have this as a consequence or be rejected. But there is no mechanism in this version of presentism to guarantee it. It is no good just appealing to the tense logic and then interpreting this in a deflationary way, since, as noted above, this semantic issue does not address the ontological issues at stake. The ontology, at least as far as Craig is concerned, is such that the present time is comprised of certain tensed facts, including those about what will happen and what has happened. We can add that all times are contingent and distinct entities. But then it follows, if we take the Humean stance, that the truth-value link cannot be guaranteed by any *necessary* connections between such times or the tensed facts of which they are comprised (at least at the ontological level of tensed facts, if not the semantic level of tense logic). Furthermore, for that matter, there cannot be *any* transtemporal relations to link times together on this view: other times don't exist; and it is hard to see how the ontological content of the present time can in itself legislate how other distinct entities, other times, can be comprised. Yet somehow it must—for if the content of the present time does not legislate this, what, according to this view, does?—since it is incredible to think that there could be missing tensed facts from, or additional tensed facts in, various times. The truth-value links have somehow to be a feature of how the facts are structured (as they are on the tenseless theory of time, for example, and as they are on the view I present below). But because no mechanism is in place to preserve the truth-value links and the possibility of a violation of them is opened up, this version of presentism should be rejected.[6]

III. Reductive Presentism

A second shot at the thought that the present itself can supply all of the required truthmakers is the Reductive version of presentism. Ludlow (1999, chp. 9) is the most recent substantial work on time that advances this view, and Le Poidevin (1991) (a tenseless theorist) recognizes this option:

> What makes a certain statement about the past true . . . is the evidence that at present exists. . . . This is possible in virtue of the fact that there are present facts which derive their character from causal connection with past states of affairs, and which determine (at least to some extent) the character of the future (37) . . . [T]his is essential to the . . . (presentist's) position. (38)

And:

> The extent to which the principle of bivalence is violated by statements about the past or future depends, for [the presentist], upon how much causal determinism he is prepared to allow. In a fully deterministic universe . . . all future- and past-tensed statements have a determinate truth-value, as this is guaranteed by present fact. But in an indeterministic universe . . . many statements about the future must for [presentism] lack a truth-value. (38)

Although it is obvious from the above discussion that, *contra* Le Poidevin, this story is not "essential to the [presentist's] position," it is a position that has been held for a variety of reasons (see, e.g., Łukasiewicz (1970, 127-28), and Wheeler (1978, 41), who thinks that the two-slit experiment in quantum physics vindicates this conception). But it is not a position which has been fully endorsed by any serious philosopher, although, of course, Dummett has played with this view for years (see Dummett (1979) and his latest (2003)). Ludlow (1999, 162) claims that although this position is not "an inevitable consequence of the A-theory," it is "a possible avenue of investigation." So, although this is only a tentative endorsement, it is important to briefly show how just a short walk down this supposed avenue reveals it to be more of a *cul de sac*.

There are two broad categories of theories of laws of nature, which I shall label "Humean" and "non-Humean." Consider the Humean theory, which states that laws consist in nothing more than: whenever there is an F, there is a G, i.e., $(\forall x)(Fx \supset Gx)$. The first issue to address here is the "whenever" quantifier. If it just ranges over times that exist (i.e., the present time!), then any present conjunction of F with G will constitute a law, which is unsatisfactory. But, more damagingly, if it is restricted to times that exist (the present time), then laws cannot do the job for which they were intended. For according to this account, they do not extend beyond the present, and thus cannot determine the truth-values of statements about the past and the future. We should, then, take it that "whenever" ranges over present times as well as those that *have* existed and *will* exist. The problem now is: what makes it true that all Fs *are*, *have been*, and *will be* Gs? For there is nothing we could appeal to in the present. We certainly can't appeal to deterministic laws of nature to ground what has happened and what will happen, since this just lands us in a circle, for the whole problem was to find something in the present that could ground what that law is. And the present state of the universe by itself cannot determine what has happened and what will happen because the present state of the universe is compatible with infinitely many different mutually contradictory pasts and futures: it is only the state of the universe at a given time *together with deterministic laws of nature* which can determine what has happened and what will happen. But there is nothing in the reductive presentist's world which can ground such laws. It is not as if a law can be seen as an extra ingredient in the present, which can help determine one course of history over another, for remember that we are talking about a Humean conception of law, and according to this conception, there is nothing more, ontologically speaking, to a law over and above what does actually happen. This leads us to consider an alternative option, namely to appeal to primitive facts: things just *did* and *will* occur in this sort of regularity. But this then renders the excursion into Reductive Presentism entirely redundant if we adopt primitive facts: we may as well be Priorian Presentists. Thus, a Humean account of laws will not give Reductive Presentists what that they would hope for.

What about non-Humean accounts? Suppose we thought that laws of nature are some kind of necessary connection between universals. According to Armstrong, the law that *F*s are *G*s is some sort of necessitation relation *N* between the universals *F*-ness and *G*-ness. $N(F, G)$ is a contingent nomic necessity that may not hold across all possible worlds, but if it actually holds, then $N(F, G) \Rightarrow (\forall x)(Fx \supset Gx)$. For various reasons, Armstrong (1983, 86-99) argues that $N(F, G)$ is a first-order relational universal that is instantiated by particular states of affairs *Fa* and *Ga*. Furthermore, according to Armstrong's Principle of Instantiation (e.g., (1983, 82)) universals must be instantiated in order to exist. It follows that given that there are only present states of affairs in the Reductive Presentist's world, any instantiation of a law can only take place in, and be true of, the present. Thus such laws of nature cannot tell us about what happens at other times, and thus cannot ground determinate truth-values for past- or future-tensed statements. (This is particularly problematic given Armstrong's view that the necessitation relation does not supervene on the state of the world, and can change in what it relates from world to world. For even if $N(F, G)$ holds of the present time, there is nothing to say it will hold across all past and future times. Yet that was what the Reductive Presentist hoped the necessitation—although it is hard to see how we can still call it "necessitation" given these considerations—would guarantee.)

This well known and well worked out non-Humean theory was worth mentioning in order to illustrate the general problem with which Reductive Presentists have to deal: *how* can the laws reach beyond the present? There are, of course, non-Humean accounts that argue that laws can exist without being so instantiated (e.g., Tooley (1977); Mellor (1980)). But a presentist who appealed to these accounts of laws of nature would either have to find something in the present to ground these laws—and it's difficult to see what this could be—or they must appeal to facts outside the present, in which case they've conceded that not all of the truthmakers can be found *in* the present, which was their initial contention. The notion of determinism will not help the presentist on the issue of securing determinateness of truth-value.

Needless to say, Reductive Presentism cannot have determinate truth-values in an indeterministic world. Yet, even in an indeterministic world, we still want to say that at least past-tensed statements have *determinate* truth-values. We want to say that it is either true or false that a certain brontosaurus had two plants for breakfast, regardless of whether there are any traces of this fact in the present, and regardless of whether the present state of the universe and the laws of nature can determine this fact. To confuse determinateness with determinism is an offense. Reductive Presentism thus violates condition (a) of section **I**.

Furthermore, Reductive Presentism, however much it initially seems to, *doesn't* in fact meet the truth-value links problem. First, consider Russell's hypothesis (1921, 159-60): what is to say that the world has not just come into existence complete with fossil records, memories, and other such causal traces? Russell intended this to illustrate the limitations of our knowledge, but pre-theoretically, it seems like a genuine possibility for how the world could be. Suppose it is true; then Reductive Presentism gives the wrong answers, for it states that certain past-tensed statements are true when they shouldn't be. Of course, nobody thinks that hypothesis is true; but Russell's hypothesis doesn't have to be actually true to cause trouble; that it is true that is it possible is enough to show that Reductive Presentism has a problem accounting for truth-value links. And we couldn't rule out Russell's hypothesis by appealing to actual laws of nature that wouldn't allow for such a complex environment in such a short space of time, for as I have argued above, the grounding of those very laws is called into question on the Reductive view. That is, in a nutshell, truth-value links are supposedly underpinned by causal links: present facts are effects of previous causes and themselves causes of later effects. But this relies on what it is in the present that grounds this causal link: for it to be true that a given fact is an effect of a previous cause, we need the present fact that it *is* the effect of a previous cause to make this true, since without this fact there is nothing to link the supposed "effect" to anything that came before, or the supposed "cause" to anything that comes after. (And we certainly need this, given the many different pasts and futures compatible with the present state of the universe.) But what sort of fact could this be? Taking it as a primitive present fact lumbers us with the Priorian Presentist's facts and all their attendant problems. We are no better off after all.

We now have an appreciation of what any successful version of presentism must hope to achieve, and the pitfalls it must avoid. The version of presentism I am about to propose meets these conditions. It also meets the challenges that some (e.g., Oaklander (2002); (2003)) throw down for presentism to tackle, namely how presentists can help themselves to the notion of *earlier than* without having to invoke real *relata*, and how presentism can distinguish the past from the future. It should also be attractive to those who see close analogies between time and modality, and who prefer ersatz theories of possible worlds (such as Adams's (1974)) over genuine modal realism (such as Lewis's (1986)), for the position I will defend is essentially this: all of us should agree that Socrates taught Plato, i.e., that the proposition that *Socrates is teaching Plato* was, at some time, true. This, to most people, is so obvious as to not be worth stating. So I ask them to keep this in mind during the more technical exposition of the position which is about to follow. For I say we should take the "i.e." seriously; that is, that *what makes it true* that Socrates taught Plato is the existence of a proposition according to which this is the case for some time in the past, where a time is a set of propositions representing the other truths about what happens at that time. For obvious reasons, I shall call this position "Ersatzer Presentism."

IV. Ersatzer Presentism[7]

a) An improvement over Prior's Presentism: e-*propositions and* u-*propositions*

I go some way with Prior ((1967, 79-82); (1968a, 122-26)) in constructing times from certain sets of present-tensed propositions.[8] However, my account differs in important respects. According to Prior, the present moment should be equated with the conjunction of all those propositions which are presently true, i.e., true *simpliciter*; and generalizing, he equates any time with a conjunction of all those present-tensed propositions which would be true at that time.

My conception differs, first, in that I distinguish present-tensed propositions that contain either **P** or **F** operators, which I shall call "embedded propositions" (*e*-propositions), from those that do not, which I shall call "unembedded propositions" (*u*-propositions). (Compare, for example, the *e*-proposition: It is now the case that it was the case that Socrates is sitting (i.e., **NP***p*), with the *u*-proposition: It is now the case that Socrates is sitting (i.e., **N***p*), or simply: Socrates is sitting (i.e., *p*).)

As we have seen, Priorian Presentism required the present to be equated with *e*-propositions in order for the present to supply all of the truths we believe there are, but that it was rather mysterious what the truthmakers were for these propositions, and neither could it account for the truth-value links between the different times. The theory to be developed avoids these difficulties.

b) Defining times, and the E-*relation*

I propose we construct times using maximally consistent sets of *u*-propositions, which intuitively we can see as those *u*-propositions that are true at that time. These propositions I take to give a complete, maximally specific description of what is true at that time. But more needs to be added: they, at least, need to be ordered by an "earlier than" relation (what I will call an "*E*-relation"), in order for the ersatz time series to be structurally similar to a real time series, so it can be taken to be a sufficient substitute. That is, we can introduce the ordered triple <**T**, *E*, *t*>, where **T** is a set, E is a relation on $\mathbf{T}$, and $t \in \mathbf{T}$.[9] Intuitively, $\mathbf{T}$ is the set of times, E is the "earlier than" relation, and t is a particular time. Times I take to be more than sets of present-tensed propositions: first, they consist of sets of *u*-propositions; second, they also contain a "date." That is, I take times (at least for the time being) to be ordered pairs of the form $t = <\mu, n \in \mathbb{R}>$, where μ is a set of *u*-propositions and $n \in \mathbb{R}$ is the date. Times can now be defined as those ordered pairs of the form $t = <\mu, n \in \mathbb{R}>$ that are members of the class of sets of ordered pairs of the form $t = <\mu, n \in \mathbb{R}>$ that are E-related. Actual times are those times that are members of the class of times that are *actually* E-related. (What this amounts to is explained below.)

Defining times in the way I have done requires us to alter our terminology slightly. Lewis (1986, 140) has complained that ersatzism concerning possible worlds commits us to saying that since worlds are, say, sets of propositions, the concrete object that we are part of—"the world," speaking with the vulgar—cannot be said to be a world. Similarly, since, according to Ersatzer Presentism, times are abstract objects, the present time is not something we inhabit. But rather than introduce a new term, or typescript, for what I am calling "times," I prefer to keep this useful terminology. The only differences it makes are that we must be said to inhabit the concrete realization of the present time, and that Ersatzer Presentism is the view not that only one time exists, but that only one time has a concrete realization. This is a mere nominal difference, equivalent to saying that presentists do not believe in any times other than the present; and all is as it should be.

The E-relation is not the genuine *earlier than* relation since it does not relate spatio-temporal objects, but it does *represent* the *earlier than* relation in the way it relates times. The properties of the E-relation match whatever we take to be the properties of the genuine *earlier than* relation. This allows presentists to have a time series related by "earlier than" without being committed to the existence of real, or rather, concretely realized *relata*, something anathema to presentism. Ersatzer Presentism thus bypasses the problems that other presentists get into when they do not take such relations as basic, and try to define them in terms of tenses (see, e.g., Oaklander (2002); Mellor (2003, 236-37)).

Various metrical and topological features of time can be represented in the structure of the abstract objects. For example, if we take time to be linear, the E-relation will have, among other things, the properties of being irreflexive, asymmetric, and transitive; if it is circular, it will be an equivalence relation, and times form an equivalence class; if time is continuous, the times are assigned real numbers as dates; if discrete, then only those times that have dates $n \in \mathbb{Z}$ will be realized; if it is infinite in both directions from the present, then the dates will have the standard order type of infinite time ($\omega^* + \omega$); if time is circular then the dates will be of the form of a periodic function (for some period with value P, $t(x+P) = t(x)$), and so on.

But despite the fact that these properties of time can be represented in this way, presentism has a very good reason for adopting a *branching* topology; that is, where the E-relation is a one-many relation in the direction from the present to future (the direction in which the dates increase in magnitude), but only a one-one relation in the direction from the present to the past (the direction in which the dates decrease in magnitude). The reason is simple: we all need a way of distinguishing the past from the future. Because presentism treats the past and the future as ontologically on a par, in the sense that it denies that there are any concrete truthmakers located there, the branching structure of times is an obvious way in which presentism can differentiate between past and future. Furthermore, it accounts for the platitude that the past is "fixed" and the future is "open," something that any good theory of time should. Thus, if we follow my proposal, presentism does have at its disposal means for distinguishing the past from the future.

c) Truth-conditions, truthmakers, truth-at-a-time, and truth simpliciter

I also distinguish truth-at-a-time from truth *simpliciter*.[10] Truth *simpliciter* is an absolute, not time-relative, notion, whereas truth-at-a-time is time-relative: all propositions at a time are true relative to it, but only those propositions which are true at the present time are true *simpliciter*. So, where the propositions involved are any atomic propositions, be they from propositional or predicate calculi, we have:

(1) *Socrates is sitting* is true-at-a-time iff there is a time, i.e., *E*-related order pair $<\mu, n\in \mathbb{R}>$ such that μ includes the *u*-proposition that *Socrates is sitting*, i.e., p is true-at-a-time $<\mathrm{T}, E, <\mu, n\in \mathbb{R}>>$ iff $p\in \mu$.

And similarly,

(2) $\sim p$ is true-at-a-time $<\mathrm{T}, E, <\mu, n\in \mathbb{R}>>$ iff $p\notin \mu$.[11]

(3) $p\&q$ is true-at-a-time $<\mathrm{T}, E, <\mu, n\in \mathbb{R}>>$ iff $p, q\in \mu$.

Whereas:

(4) *Socrates is sitting* is true *simpliciter* iff Socrates (i.e., an actual, concrete, flesh and blood Socrates) is presently sitting.

Furthermore, it is not only present-tensed *u*-propositions that can be true *simpliciter*; past-tensed propositions can be too. The difference is that whereas present-tensed *u*-propositions are made true *simpliciter* iff corresponding actual concrete facts are presently realized, past-tensed propositions are made true *simpliciter* by something entirely different.

> *It was the case that p* is true *simpliciter* iff p is a member of a set μ of *u*-propositions that is the first element of an ordered pair $<\mu, n_i\in \mathbb{R}>$ actually

E-related to the presently instantiated ordered pair $<\nu, n_j \in \mathbb{R}>$, where ν is the set of u-propositions that is true *simpliciter*, and $n_i < n_j$.[12]

Note that the "actual" here is not superfluous: it has a technical use. Since time has a branching structure, then time *actually* has a branching structure and all possible times on each of these branches are *in some sense* of "actual" actually E-related. Yet I require only *some* times as truthmakers for past- and future-tensed statements, namely those that correspond to what we would ordinarily want to call the actual history of the world. Thus, when I use the phrase "actually E-related" I mean that there is a time which is E-related to the time that is true *simpliciter*, i.e., that is actually concretely realized. This grounds which E-related branch we should use for finding truthmakers for past-tensed statements, since only one E-related branch is accessible from the concretely realized time in the direction towards the past.

In short we have the following story about truthmakers:

- What makes u-propositions true *simpliciter* are actually realized concrete facts.
- What makes past-tensed propositions true *simpliciter* are actually E-related ordered pairs of u-propositions and dates.
- What makes p true-at-time-t is the fact that $p \in \mu$, where $t = <\mu, n \in \mathbb{R}>$.

It now remains to complete the story by dealing with quantification. As noted in n. 2, presentists should read the existential quantifier tenselessly. I also take it that "$(\exists x)Fx$" should be read in the standard objectual way. The quantified propositions $(\exists x)Fx$ and $(\forall x)Fx$ themselves are to be given their standard truth-conditions in terms of the satisfaction of the incomplete proposition Fx. Despite being taken tenselessly, the proposition that $(\exists x)Fx$ might change truth value from time to time because of the continually changing concrete facts and subsequent change in the domain of quantification. Thus:

$\mathbf{P}(\forall x)Fx$ is true at $\langle \mathbf{T}, E, t\rangle$ (and is true *simpliciter* iff the set of propositions at t is true *simpliciter*) iff $\forall xFx$ is true at some time E-related to t, where $(\forall x)Fx$ is true-at-a-time $\langle \mathbf{T}, E, \langle \mu, n \in \mathbb{R}\rangle\rangle$ iff $(\forall x)Fx \in \mu$..

$\mathbf{P}(\exists x)Fx$ is true at $\langle \mathbf{T}, E, t\rangle$ (and is true *simpliciter* iff the set of propositions at t is true *simpliciter*) iff $(\exists x)Fx$ is true at some time E-related to t, where $(\exists x)Fx$ is true-at-a-time $\langle \mathbf{T}, E, \langle \mu, n \in \mathbb{R}\rangle\rangle$ iff $(\exists x)Fx \in \mu$..

d) The future

An interesting question that arises for Ersatzer Presentism is accounting for future-tense propositions. Can the proposition that *It will be the case that p* be true *simpliciter*? According to this theory,

> *It will be the case that p* is true *simpliciter* iff *p* is a member of a set μ of *u*-propositions that is the first element of an order pair $<\mu, n_i \in \mathbb{R}>$ actually *E*-related to the presently realized ordered pair $<\nu, n_j \in \mathbb{R}>$, where ν is the set of *u*-propositions that is true *simpliciter*, and $n_i > n_j$.

However, *it will be that p* is neither determinately true *simpliciter* nor determinately false *simpliciter*, when *p* is contingent, given that, as of the present, no one future branch is actual. Furthermore, consider the truth-conditions for future-tensed quantified statements:

> $\mathbf{F}(\forall x)Fx$ is true at $<\mathbf{T}, E, t>$ (and is true *simpliciter* iff the set of propositions at *t* is true *simpliciter*) iff $(\forall x)Fx$ is a member of some μ element of times later than *t* (understood in the ersatz way), where $(\forall x)Fx$ is true-at-a-time $<\mathbf{T}, E, <\mu, n \in \mathbb{R}>>$ iff $(\forall x)Fx \in \mu$.

> $\mathbf{F}(\exists x)Fx$ is true at $<\mathbf{T}, E, t>$ (and is true *simpliciter* iff the set of propositions at *t* is true *simpliciter*) iff $(\exists x)Fx$ is a member of some μ element of a time later than *t* (understood in the ersatz way), where $(\exists x)Fx$ is true-at-a-time $<\mathbf{T}, E, <\mu, n \in \mathbb{R}>>$ iff $(\exists x)Fx \in \mu$..

Given the branching future, $\mathbf{F}(\forall x)Fx$ is only true *simpliciter* when the proposition that $(\forall x)Fx$ is true at all times later than the present, i.e., on all possible branches. And $\mathbf{F}(\exists x)Fx$ is only false *simpliciter* when the proposition that $(\exists x)Fx$ is true at no time later than the present, i.e., on no possible branches. That they have determinate truth-values only in these extreme circumstances, however, must only really be treated as a consequence of the branching view, not a criticism of it. But I should discuss here Lewis's (1986) argument against branching time, since it is based on a common misconception of it. Lewis writes:

> The trouble with branching exactly is that it conflicts with our ordinary presupposition that we have a single future. If two futures are equally mine, one with a sea fight tomorrow and one without, it is nonsense to wonder which way it will be—it will be both ways—and yet I do wonder. The theory of branching suits those who think this wondering makes sense only if reconstrued: you have leave to wonder about the sea fight, provided that really you wonder not about what tomorrow will bring but about what today predetermines. (207-8)

Lewis is wrong, however, to conclude that a branching future commits one to holding "it will be both ways." He is confusing the fact that there *are* many future branches to be realized as of the present time, with the fact that only one of them *will* be realized. After all, that is the only way to make sense of the claim that future contingent statements have indeterminate truth-values[13]—if it will be both ways, then both of the statements are true. But according to the branching view, future contingents are presently indeterminate precisely because only *one* branch of the possible branches *will* get realized, but that there *is* no branch in particular that is presently that branch. Thus it makes good sense to wonder now which of the unrealized branches will become *the* future. Thus branching does not conflict with our ordinary presupposition that we *will have* a single future, although it does conflict with the idea that we *have* a single future. The former presupposition is something we surely do not want to reject—it is close to being a platitude. But it is far from clear that the latter presupposition has such a status, and arguably is contrary to most people's commonsense view of the future: many people think, in a clear-headed way, that *pace* Lewis, we don't *have* a single future, but we *will* have one. And, moreover, this has nothing to do with what is "predetermined": as already noted above (§III), *determinism* is neither here nor there on the issue of the *determinateness* of the future. Lewis's argument is thus far from compelling.

V. Branching Time for Presentists

It just remains to fill out the story concerning the branching structure and show how Ersatzer Presentism can account for the truth-value links between times. Consider:

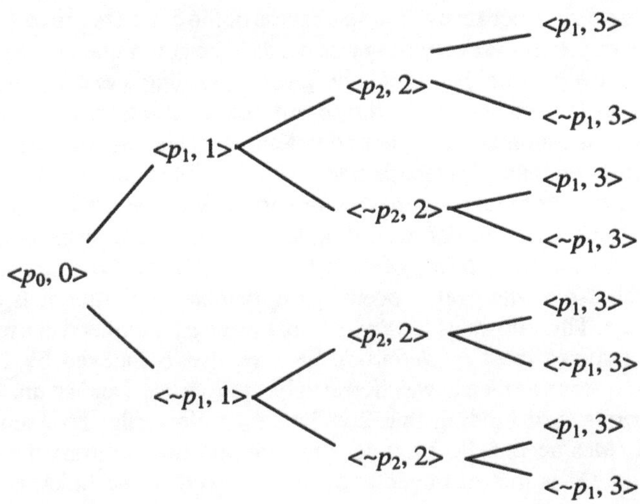

At the present, there is only one past accessible to it and no unique future. The set-up is such that all possible futures are represented and accessible from each node on a particular branch, so whichever time gets realized next will be represented somewhere on one of the future branches accessible from it: the branching structure gives us completeness.

a) A problem with truth-value links?

But a problem arises. Suppose that p_0 is true *simpliciter*, i.e., that time t_0 is presently realized. Now suppose that the proposition p_1 is made true *simpliciter* next. What guarantee have we that it is p_1 at t_1 that has been realized rather than, say, p_1 at t_3? If there is nothing to constrain which time is realized, it seems that we are committed to saying that the concrete facts that have just "become" realize all of these times equally. But if this is the case, then there are branches that state

$$<<p_1, 1>, <p_0, 0>>$$

is the history of the world, whereas others state

$$<<p_1, 3>, <p_2, 2>, <p_1, 1>, <p_0, 0>>$$

is, whereas others state that

$$<<p_1, 3>, <\sim p_2, 2>, <\sim p_1, 1>, <p_0, 0>>$$

is, and if p_1 is true *simpliciter*, it seems that there is nothing to choose between these options. Something needs to be done to preserve the truth-value links.

The problem here arises because of the sparseness of the concrete present: it is not sufficiently rich to determine everything that needs to be determined in order to defend a plausible view of time. But this is no great news; that's exactly what I have been arguing so far, and precisely the reason for invoking more than the concrete present facts as truthmakers for tensed statements. This was, after all, what confined the previous versions of presentism to an implausible position. But once we remember that Ersatzer Presentists have more artillery at their disposal, this problem has an obvious solution. For according to the theory here proposed, the sets of propositions follow the *ordering* of the dates; so although the content of the proposition p_1 is the same wherever it occurs as a member of a time, it is also associated with a date. This allows us to *stipulate* that when p_0 is realized at time 0, the next set to be realized must *by definition* be a set that is indexed by 1 (or whatever number sequence or metric we choose to be appropriate), and so on. This is no more mysterious than holding that 2nd January follows the 1st January, something that all sides accept. So there is a natural and non-arbitrary way of specifying which time it is that has been realized. The truth-value links remain intact.

b) A further problem with truth-value links for branching time presentists?

However, there is another potential objection. Suppose that it is $<p_1, 1>$ that is realized after $<p_0, 0>$, rather than, say, $<p_1, 3>$, due to the fact that $<p_1, 1>$ is indexed by 1 and $<p_1, 3>$ is indexed by 3, and 1 is the least available number greater than 0. Now suppose that proposition p_2 is made true *simpliciter* by the next concrete facts to "become." What guarantee have we that it is the p_2-set in the history

$$<<p_2, 2>, <p_1, 1>, <p_0, 0>>$$

that is realized, rather than the p_2-set in the history

$$<<p_2, 2>, <{\sim}p_1, 1>, <p_0, 0>>?$$

Indeed, since there is nothing in the present to determine which one is realized (both p_2-sets are, after all, indexed by 2), it seems that we are committed to saying that both have equal claim to have been realized, and our truth-value links again break down.

This objection again relies on there not being enough in the present to determine which of the sets is realized. But, again, this is not the theory that I am putting forward; there *are* more facts available to determine which p_2-set is realized. For once we have in place the fact that it is $<p_1, 1>$ that is realized after $<p_0, 0>$, there is a fact of the matter as to which branch has been realized, namely the $<<p_1, 1>, <p_0, 0>>$ history rather than the $<<{\sim}p_1, 1> <p_0, 0>>$ history. It follows that it must be the p_2-set accessible from the $<p_1, 1>$ that is the next to be realized, since it is a fact that that branch is realized and not the $<{\sim}p_1, 1>$ branch.

This can be represented formally, rather than diagrammatically, by adding an accessibility relation defined over times, and adding a "positional" element to times. Times are now ordered triples of the form $<\mu, n \in \mathbb{R}, \sigma>$, where σ is the position of the time in the branching structure, something given by specifying the sequence of times along its past. For discrete time, this can be achieved simply by including the last time realized as the third element (since this time will specify its past, and so on). Thus, for the case of discrete branching time, time $n+1$ has the form $t_{n+1} = <\mu, n+1, t_n>$, where $t_n = <\nu, n, t_{n-1}>$, and so on. The accessibility relation R for $<\mathbf{T}, E, t, R>$ can be defined as follows: for all times, t_{n+1} is accessible to t_n iff t_n is a member of t_{n+1}, i.e.,

$$(\forall t)(R\ (t_{n+1}, t_n) \equiv t_n \in t_{n+1})$$

We can also generalize to continuous time. The reason that we must say something slightly different for continuous time is that there is no "next available date" or "last time realized" to be the positional element. But all that needs to be said to constrain the order of realization is first to say that the order of realization must follow the order of the real numbers as given by the *greater than* relation. Second, it may not be possible to specify a particular time as the positional element of a time, but it *is* possible to specify the position using the *sequence* of times leading up to that time (which I represent as "$\sigma_{<a}$"), and then represent times as follows:

$$t_a = <\mu, a \in \mathbb{R}, \sigma_{<a}>$$

So generalizing to include continuous time, the accessibility relation R for $<\mathrm{T}, E, t, R>$ can now be defined as follows: for all times, t_b is accessible to t_a iff t_a is included in the third element of t_b, i.e.,

$$(\forall t)(R\ (t_b\, , t_a) \equiv ((t_a \subset \sigma_{<b})\ \&\ (t_b = <\nu, b \in \mathbb{R}, \sigma_{<b}>))).$$

c) Tying up loose branches: one last objection met

Now that the metaphysics and semantics have been given for Ersatzer Presentism, I should answer one possible objection: *Why* is it that a certain atomic proposition p appears in a given E-related time if not because *it was the case that* p? But if this is the explanation, if this is how the story ultimately bottoms out, then how is Ersatzer Presentism any better off than Priorian Presentism?

There are a few things to say in response. First, whether a given proposition appears in a time is simply a brute fact. Or rather, since all possible sets of propositions appear on the branches, it is a necessary truth that such propositions appear in some of the times, and it is a brute fact that one of them gets realized. This is no objection since it is no more mysterious than concrete facts being realized at the times they are according to tenseless theorists. The potential threat of the question comes from explaining why the various times are E-related. But on reflection it is more of a toothless suck than a vicious bite. For according to Ersatzer Presentism, what makes *It was the case that p* true is an actually E-related ordered triple, whereas according to the tenseless theory, what makes it true is an actually *earlier than*-related concrete fact. Now to ask why these ordered triples are actually E-related is about as fair as asking why the concrete facts are actually *earlier than*-related in the tenseless theory, i.e., not at all—they just *are.* In this sense, then, all theories take it as a brute fact that it was the case that p; the advantage over Priorian Presentism is that these other theories have an account of what this fact looks like, be it an E-related abstract structure or an *earlier than*-related concrete one.

VI. The Advantages of Ersatzer Presentism

Ersatzer Presentism, then, can meet the three conditions that any satisfactory account of time should: it allows us to state truths about the past; it wears its ontological commitments on its sleeve; and it ensures that truth-value links are preserved. There are two further advantages. First, it is often argued that there is no substance to the tensed-tenseless debate—that once we recognize the confusion over tensed and tenseless readings of various quantifiers and copulas, the tenseless theorist's "does exist in the past" is not substantially different from the tense theorist's "existed"—they are the same theory under different descriptions. I need not show here all the ways in which I think this line of argument is confused, for the charge that the debate is insubstantial certainly cannot be levelled against Ersatzer Presentism, whether or not it affects other versions of presentism. According to Ersatzer Presentism, the constituents of past and future times are not spatio-temporally or causally related objects, unlike past and future times according to the tenseless theory. The constituents of ersatz past and future times are propositions, numbers, and sets of propositions and numbers, and so can only have those relations that can hold between sets, propositions, and numbers. These conceptions of past and future times are so radically different that it would stretch the meaning of the word "same" to claim that they are essentially the same theory under different descriptions! Unlike the tenseless theory, according to Ersatzer Presentism, there really is no time like the present.

The second advantage is that many have complained that there is no good motivation for becoming a tense theorist, let alone a presentist. Tooley (1997, 250) has argued that before we do commit ourselves to a tensed theory, there is a "need for metaphysical argument." Here I claim the argument in favor of Ersatzer Presentism over its rivals is *ontological parsimony*. The tenseless theory, for instance, postulates appropriately related past and future concrete objects as truthmakers for various tensed statements; but this is something Ersatzer Presentism can discard. Neither does Ersatzer Presentism postulate mysterious facts *in* the present to make past-tensed statements true. Indeed, the discussion indicates that a theory that just rests solely on concretely realized present facts will not adequately account for all the truths we need. So the appeal to more objects other than concretely realized present facts is the necessary concession presentism must make to be viable. But it is still a highly attractive theory if you *already* believe in abstract objects (propositions, numbers, and sets), which many of us do. Thus it isn't an issue of quantitative versus qualitative parsimony (see Bacon (1995, 87); Lewis (1973, 87) for this distinction). The argument is that Ersatzer Presentism is *quantitatively* more parsimonious than the tenseless theory; that is, the set of objects that Ersatzer Presentism postulates is a proper subset of those objects that the tenseless theory postulates. But quantitative savings are still significant: everyone agrees that one *can* have too much of a good thing. Thus, if you prefer ontologically economical theories without having to live on a shoestring, then Ersatzer Presentism is the theory to adopt.[14]

Notes

1. There is a clear partition between tensed and tenseless theories of time: essentially, tensed theories of time assert that in some objective, mind-independent sense or other, the present is privileged, whereas tenseless theories of time assert that all times are real, no one of which is privileged. Thus all tenseless theorists (e.g., Mellor (1981) and (1998); Oaklander (1984); Le Poidevin (1991)) are realists about the past, but tensed theorists

needn't be, although some are, e.g., Broad (in some of his writings); Gale (1968); Schlesinger (1980); Smith (1993); McCall (1994); and Tooley (1997).)

2. Throughout this paper, "exists" is meant *tenselessly*, for to equate "exists" with "exists now" renders presentism trivial. The tenseless reading is the only reading that makes presentism a substantial alternative to the tenseless theory, which currently dominates the philosophy of time.

3. As is widely known, relativity theory creates problems for all tensed theories of time, not just presentism. How to reconcile the two cannot be addressed in this paper, but see my (2006) for a proposed solution. Note that this solution is perfectly compatible with the version of presentism I present below.

4. For ease, I shall talk throughout of facts and their constituents as truthmakers. Those sympathetic to particulars as truthmakers can make the relevant changes.

5. I am here assuming the truth-value links generated in linear time. Those who don't accept that *p* also generates **PF***p* (because they hold that the future branches, say) would still want to hold that *p* generates **FP***p* and the truth-value links associated with it. My point is that this version of presentism cannot even guarantee that these truth-value links are preserved.

6. The versions of presentism proposed by Bigelow (1996) [chapter 8 in this volume—*Editors*] and Craig (1997) [chapter 4], which also pack all of their proposed truthmakers for past- and future-tensed statements into the present, are subject to such criticisms. See Oaklander (2002) for specific criticisms of these versions of presentism, and Oaklander (2003) [chapter 5] for criticisms of Craig's later (2000a), (2000b) attempts at defending presentism.

7. Lewis (1986, 204) briefly mentions the possibility of holding some kind of ersatz position concerning time. His reason for dismissing it is that "No man . . . believes that he has no future; still less does anyone believe that he has no past." This is ambiguous. On one reading it is plain false: no presentist believes they have a future in the sense of real times that exist later than now. On the other reading, it is true but harmless, for presentists (can) believe that there *will* be more times to come, even though there *are* no real times later than our present time (see §**IV** (*d*)).

8. For the purposes of this paper, I shall take propositions to be primitive abstract objects. However, this is not to say that nothing more can be said about them: they are entities capable of being either true or false, entities to which we can have attitudes (such as beliefs, desires, hopes, and fears), and they are capable of representing states of affairs. How they manage to do this is not within the scope of this paper. Being abstract entities, they may raise the understandable question of how we come to grasp them. But I think this worry is misguided. It is no more mysterious than coming to know any truth. Grasping the content of propositions about the concrete world requires interaction only with the concrete world and not with an abstract object.

9. À la Kripke (1963) and also McArthur (1976).

10. I take my cue here from Adams (1974) in his analogous discussion of truth *simpliciter* and truth-at-a-world in the possible worlds debate.

11. I use this truth-condition since μ is defined as a set of *u-propositions* that is true at that time, thus it is not the case that "~*p*" is true iff "*p*" is false, since *p* does not even appear in that time if "*p*" is false at that time.

12. Thus Russell's hypothesis is harmless to Ersatzer Presentism. The *E*-relation simply does not hold if the world has just come into existence; so no past-tensed propositions are true, as required.

13. See Bourne (2004) for a treatment of future contingent statements with indeterminate truth-values which is compatible with the laws of non-contradiction and excluded middle.

14. Many thanks to Oren Goldschmidt, Katherine Hawley, Hugh Mellor, Peter Smith and one of the anonymous referees for helpful comments. This paper was first given at the Cambridge Graduate Seminar (April 2001) as "The Presentist Manifesto," then at the Moral Sciences Club, Cambridge (March 2002) as "Between Timid and Timbuktoo" and then at Leeds University (February 2004) as "Real Time III: The Presentist Strikes Back." Many thanks to those audiences.

References

Adams, R. M. "Theories of Actuality," *Noûs* 8 (1974): 211-31.

Armstrong, D. M. *What is a Law of Nature?* Cambridge: Cambridge University Press, 1983.

———. *A World of States of Affairs*. Cambridge: Cambridge University Press, 1997.

Bacon, J. *Universals and Property-Instances: The Alphabet of Being*. Oxford: Basil Blackwell, 1995.

Bigelow, J. "Presentism and Properties," *Philosophical Perspectives* 10 (1996): 35-52. Chapter 8 in this volume.

Bourne, C. P. "When Am I? A Tense Time for Some Tense Theorists?" *Australasian Journal of Philosophy* 80 (2002): 359-71.

———. "Future Contingents, Non-Contradiction, and the Law of Excluded Middle Muddle," *Analysis* 64 (2004): 122-28.

———. *A Future for Presentism*. Oxford: Clarendon Press, 2006.

———. "Numerical Quantification and Temporal Intervals: A Span-er in the Works for Presentism?" *Logique et Analyse* 50 (2007).

Craig, W. L. "Is Presentness a Property?" *American Philosophical Quarterly* 34 (1997): 27-40. Chapter 4 in this volume.

———. *The Tensed Theory of Time: A Critical Examination*. Dordrecht: Kluwer, 2000a.

———. *The Tenselesss Theory of Time: A Critical Examination*. Dordrecht: Kluwer, 2000b.

———. "In Defense of Presentism." Pp. 391-408 in *Time, Tense, and Reference*, edited by A. Jokić and Q. Smith. Cambridge, Mass.: MIT Press, 2003.

Dummett, M. "The Reality of the Past." Pp. 358-74 in *Truth and Other Enigmas*, by M. Dummett. London: Duckworth, 1978.

———. *Truth and the Past*. Columbia University Press, 2003.

Gale, R. M. *The Language of Time*. London: Routledge and Kegan Paul, 1968.

Kamp, H. "Formal Properties of 'Now,'" *Theoria* 37 (1971): 179-87.

Kripke, S. A. "Semantical Considerations on Modal Logic," *Acta Philosophica Fennica* 16 (1963): 83-94.

Le Poidevin, R. *Change, Cause and Contradiction*. London: Macmillan, 1991.

Le Poidevin, R., ed. *Questions of Time and Tense*. Oxford: Oxford University Press, 1998.

Lewis, D. K. *Counterfactuals*. Oxford: Basil Blackwell, 1973.

———. *On the Plurality of Worlds*. Oxford: Basil Blackwell, 1986.

———. "Forget about the 'Correspondence Theory of Truth,'" *Analysis* 61 (2001): 275-80

———. "Things *qua* Truthmakers." Pp. 25-38 in *Real Metaphysics: Essays in Honour of D.H. Mellor*, edited by H. Lillehammer and G. Rodriguez-Pereyra. London: Routledge & Kegan Paul, 2003.

Ludlow, P. *Semantics, Tense, and Time: An Essay in the Metaphysics of Natural Language*. Cambridge, Mass.: MIT, 1999.

Łukasiewicz, J. *Selected Works*, edited by L. Borkowski. Amsterdam: North Holland, 1970.

McArthur, R. P. *Tense Logic*. Dordrecht: Reidel, 1976.

McCall, S. *A Model of the Universe*. Oxford: Clarendon Press, 1994.

Mellor, D. H. "Necessities and Universals in Natural Laws." Pp. 105-25 in *Science, Belief and Behaviour*, edited by D. H. Mellor. Cambridge: Cambridge University Press, 1980.

———. *Real Time II*. London: Routledge, 1998.

———. "Replies: Change and Time." Pp. 235-37 in *Real Metaphysics: Essays in honour of D. H. Mellor*, edited by H. Lillehammer and G. Rodriguez-Pereyra. London: Routledge, 2003.

Oaklander, L. N. *Temporal Relations and Temporal Becoming*. Lanham: University Press of America, 1984.

———. "Presentism, Ontology and Temporal Experience." Pp. 73-90 in *Time, Reality and Experience*, edited by C. Callender. Cambridge: Cambridge University Press, 2002.

———. "Presentism: A Critique." Pp. 196-211 in *Real Metaphysics: Essays in Honour of D. H. Mellor*, edited by H. Lillehammer and G. Rodriguez-Pereyra. London: Routledge, 2003. Chapter 5 in this volume.

Plantinga, A. *The Nature of Necessity*. Oxford: Clarendon Press, 1974.

Prior, A. N. *Time and Modality*. Oxford: Oxford University Press, 1957.

———. *Past, Present and Future*. Oxford: Oxford University Press, 1967.

———. *Papers on Time and Tense*. Oxford: Oxford University Press, 1968a.

———. "Now," *Noûs* 2 (1968b): 101-19.

———. "The Notion of the Present," *Studium Generale* 23 (1970): 245-48.

———. *Objects of Thought*, edited by P. T. Geach and A. J. P. Kenny. Oxford: Clarendon Press, 1971.

———. "A Statement of Temporal Realism." Pp. 45-46 in *Logic and Reality: Essays on the Legacy of Arthur Prior*, edited by J. Copeland. Oxford: Oxford University Press, 1996.

Russell, B. *The Analysis of Mind*. London: Allen and Unwin, 1921.

Schlesinger, G. N. *Aspects of Time*. Indianapolis: Hackett, 1980.

Smith, P. "Deflationism: The Facts." Pp. 43-53 in *Real Metaphysics: Essays in Honour of D. H. Mellor*, edited by H. Lillehammer and G. Rodriguez-Pereyra. London: Routledge, 2003.

Smith, Q. "Problems with the New Tenseless Theory of Time," *Philosophical Studies* 52 (1987): 371-92.

———. *Language and Time*. Oxford: Oxford University Press, 1993.

Tooley, M. "The Nature of Laws." *Canadian Journal of Philosophy* 7 (1977): 667-98.

———. *Time, Tense, and Causation*. Oxford: Oxford University Press, 1997.

Wheeler, J. A. "The 'Past' and the 'Delayed Choice' Double-slit Experiment." Pp. 9-48 in *Mathematical Foundations of Quantum Theory*, edited by A. Marlow. New York: Academic Press, 1978.

10
A Defense of Transient Presentism

M. Oreste Fiocco

Introduction

Presentism is a controversial and much discussed position in the metaphysics of time. The position is often glossed as simply the view that *everything that exists is present*.[1] This gloss, however, does not in itself characterize a single view. There are different interpretations of "to exist at present," so that one who attempts to provide an account of the nature of temporal reality must—if he is to address such relevant issues as the ontological status of entities before and after this moment, the truth-value of statements about such, and the semantic grounds for these statements—adopt more general metaphysical theses to complement any bald claim about present existence. Since the adoption of these more general theses yields distinct views, there are several incompatible views of the nature of temporal reality consistent with this familiar gloss.

Traditionally, many objections have been leveled against presentism. Given the variety of views that can legitimately be classified as presentist, it is futile to consider these objections or, more generally, to assess presentism without first examining the differences among these views. In this paper, the variety of presentist views is first propounded, with a characterization of the primary dimensions along which the views differ. Then the version of presentism plausibly judged to be optimal is presented. The variety among presentist views is so great that this version—*transient presentism*—is not even characterized by the standard gloss given above. Next, all of the familiar objections to presentism are discussed, with consideration of the relevance of each to transient presentism. Also considered, though in less detail, is the relevance of these objections to other versions of presentism in the contemporary literature and the means these views have to counter them. The conclusion is drawn that transient presentism provides the most satisfying responses to the traditional objections and, furthermore, captures very well the motivation for adopting a presentist position in the first place. The grounds for these two conclusions constitute the defense of transient presentism.

Part I. The Varieties of Presentism

Because the gloss on presentism as the view that everything that exists is present is too vague to be practicable and, moreover, is inconsistent with the version of presentism that is plausibly judged to be optimal, one should regard presentism as a generic position, subsuming many incompatible, specific accounts of the nature of temporal reality. In particular, presentism should be regarded as the position according to which there is something metaphysically distinctive—be it qualitative or ontological—about this moment, now. A variety of presentist view, that is, a version of presentism, is any view that conjoins with this principle such other metaphysical and ontological theses as may be necessary for a substantive account of the nature of temporal reality.

The two distinctions that provide the most perspicuous means of classifying the versions of presentism are (i) whether, according to the view, *there is but a single moment* or *there is more than one moment* and (ii) whether, according to the view, *there is but one way to be real*, that is, to exist *simpliciter* or whether *there is more than one way to be real*. Hence, for example, there are varieties of presentist view according to which there are many moments, though this moment, now, has some significant feature that sets it apart from all others; and there are varieties according to which there is only one moment, though there are also past and future entities whose ontological status differs from this moment (and anything that exists at it). The assortment of views is increased with the adoption of independent metaphysical theses, such as the principle that non-existent entities can bear properties.[2]

§1.1 Constraints on the Optimal Version of Presentism

Elsewhere,[3] five constraints on the optimal version of presentism are proposed: that (i) there be no distinction among ways or levels of being *simpliciter*; (ii) an entity must exist in order to bear any property whatsoever; (iii) there exist but a single moment (where a *moment* is a unique sort of entity, irreducible to any other); (iv) no obvious, uncontroversial truth regarding the structure in temporal reality (e.g., *that Socrates was a philosopher*, *that dinosaurs existed*) need be denied; and (v) grounds for all truths regarding this structure be provided.

The reasons for proposing four of these constraints ((i), (ii), (iv), and (v)) are intuitive. It seems that either a thing exists or it does not—that there is no intermediate ontological status between existence and non-existence—and hence no genuine distinction between *existing* and *subsisting* (or *existing* and *having being*). Furthermore, it seems intuitively obvious that in order for a thing to bear any property, it must first exist. These claims are so basic, that it does not seem possible for any arguments that are not question-begging to be made in support of them.

Similarly, an account of the nature of temporal reality that requires one to deny obvious truths about the temporal structure of the world does not seem to be an account worth having. If one cannot be confident in such trivialities as that it is literally true that Socrates was a philosopher and that dinosaurs once existed, then it seems one is completely without guidance when considering temporal reality and, consequently, that any attempt to gain insight into its nature is futile. The assertion that these truths have grounds—that they are made true in virtue of some feature of reality—is another intuitive claim that seems too basic to be made compelling, via argument, to one who does not find it so.

The reasons for proposing constraint (iii) are discussed in detail elsewhere.[4] It is hoped that those who do not share the intuitions or reasons for proposing this constraint will nonetheless find the present discussion interesting, if only because it shows that there is a coherent view in the metaphysics of time that meets such stringent constraints.

§1.2 The Optimal Version: Transient Presentism

Transient presentism is incompatible with the principle that everything that exists is present (regardless of how "to exist at present" is construed) for the reason that the view is explicitly committed to the existence of *atemporal* entities, that is, entities that do not exist in time. Examples of atemporal entities are numbers, properties, and facts.[5] Insofar as such entities exist outside of time, they do not bear any temporal properties and so, in particular, they do not bear the property of being present.

Thus, in order to appreciate transient presentism, one must have a sense of what *time* is in order to distinguish atemporal from temporal entities. One might feasibly regard time as the entity, abstract and sui generis, that renders change possible. Change occurs when an entity bears a property at one moment and an incompatible property at a distinct moment. Time renders change possible by bringing forth[6] the moments at which (temporal) entities bear properties, the moments that are essential to change. On this simple account of time, the essential mark of a temporal entity is its capacity to change. Hence, an atemporal entity is an entity that does not and could not possibly change. The atemporal entities crucial to transient presentism are *facts* (of a particular sort).

According to the metaphysics of time that underlies this version of presentism, there is but a single moment; specifically, the one that exists now (this is the only moment that has any being whatsoever). A moment, however, is an entity that exists instantaneously, then ceases to be. It is replaced by a distinct moment.[7] Although a moment exists only instantaneously, the states of the world at that moment indelibly mark the world by yielding facts.

One can construe facts as simple, that is, non-structured entities that, being immutable, do not exist in time. A specific fact would not have existed without the moment it is about; to this extent, facts depend on moments. Since, however, the facts about a moment, *m*, do not exist *at* m, they can remain despite *m*'s ceasing to be. Despite their not existing at a moment, this kind of dependence between facts and moments enables facts to underpin the structure of temporal reality. Atemporal facts of this sort are the basis of truths about the past; as such they are the *truth-makers* of past-tensed sentences. Truth-makers are non-representational (or non-intentional) entities, namely, those entities, abstract or concrete, correspondence with which renders a representation true.[8]

Note that being atemporal does not preclude coming into existence. The mark of a temporal entity is capacity to change. As noted above, change occurs when an entity bears a property at one moment and an incompatible property at a distinct moment. If moment, *m*, is the first at which an entity, *e*, exists, then there is no moment distinct from *m* at which *e* bears incompatible properties. Coming into existence, therefore, is not change. Hence, an atemporal entity—the sort of thing that could not possibly change—can indeed come into existence. It does not, however, come into existence at a moment. Since it does not come into existence at a moment, nor exist at any moment, it neither bears temporal properties, nor stands in temporal relations. An atemporal entity simply exists and remains. It might, nonetheless, be an entity with an origin. Indeed, the facts that provide the basis of the structure in temporal reality have origins—despite not existing in time.

It should be clear that transient presentism meets three of the proposed constraints on the optimal version of presentism, namely the constraints regarding the existence of a single moment, that there be a metaphysical basis for the structure in temporal reality and, in virtue of the latter, that no uncontroversial truth about the nature of temporal reality need be denied. To ensure that it meets the other two constraints, transient presentism simply adopts explicitly the theses that there is but one way of being *simpliciter* and that no entity can bear any property whatsoever without existing. The coherence of the view is preserved by acknowledging the atemporality of the facts that serve as the basis of the structure in temporal reality.[9]

On transient presentism, then, the past does indeed exist. The past has the same ontological standing as the present and, hence, is just as real; however, it is a completely different sort of thing. The present is the world at this moment; as such, it is in some sense constituted by the great variety of familiar (and exotic) mutable things that now exist: desks, trees, bicycles, persons, minds, etc. The past is not the world at each moment prior to this one, now; there are no past moments or past entities. Rather, the past is an indefinitely large collection of simple facts.

Therefore, both the past and the present exist; the future, on the other hand, does not exist in any sense. It is plausible to think there are no facts about the future and, furthermore, that there could not be such facts, given the foregoing account of the genesis of such things. Indeed there is compelling reason to deny that there ARE (tenselessly) or are now facts about the future. The reason is that if there were such facts, the future would not be open; it would be as fixed as the past and necessary in the same sense the past is (in whatever sense, exactly, this is). Hence, if there were facts about the future, agents would not ever be free in the sense of being able to do otherwise than what they in fact do and no truths about the future would be contingent. One can find detailed discussion of this problem for facts about the future elsewhere,[10] and so it is not considered further here; the problem is mentioned in passing simply to elucidate the nature of temporal reality as it is according to transient presentism.

Part II. Objections to Presentism

There have been many objections directed at presentism. But presentism is not a single view. In order, therefore, to evaluate how problematic these objections are, it is best to choose a specific version of presentism to evaluate in light of them. Although the primary purpose here is to consider how transient presentism fares against these objections to presentism in general, some other versions of presentism are introduced, and some consideration is given to how one might respond to these objections on a different presentist view. In the course of such discussion, however, these other responses are contrasted with the one available to the proponent of transient presentism and the preferability of the latter is maintained. The reason for this is always the same: the transient presentist's response is the only one that respects the five constraints propounded above.

The problems raised by objections to presentism are varied. One of the objections in the contemporary literature poses, in itself, no problem at all. Some objections are pertinent only to specific versions of presentism and are idle when directed at others. Yet another objection raises philosophical problems so general, it should not be taken as an objection to presentism per se. Nevertheless, consideration of these objections is illuminating because it does clarify the distinctions between presentism in general and other accounts of the nature of temporal reality (and between transient presentism and other versions of presentism). More specifically, it focuses attention on some of the metaphysical and, in particular, ontological consequences of holding the view that there is something distinctive—be it qualitative or ontological—about this moment, now, while respecting the aforementioned constraints.

Below are considered six more and less familiar objections to presentism and the putative problems to which they give rise.

§2.1 The Problem of There Being Only One Moment

Ned Markosian maintains that an embarrassing problem for presentism is that the position *seems* to entail that there is no moment except the present one. The embarrassment for the presentist is supposed to arise because, "It is very natural to talk about times. We often speak as if times are genuine entities, and we often appear to express propositions about times."[11] In fact, Markosian does *not* believe that presentism entails that there is but one moment. And one should concur with this claim. It is consistent with the general statement of presentism that there is any number of moments—just so long as the present one bears some metaphysical mark that distinguishes it from the others.

Markosian propounds a version of presentism closely modeled on a position in the metaphysics of modality, namely, *actualism*; for this reason, it is apt to call Markosian's view, *actualist presentism*.[12] According to this view, there are many (presumably infinitely many) abstract moments; an *abstract moment* is supposed to be a *maximal, consistent proposition that has been, is, or will be true*, that is, an entity that represents everything that has been the case, is the case, or will be the case. In addition to the many abstract moments, there exists a *concrete moment*, which is supposed to be the *sum total of all present facts*, the totality of all things that now exist and the properties they now bear.[13]

One abstract moment is distinguished from all others in that it corresponds to the unique concrete moment. Markosian believes that this view avoids the sort of embarrassment characterized above, because it acknowledges the existence of more than one moment and, hence, provides a subject matter for one's claims about moments other than this one, now.

There are reasons for being dissatisfied with actualist presentism. First of all, it does not seem justifiable to think the ontological reduction of moments to propositions is feasible or at all helpful, for the notion of a moment is indispensable to articulating how exactly a particular proposition represents the world as being.[14] Similarly, it does not seem natural to construe one's talk about moments as being about *representations* of the world, rather than about features of the world themselves. Moreover, it is not at all clear how things so different in nature as the concrete moment and the myriad abstract moments are both appropriately regarded as *moments* or how it is plausible to regard a past (abstract) moment as being related—causally or otherwise—to the present (concrete) moment.

These objections to actualist presentism are significant, but are perhaps less so compared to the following: According to this variety of presentist view, in addition to the maximal, consistent proposition, i.e., moment, that represents how things now are, there are countless propositions that truly represent how things have been and how things will be. Whereas the basis of the truth of the one proposition that correctly represents how things are at present is the unique concrete moment, there is nothing in the world that grounds the truth of those propositions that are supposed to represent how things were and how they will be. In other words, there is nothing whatsoever that makes it true that Socrates was a philosopher or that there will be a woman president in 2016 and, hence, no explanation for why or how these propositions are true. This objection indicates the *problem of the basis of temporal reality* (also known as the *problem of truthmakers*): it is obvious that there is some structure in temporal reality; it is true that the world has been some ways (and not others), is now a certain way and, perhaps, will be certain ways. As developed by Markosian, actualist presentism has no way to account for this structure—despite its commitment to propositions that accurately portray it. Since the idea of truths without bases is unacceptable, one should reject Markosian's version of presentism. (The problem of the basis of temporal reality is taken up again below.)

If the objection that Markosian raises, to wit, that it is unacceptable for an account of temporal reality to posit but one moment, were cogent, it would be appropriate at this point to present a different version of presentism, one that assumes that there are many moments. One should not, however, find this objection at all compelling. Transient presentism entails that there is but one moment (of any kind), yet this feature of the view is not embarrassing or problematic in any way. The question whether an account of temporal reality posits one moment or many is, in itself, not at all relevant to the adequacy of the account. Whether an account is adequate (and appealing) is determined by the success it has in preserving one's basic intuitions regarding temporal reality and more general metaphysical principles. It does not seem that one's intuitions regarding the number of moments—if one has any at all—are appropriately basic.

Whereas one should agree with Markosian that it is natural and meaningful to talk about other moments, such moments nevertheless need not exist. (Of course, an adequate semantic account of such talk is needed and, fortunately, is available. More on this below.) Nor does the fact that only one moment exists entail that there is neither structure to temporal reality nor a basis of the truth of one's claims about it. Therefore, one can conclude that its being a consequence of transient presentism that there is but one moment presents no problem for the view.

§2.2 The Problem of McTaggart's Paradox

L. Nathan Oaklander contends that "in one way or another [the versions of presentism] are all susceptible to McTaggart's paradox" [chapter 5 in this volume—*Editors*].[15] McTaggart's Paradox[16] holds a vexing position in modern discussions of the metaphysics of time. Some take it to be of the utmost importance in understanding the nature of temporal reality and presenting an adequate account thereof;[17] others dismiss it as a "howler."[18] The paradox does indeed show something, but only about one unpromising variety of presentist view that few—if anyone—explicitly hold.

There are different ways of interpreting McTaggart's obscure presentation of the paradox. The simplest way of appreciating it is to view the paradox as a special case of the problem of temporary intrinsics:[19] On certain accounts of the nature of temporal reality, entities that exist in time are supposed to be future at one moment, then present, and then past at subsequent moments. On another account of the nature of temporal reality, all moments of time—and, hence all things existing at these moments—have the same ontological status and so are equally real. When one attempts to combine these two sorts of account, one is left with a view according to which an object, *o*, is future at *m1*, present at *m2*, and past at *m3*; however, since all moments of time (and the things that exist thereat) have the same ontological status, *o* IS (tenselessly) past, present, and future. The fact that *o* is future at one moment, then present at another, then past is insignificant, for all moments of time have the same ontological status and, hence, so does *o* as future, *o* as present, and *o* as past. The problem, obviously, is that no one thing can exist bearing these three incompatible properties.

There are a number of ways one might respond to this problem. Perhaps the simplest, given its presuppositions, is to adopt some variety of *temporal parts theory*, according to which there is not literally a unique object that is past, present, and future, but rather distinct parts of a composite object bearing these temporal properties. More important to present purposes, though, than consideration of the means available to respond to McTaggart's Paradox, is recognition of the fact that if one is to be vulnerable to the paradox in the first place, one must hold very specific views regarding the nature of temporal reality. The paradox is generated only if one assumes both that *more than one moment of time has the same ontological status*[20] and that *there is a certain kind of dynamism in temporal reality*, to wit, (tenselessly) EXISTING moments taking on, then losing temporal properties. There is, however, only one version of presentism that conjoins these two assumptions; it is aptly called *flowing presentism*. On this view, all moments of time have the same ontological status; nevertheless, the view is genuinely presentist because it is assumed that one of these moments, to wit, this one, now, bears the qualitative distinction of *being present* (a property it will lose in just a moment).

Flowing presentism is of dubious coherence, as it attempts to combine the stasis of a universe in which all moments EXIST with the dynamism that is naturally and intuitively thought to be a feature of temporal reality. It is quite dubious that the view is, in itself, defensible. McTaggart's Paradox merely shows that if one endorses flowing presentism, one most forgo the view that objects persist by enduring (or adopt some other means of solving the putative problem of temporary intrinsics).[21] However, if one rejects the principle that all moments of time have the same ontological status—or the assumption that moments sequentially take on and lose temporal properties—McTaggart's Paradox does not even arise. All varieties of presentist view—other than flowing presentism—deny that all moments have the same ontological status; according to transient presentism, in particular, there exists, in any sense, but a unique moment. Therefore, Oaklander's claim that all versions of presentism are confronted by McTaggart's Paradox is not true and, more importantly, the paradox presents no problem for transient presentism.

§2.3 The Problem of the Basis of Temporal Reality

Perhaps the most significant and most accessible objection to presentism in general is the claim that undeniably there are truths about moments other than this one, now; certainly some accounts of the ways the world was are accurate, whereas others are not. (Moreover, some believe that at present there are truths regarding the ways the world will be.) The problem for the presentist is to provide a satisfying account of what in the world makes some claims about moments distinct from this one, now, true and other claims false. What is needed, in other words, is an adequate account of the basis of temporal reality.

Objections very similar to the one just presented have received much attention in recent discussions of the metaphysics of time. This sort of objection, variously formulated, can be called the *truth-maker objection*.[22] Although one need not accept all formulations of this objection,[23] one should accept the principle on which it is premised, namely, that *all truth supervenes on reality*—there can be no difference in which representations are true without some difference in what exists and how it is.

There are different ways a presentist can respond to this objection and, hence, different versions of presentism that purport to resolve it. One response is simply to deny the foregoing principle, that is, to acknowledge that there are indeed truths about moments other than this one, now, and yet maintain that these truths supervene on no features of reality. Here resolution comes simply by denying the problem. This sort of response is not uncommon. One who responds in this way seems to assume that the structure in temporal reality is primitive; hence, the variety of presentist view based on this assumption is aptly called *primitive presentism*.

The proponent of primitive presentism freely uses tensed expressions—thereby preserving ordinary temporal discourse—while denying that such use raises any metaphysical difficulties.[24] For example, one claims that a past (or future) event or individual that is supposed *not to exist* at present, nevertheless is aptly characterized or truly described in certain ways, because it *was* (or *will be*) that way. Thus, although Socrates does not now exist he nonetheless *was*. In this way, one can maintain the truth of a platitude, to wit, that Socrates existed or that Socrates was a philosopher, without having to maintain that Socrates exists to bear the property of being past or having been a philosopher or any property whatsoever.[25]

Primitive presentism preserves the truth of many intuitive claims about temporal reality, but it does so merely by reiterating them (or by providing a bit of formalism, employing primitive sentential operators, that is supposed to express these claims). It provides no explanation of the metaphysical basis of these truths and, hence, although it attributes structure to the nature of temporal reality while being consistent with the claim that but one moment—the present one—exists (or has any being at all), it simply takes this structure for granted. Of course, one's account of the nature of temporal reality must stop at some point; nevertheless, it seems that such an account is objectionably incomplete in that it fails to provide any sort of explanation of how or why statements about temporal reality are true, that is, it makes no attempt to ground the structure of temporal reality.

Other responses to the truth-maker objection do not simply take the structure in temporal reality as primitive. However, in order to respond to the objection in a way consistent with presentism in general, some philosophers have claimed that there is more than one way to be real, that is, that there is more than one mode of existence *simpliciter*. There is a view according to which only the present exists, but the past is real (has being) in that it *was* and the future is real (has being) in that it *will be*. On this view, there are three distinct modes of existence *simpliciter*: *having been*, *existing now*, and *going to be*. Thus, there could be an individual intrinsically exactly similar to one existing now, differing only in its mode of existence—it *has been*, rather than *is now*.[26] Hence, the truth that Socrates existed is explained by noting that although Socrates does not now exist, he nonetheless *was*. Here, though, this claim is not supposed to describe a primitive feature of temporal reality but to relate the special mode of existence that the individual, Socrates, enjoys—he *was*—and this is taken to ground the present truth that Socrates existed.

There is also the *degree presentism* defended by Quentin Smith.[27] On this view, only the present moment and the things that exist thereat have the maximal degree of existence; everything else exists to some lesser extent, depending on its temporal distance from this moment, now. Consequently, Smith believes there is an infinite number of ways to be real. This array of things and moments, with their different degrees of existence, is supposed to be the basis for the structure in temporal reality.

According to these views, only this moment now exists (or exists to the maximal degree); the structure in temporal reality, however, can be accounted for in terms of moments and things which although they do not exist, nevertheless are real.[28] One of the constraints on the optimal version of presentism enumerated above is that it make no distinction among ways or levels of being *simpliciter*. Since the foregoing views do not satisfy this constraint they are not acceptable. The notion of being is, it seems, so fundamental that it can only be construed as simply the contrary of nonexistence.

Transient presentism respects this simplicity in the notion of existence; it is a version of presentism that makes no distinction among ways of being or modes of existence. As noted below,[29] one of the strongest motivations for accepting some variety of presentist view is the belief that objects come into and go out of existence *simpliciter*. If this is indeed the case, then, insofar as there are true claims about an object that has ceased to be, such claims must be substantiated by something other than that object.[30] It seems that this motivating belief and the position on which it rests cannot be defended without some addition to one's ontology. Hence, the posit of the existence of simple, atemporal facts. The structure in temporal reality is explained by these facts, for true claims about moments preceding this one, now, are made true in virtue of them. These facts exist—they have the same ontological status as any other existent thing—but, insofar as they do not exist in time, they bear no temporal properties and so, in particular, are not present. Furthermore, insofar as they are simple, they do not have constituents of dubious existential status.

Whereas on primitive presentism, there is no explanation of the truth that Socrates was a philosopher—the claim is simply true—on transient presentism the truth of the claim is explained by the existence of a fact, viz., the fact that Socrates was a philosopher. There are significant differences between a view that maintains that a claim is *simply true* and one that maintains that a claim is true *simply in virtue of the existence of a fact*. Most obvious is the ontological commitment to facts made by the latter view. It is by dint of this commitment, and the accompanying account of facts as the simple, atemporal, effluvia of a world in time, that transient presentism can provide some explanation of the metaphysical basis of tensed truths.

Facts, on this account, are essentially—and nothing but—truth-makers. The fact that Socrates was a philosopher makes true the claim that Socrates was a philosopher (and not some other claim) because it is this fact that exists as a consequence of Socrates' being a philosopher when he existed. Similarly, the fact that Socrates was an Athenian makes true the claim that Socrates was an Athenian because it is this fact that exists as a consequence of Socrates' being an Athenian when he existed. (The relationship between true claim and fact is so intimate as to stultify further explanation—but not so intimate as to obliterate the distinction between the two.) These two facts are not distinguishable in terms of their structure or constituents, for, as simple, they have neither. They can, however, be distinguished by means of their properties, e.g., one fact pertains to philosophy and the other to Athens and one fact would not have existed had Socrates been a fisherman and the other would not have existed had he been from Sparta. Just as simple, unstructured entities like numbers and colors (and the properties they bear) can be the basis of the structure in mathematical reality and the color wheel, the properties of simple facts underpin the structure in temporal reality. It is they that determine that some accounts of the ways the world was are accurate, whereas others are not.

Given the account of the genesis of facts presented above (in §1.2), there are no truths about moments subsequent to this one, now. One should find this result neither implausible nor lamentable. It seems that one of the strongest intuitions regarding the nature of temporal reality is that of a significant asymmetry between the existential status of the past and future.[31] Thus, the proponent of transient presentism can resolve the problem posed by the truth-maker objection through recourse to atemporal, simple facts. It is the existence of such facts that renders the view compatible with the very plausible principle underlying the truth-maker objection, to wit, that all truth supervenes on reality.

§2.4 The Problem of Cross-Temporal Relations

Another putative problem commonly associated with presentism arises in light of claims that seem to attribute relations among entities that do not exist at the same moment(s). Claims such as: *I am of the same political party as FDR*, *Abraham Lincoln was taller than Napoleon Bonaparte*, *WWI preceded WWII*, and *I admire John Lennon* are not unfamiliar and, in some cases, obviously and literally true. The problem emerges given the assumption that *in order for a relation to hold, all relata must exist*. FDR does not exist and so cannot stand in the *same-political-party-as* relation, neither Abe Lincoln, nor Napoleon (nor either World War) exists to stand in any relation whatsoever and Lennon does not exist to be admired. This problem has been called the *problem of relations* but is more perspicuously called *the problem of cross-temporal relations*.[32]

In order to address this problem satisfactorily, one must get clear on what kind of difficulties it is supposed to pose for the presentist. These difficulties are either more or less semantical or more or less metaphysical. In the former case, the primary objection is presumably that the presentist is unable to provide an account of the *meanings* of claims of the sort presented above, on the assumption that the only people, wars, etc. that exist are those that exist at this moment, now. In the latter case, the concern seems to be that given the meaningfulness of such claims, the presentist has not the means to account for the obvious truths among them.[33]

Regarding the problem in this way, it is indeed incumbent upon the presentist to say something about the semantic subtleties of talk about entities that do not exist. This issue is considered in the section below. The more metaphysical concerns, however, seem to be little more than a special case of the problem of the basis of temporal reality considered in the prior section. Hence, what makes the claim *Abraham Lincoln was taller than Napoleon Bonaparte* true is that it is simply an atemporal fact that Abe Lincoln had a certain height, it is simply an atemporal fact that Napoleon had a certain height and the former height is greater than the latter. The claim *WWI preceded WWII* is true in virtue of the existence of the simple, atemporal facts that WWI existed in 1914 and WWII existed in 1939 and 1914 existed in the absence of any fact about 1939. Similarly, *I admire John Lennon* is made true by the simple, atemporal fact that John Lennon had certain attributes and the fact that one admires those attributes. An essentially similar account can be given, it seems, for any claim that appears to attribute a relation between an individual that exists at this moment, now and some entity that no longer exists and, more generally, for any claim that seems to attribute a relation between two entities that no longer exist.[34]

Some have focused attention on a specifically causal version of the problem of cross-temporal relations.[35] Thus, this particular version of the problem is worthy of comment. The problem is supposed to be that the presentist is unable to provide any account of how the ways the world is at one moment cause the world to be as it is at another moment; that is, the presentist is unable to provide any account of how a cause at one moment has effects at another moment. According to transient presentism, though, there is but one instantaneous moment, namely, this one, now. There are, therefore, no causal connections between moments. Nevertheless, the states of the world at this moment are not independent of the ways the world was. To the extent that those ways, which no longer exist, *cause* the world to be as it is now, causation must be construed in terms of atemporal facts about those ways. If the streets are now wet, one cause of this is the fact that rain fell at a moment that no longer exists; other causes pertain to general, atemporal features of water and asphalt and how they interact. Similarly, assuming it is an atemporal fact that the streets were wet at *m* (some moment that has ceased to be), this fact is presumably caused by the fact that it had rained at *m-1* (some moment that had existed in the absence of any fact about *m*). The notion of causation employed here is not alien, for it seems amenable to analysis via consideration of counterfactuals. More can certainly be said about these issues, but further discussion must be deferred. Given these comments, however, it does seem that causal relations are susceptible to the proposed account of how to understand what appears to be an attribution of a relation between entities (such as conditions) that do not exist at the same moment.

On this account, all relations hold between existing entities, for example, heights, dates, a person, and certain attributes. Claims that seem to attribute relations between entities that do not exist at the same moment(s) are made true in virtue of relations that hold between existent entities (some of which are plausibly regarded as not existing in time) and, furthermore, certain atemporal, simple facts. On this sort of response, though, it is clear that the *logical form* of a claim, that is, the form that best represents the ontological commitments and inferential cache of that claim, might be more complex than it appears to be. For instance, the logical form of the claim *Abraham Lincoln was taller than Napoleon Bonaparte* is not, despite syntactic appearances, one correctly regarded as merely representing two persons standing in the *was-taller-than* relation.

This sort of response to the problem of cross-temporal relations is, presumably, somewhat familiar; it is discussed in the literature, but not attributed to any particular source. Matthew Davidson classifies it as a *Hidden-logical-form* solution to the problem and criticizes such solutions on a few different rounds [chapter 11].[36] He maintains that one should be concerned that the logical forms attributed to claims of this sort quantify over "all sorts of questionable entities, and it's not clear that there is any systematic, or even piecemeal method of eliminating these entities from one's ontology."[37] This line of criticism is not compelling. For it does not seem implausible that there are such things as political parties, heights, attributes, etc. Moreover, a commitment to such abstracta is not obviously avoidable simply by rejecting transient presentism (or presentism in general).

Consider two other sorts of response to the problem of cross-temporal relations: (i) *anti-existentialist presentism*, a variety of presentist view whose distinguishing feature is that it countenances non-existent entities bearing properties and standing in relations, and (ii) a view on which temporal reality is *ontologically homogenous*, so that all entities that have existed, exist, or will exist have the same ontological status. Even assuming that John Lennon can stand in the relation of being admired *despite not existing*, it is not Lennon per se who is admired, but Lennon in some respects—it is plausible to think that these respects are certain attributes Lennon possessed. Similarly, even assuming that both Abe Lincoln and Napoleon EXIST (in a tenseless sense), it is not that the former is, in himself, taller than the latter; rather, it is in virtue of some particular feature that Abe Lincoln POSSESSES that he SURPASSES Napoleon in height; it is plausible to think that this feature is the *height* he POSSESSES. Hence, even on these alternatives, a commitment to attributes and heights and abstracta of this ilk is still plausible and not obviously (or credibly) avoided.

Davidson also criticizes hidden-logical-form solutions to the problem of cross-temporal relations on the grounds that there is no syntactic evidence for supposing a hidden logical form in the disputed relational sentences and, hence, the supposition is entirely theory driven, "motivated by an attempt to avoid a seemingly significant problem for presentism."[38] This, however, presumes that the logical form of a claim attributing a relation between existent entities in fact differs from that of one of the disputed claims. But this presumption is precisely what is challenged above: the logical form of *the bookcase is taller than the table* (assuming both pieces of furniture exist at this moment, now) is *not*, it seems, simply one that represents a relation holding between two concrete individuals. Rather, the claim is most perspicuously analyzed as attributing heights to the two pieces of furniture and stating that the height had by the bookcase is greater than that of the table. Thus, the logical form of this sentence is not so different from one seeming to attribute a relation between entities that do not exist at the same moment.[39]

Davidson has a third objection to hidden-logical-form solutions, but this objection seems misdirected, for it appears to be based merely on an infelicitous example.[40] Davidson believes that the two claims, *I am of the same political party as FDR* and *I am of the same political party as Russ Feingold*, provide an example of a pair of true propositions that certainly seem to have the same logical form, but which must be interpreted, by the proponent of a hidden-logical-form solution, as having different logical forms because of the non-existence of FDR. He then asserts: "Surely this can't be right; surely the logical form of the proposition I express doesn't change simply because one of the objects doesn't exist because it's not present. The truth of the relational statement may change, but surely the logical form of the proposition can't vary only with the existence or non-existence of the 'related' objects."[41] However, given these two claims, there is no reason for the presentist to deny that the two propositions have the same logical form. Of course, if they do have the same form, the former proposition is, pace Davidson, false. If one wants to express a truth regarding one's membership in the Democratic Party—the party of which FDR was once a member—one might utter the sentence "I am of the same political party as FDR *was*." Here, though, any difference in logical form between the proposition expressed and that of *I am of the same political party as Russ Feingold* can be attributed to the difference in tense and, so, is not simply in virtue of the non-existence of FDR.

Therefore, for these reasons, Davidson's criticisms do not seem cogent. The metaphysical difficulties presented by the problem of cross-temporal relations are simply special cases of the problem of providing an adequate basis for temporal reality. In light of this, the means of responding to the more general problem are apposite in this case. By exploiting simple, atemporal facts and by recognizing that attributions of relations—holding among both existent and nonexistent entities—have logical forms different from their syntactical forms, one can provide adequate resolution to the problem of cross-temporal relations. Hence, the problem does not undermine transient presentism.

§2.5 The Problem of Singular Propositions

The two preceding sections attempt to make the case that at least one version of presentism—transient presentism—can account for the basis of obvious truths about moments other than this one, now, and can account for those claims that appear to attribute cross-temporal relations. Some, however, object to presentism on more directly semantical grounds, maintaining that the general view is incapable of providing an adequate account of the accepted *meanings* of many mundane claims. The problem is supposed to arise over the issue of singular propositions.[42]

A singular proposition is expressed by means of a term that is *directly referential*; a directly referential term is one that refers to a particular object (or kind) and yet does so without determining that thing by way of some conceptualization or by way of the properties that thing bears. A vivid way of understanding this notion is to think of a singular proposition as one that has as a constituent the very thing that proposition is intuitively about. Rather than representing its subject via proxy—by some property or set of properties that uniquely characterize that subject—a singular proposition represents, as it were, without intermediary. Such a proposition is supposed to be the content of one's thought about an object, *o*, when one is regarding *o* not in terms of any particular feature it has, but as the very object *o* is. It has been argued that all simple singular terms, such as one-word proper names, indexicals (such as "I," "here," "now"), demonstratives (such as "this" and "that"), and pronouns, are directly referential and, thus, that the contents of sentences making use of these terms are singular propositions. These arguments are in large part based on a consideration of the inferences supported (and undermined) by use of these terms in various modal and propositional attitude contexts.

If the meaning or content of a sentence using such commonplace terms as proper names, indexicals, and the like is a singular proposition, and if, as seems plausible, a singular proposition cannot exist unless each of its constituents does, then an ostensible problem arises for presentism. Insofar as the presentist maintains that the only temporal entities that exist are those that exist at this moment, now, he must deny that there are singular propositions about past (and future) entities. Consequently, he must deny a prevalent account of the semantics of sentences seemingly about past entities, that contain simple singular terms. Yet it seems clear that one is able to speak meaningfully about entities that no longer exist.[43]

Ned Markosian claims that this issue of singular propositions creates "one of the most obvious problems facing Presentism," for the position has the "very counterintuitive consequence" that there are no singular propositions about, for instance, objects that existed prior to this moment, now.[44] But, it seems, Markosian overstates the difficulty the issue poses for this class of views. When one recognizes that by grasping a singular proposition one is put into direct cognitive contact with the subject of that proposition, it is not at all farfetched to think that there are no such propositions about entities that do not exist. One can hardly regard an object as the very thing it is, if there is nothing there to be so regarded. Similarly, one can hardly refer directly to something, that is, refer without recourse to some sort of intermediary—conceptual or otherwise—if there is simply nothing there to be referred to in this manner. This is not to say that one cannot think about or speak meaningfully about objects that do not exist and, in particular, objects that once existed but no longer do. But, insofar as one can do so, it is by some mechanism other than the one used to think or speak about directly presently existing objects.

Hence, if there is a problem here for the presentist, it really has little to do with the intuitiveness or unintuitiveness of denying that there are singular propositions about nonexistent entities. Rather any problem would have to do with whether there is an acceptable account of the semantics of sentences with non-referring singular terms, consistent with the general position of presentism. A natural assumption (and one with much historical precedent) is that a non-referring singular term is associated with some definite description whose content is the contribution made by that term to the proposition expressed by a sentence containing it. For example, it is perhaps plausible to think that the non-referring singular term "Socrates" is associated with something like the definite description "the man called 'Socrates,' the most-esteemed teacher of the author of *The Republic*, sentenced to death for corrupting the youth of Athens" and that what is conveyed when one uses the term "Socrates" is the content of this definite description.

Since the concepts employed by such a description exist and are easily grasped and since, moreover, it is obvious that one can, via conceptualization, represent something that does not exist, it might be thought that this account of the content of non-referring singular terms is both appealing and compatible with presentism. In fact, Theodore Sider has claimed that this account would seem to be the *only* one available to the presentist. However, he then points out: "[b]ut now the problem is that Saul Kripke and other antidescriptivists have argued powerfully that names do not abbreviate descriptions" and, hence, that the account should be rejected.[45]

If this descriptivist account of the semantics of non-referring singular terms were indeed the only one available, then the presentist might be in a predicament.[46] Fortunately, however, this is not so. There is at least one account of the semantics of non-referring singular terms that takes seriously the arguments of direct-reference theorists and also appears perfectly compatible with presentism. In his recent book,[47] R. M. Sainsbury propounds a theory intended to arrest the oscillations between descriptivist and direct reference theories that mark so much of contemporary discussions of the semantics of natural language. Sainsbury's theory[48] exploits the insights and benefits of both kinds of theory in providing an account of how a non-descriptivist referring expression can be intelligible despite lacking a referent. Crucial to his approach is a rejection of the notion that a semantic theory must regard meanings as entities that are "grasped" in understanding.

Sainsbury prefers a broadly truth-theoretic approach to semantics on which sentences are associated with truth-conditions, rather than entities such as propositions; referring expressions are associated with reference conditions, rather than referents. Thus, a name, such as "Socrates," which appears to refer to something which no longer exists is not meaningful by dint of reference to an individual, viz., Socrates, nor by means of some associated description. Like all names (empty or referring), "Socrates" is associated with a reference condition, in this case:

for all x ("Socrates" refers to x iff x = Socrates).

The meaningfulness of the term consists in the ability of speakers to master this condition and to employ the term in successful communication. Further constraints on success are imposed by the use of tense (in sentences such as "Socrates was a philosopher"), but, judging by the ease with which tensed language is employed, these constraints do not seem to be onerous. On this approach, singular propositions simply "have no role to play."[49]

On Sainsbury's theory, a referring expression without a referent can occur in a truth. Given that there is a metaphysical basis for truths regarding temporal reality (as argued in the two preceding sections), it seems that by combining the insights of Sainsbury's theory with a proper appreciation and assessment of the role of linguistic tense, one should be able to characterize the semantic mechanisms at work when uttering and understanding a claim about the past. Of course, doing so would undoubtedly require much work and ingenuity (certainly more than what is represented by the brief remarks here)[50] but there is no reason in principle to think it cannot be done. Hence, the problem of singular propositions does not seem to present an insuperable difficulty to presentism in general, nor to transient presentism in particular.

§2.6 The Problem of the Special Theory of Relativity

The supposed problems for presentism considered above are disparate, yet similar in that each is thought to pose some sort of *conceptual* or *philosophical* difficulty to the position. The problems are propounded by opponents of presentism with the intention of showing that the generic position is untenable on theoretical grounds, that it has some untoward or otherwise unacceptable metaphysical (or semantical) consequence. Presentism, however, has been challenged in another way. It has often been claimed that the position must be rejected on *empirical* grounds, for it is inconsistent with a highly-confirmed physical theory, the Special Theory of Relativity. Thus, even if one is entirely congenial to the responses presented in the preceding sections to the philosophical difficulties, one might have lingering concerns about the viability of any version of presentism based on some awareness of this other kind of challenge.

This challenge based on the Special Theory of Relativity (STR) is perplexing. It seems that a theory (e.g., transient presentism) could conflict with a well-confirmed account of how things behave in the physical universe (e.g., STR) only if the theory has empirical consequences inconsistent with data explained or predicted by the account. Setting aside the deep and interesting question of whether *any* philosophical theory has empirical consequences, it is clear that transient presentism, as characterized above, does not. The theory adopts some very general principles about the nature of reality—for instance, that there is but one way of being *simpliciter* and that no entity can bear any property without existing and, making use of atemporal entities, codifies a natural and intuitive view of the nature of temporal reality. But the philosophical theory makes no claim whatsoever about the observable behavior of anything in the physical universe and, thus, has no empirical consequences. Transient presentism is not distinguished from other versions of presentism in this respect.

Hence, with this challenge to presentism on empirical grounds comes a certain problem of engagement: the natures and subject matters of the theories supposed to be in conflict are so different that it is not clear that conflict is possible and, assuming it is, where exactly it arises. This point is recognized by Simon Saunders: "Of course special relativity is an empirical theory. One might take the view that presentism is concerned with a level or reality that is beyond the reach of experimental methods. But I do not grant that physics is so limited in scope, or that metaphysics can find anything deeper. But I will not argue for either of these claims here."[51] Once this is recognized, however, and the abstract—the "deeper"—nature of the variety of presentist views is articulated, the burden of proof seems to be on the one who maintains that the physical and the metaphysical can conflict.

This problem of engagement is reflected in the literature, for there is stark disagreement regarding the compatibility of STR and presentism and even regarding the relevance of the latter to the former. Thus, Theodore Sider asserts, in his extended argument against presentism, that Special Relativity provides the "final blow" against the position, for presentism "is inconsistent with special relativity."[52] Yet Lawrence Sklar asserts, when evaluating the metaphysics of time in light of relativity theory: "One thing is certain. Acceptance of relativity cannot force one into the acceptance or rejection of any traditional metaphysical views about the reality of past and future."[53] Some defenders of presentism tacitly concede the relevance of STR to the metaphysics of time by making or responding to arguments couched in the vocabulary (of inertial reference frames, Minkowskian spacetime, time-like separation, etc.) in which the physical theory is frequently presented; in this way, they admirably try to confront the empirical challenge directly.[54] However, until this more basic issue of engagement is settled, it is unclear how best to view these arguments and whether they are really necessary for a defense of presentism.

Of course, the crucial feature of any version of presentism is a commitment to the idea that there is something *metaphysically* distinctive about this moment, now. And a natural understanding of STR indicates that there is nothing *physically* distinctive about this moment, now. However, even staring at this bald feature of presentism on the one hand and STR on the other, the conflict is not obvious. What is needed is some account of the relationship between the metaphysical and the physical. One might simply read off, as it were, a metaphysics from the physical theory, a metaphysics that would render presentism and STR patently incompatible. One must recognize, though, that such a metaphysics would not be justified by the empirical evidence that confirms the physical theory; the assumption of the appropriateness of the metaphysics is of an entirely different order and, as such, requires justification of a different kind.

It seems that a metaphysical theory provides the means of interpreting empirical data, that is, of understanding those data in a more abstract, grander context. If this is so, any theory codifying empirical data—any physical theory—is consistent with any number of interpretive metaphysical theories. (What constraints, if any, the empirical data could place on an acceptable metaphysics is a profound question, but one that cannot be resolved here.) In the case of STR, there are, for instance, Minkowskian interpretations, on which matter-energy is spread out over a four-dimensional manifold, as well as "neo-Lorentzian" interpretations, on which matter-energy does not have this arrangement. Despite differences in the metaphysical presuppositions of the two sorts of interpretation, they are empirically equivalent.[55]

Insofar as the two interpretations are empirically equivalent, the only thing to guide one's choice between them is the metaphysical presuppositions and consequences each has. The behavior of things in the physical universe is not relevant at this point. One's judgment must be based on abstract considerations, such as the nature of temporal reality (e.g., whether the past and future are ontologically similar) and the nature of things per se (e.g., whether ordinary objects persist through time by enduring or perduring). Although Minkowskian interpretations are currently predominant, there are good reasons for rejecting the existence of the four-dimensional manifold this interpretation requires.[56] Regardless of one's choice of interpretation, though, it is clear that the empirical data underlying the acceptance of STR can provide no challenge to presentism, for this data, and the mathematical formalism that represents it, do not in themselves constrain one's choice of metaphysical theory. One can conclude, therefore, that the empirical challenge to presentism based merely on the truth of STR is misguided.

Nonetheless, one might hope (or expect) to find, in this defense of transient presentism, a novel interpretation of STR,[57] in order to evaluate the appeal or feasibility of an interpretation of the mathematical formalism of this physical theory in terms of this version of presentism. Such an interpretation, however, would be beyond the scope of this paper. Although an interpretation of STR in terms of the theory of transient presentism might be a valuable and satisfying heuristic, given the loose relation between the physical and metaphysical—on which the physical provides no obvious constraints on the metaphysical—such an interpretation could have no real justificatory power. Thus, it would not contribute to a defense of transient presentism.

Part III. Conclusion

Presumably, one adopts a presentist view because one thinks that there is something metaphysically distinctive about this moment and one thinks that the distinction is ontological: if anything exists in time, then it exists *now*. Accompanying and perhaps inseparable from this thought is the sense that objects come into and go out of existence *simpliciter*. That is, one has the sense that objects do not merely come into being at some specific moment, *m*, and cease to be at some other moment, *m'*. This sort of "coming and going" is compatible with the view that there are infinitely many moments, all with the same ontological status—rather, one has the sense that objects can come into being when once they did not exist at all (at *any* moment) and can entirely cease to be. Insofar as things really can come and go simpliciter, the view that there EXISTS an infinity of moments, all with the same ontological status must be rejected; once this view is eschewed, one is a de facto presentist. Transient presentism accommodates very well—and captures vividly—this motivation for presentism.

A succinct account of transient presentism, which is plausibly regarded as the optimal version of presentism, is that it is the view that all temporal entities, entities that do or must exist in time, exist at the present moment. All temporal entities now exist, but not all entities whatsoever do, for there are many sorts of things, in particular, facts, that do not exist in time. Nevertheless, these atemporal facts are the basis of the intuitive structure in temporal reality. In order to present a substantive account of the nature of temporal reality, transient presentism includes, in addition to the principle that there is a metaphysical basis for the structure in temporal reality, the theses that there exists but a single moment (where a moment is an instantaneous, irreducible entity yielded by time), that there is but one way of being *simpliciter*, and that an entity must exist in order to bear any property whatsoever.

In the preceding discussion, this version of presentism is examined in light of six problems traditionally leveled at the generic position of presentism. One can reasonably conclude that none of these problems presents a reason for rejecting transient presentism and that, in several cases, this version of presentism provides more satisfactory resolutions to the problems than other versions. Hence, insofar as one finds presentism in general appealing and finds intuitively compelling the strict metaphysical constraints motivating transient presentism, one should find this variety of presentist view an attractive account of the metaphysics of time.[58]

Notes

1. For those who gloss presentism in these terms, or very similar ones, see: Markosian 2004: p. 47; Keller 2004: p. 84 [chapter 15]; Craig 2003: p. 391; Davidson 2003: p. 77; Saunders 2002: p. 278; Keller and Nelson 2001: p. 333; Bergmann 1999: p. 125; Sider 1999: p. 325; Mellor 1998: p. 20; Tooley 1997: p. 232; Bigelow 1996: p. 35 [chapter 8]; Hinchliff 1996: p. 123; Merricks 1995: p. 523.

2. In "The Varieties of Presentism" (under review), one finds distinguished and discussed flowing, actualist, ersatzist, anti-existentialist, subject, degree, classical (or austere), tensed-facts, and primitive presentisms. Also distinguished are two specious varieties of presentism—eternal "presentism" and indexical "presentism." In that paper, as well, the version of presentism that is plausibly regarded as optimal, to wit, transient presentism is introduced and motivated.

3. Ibid.

4. These reasons are presented explicitly or can be inferred from the discussions in "The Varieties of Presentism," "Temporary Intrinsics and Relativization" (under review), "The Problem of Fixity" (under review), and "Passage, Becoming, and the Nature of Temporal Reality" (forthcoming in *Philosophia*).

5. One must distinguish between *atemporal* entities, which do not exist in time, and *eternal* entities, which exist at each moment of time. One might be inclined to classify some or all of the entities cited above as eternal rather atemporal; to do so, however, seems mistaken. As becomes clear in the next paragraph of the text, time is the entity that renders change possible. Thus the mark of a temporal entity is capacity to change. Since it does not seem possible for a number or property to undergo change per se, one should take this as evidence that these entities do not exist in time at all.

6. The exact mechanism is, it seems, certainly ineffable.

7. It is an underlying assumption of the presentist view of temporal reality propounded in this paper that there is no difference among the instantaneous moments that come into being and cease to be *with respect to their being moments*. Consequently, one should assume that the apt account of temporal reality at any moment—presentist or otherwise—is the apt account at every moment.

8. Since it is common to speak of recognizing logical (or semantical) relations between *facts*, the notion of a non-structured fact might seem a significant doctrinal departure, for no such relations could exist if facts were simple. However, to speak of logical relations between facts seems to neglect the distinction between propositions (and other intentional

entities) and facts. The former are the means by which conscious minds represent and conceive the world; the latter are what is being conceived, the basis of reality.

9. Quentin Smith argues that all entities exist in time (see Smith 1993: chap. 6; Smith 1998: §6). There is not space here to consider Smith's arguments; however, it seems they can be successfully challenged.

10. In "The Problem of Fixity."

11. Markosian 2004: p. 51.

12. See "The Varieties of Presentism": §1.2.1. For Markosian's explicit statement of the similarities between his accounts of the nature of temporal reality and the metaphysics of modality, see Markosian 2004: 76. For a classic statement of actualism in the metaphysics of modality, see Plantinga 1976.

13. It should be noted that Markosian's notion of a *fact* is very different from the one proposed and developed in the present paper.

14. For more on this see "The Varieties of Presentism": §1.2.1.

15. Oaklander 2003: p. 196.

16. The paradox was originally presented in McTaggart 1908 in an attempt to show that time is not real.

17. See, for example, Oaklander's work and Mellor 1981, 1998.

18. See, for example, Broad 1938: pp. 309–17 and Sider 2001: p. 35, n. 19.

19. For the *locus classicus* of the statement of this putative problem, see Lewis 1986: pp. 202–5. For more discussion of the relation between this problem and McTaggart's Paradox see Craig 1998, Oaklander 1999, and "Temporary Intrinsics and Relativization."

20. One can articulate this assumption with more precision. The two most natural ways of doing so are to express it either as the assumption that (i) all moments of time have the same ontological status or that (ii) the present moment and all moments prior to it have the same ontological status.

21. In "Temporary Intrinsics and Relativization," one finds the argument that the metaphysical assumptions underlying the problem of temporary intrinsics counteract any resolution that attempts to preserve the view that objects endure. Since there are reasons for thinking that objects do in fact endure, one should reject these assumptions.

22. See Keller 2004; Lewis 1999: p. 207; Bigelow 1996: p. 38; Adams 1989: pp. 32–33. The objection is discussed somewhat in Sider 1999 and is discussed explicitly in Sider 2001: p. 35ff. Surprisingly, Markosian does not address the truth-maker objection in his defense of presentism in Markosian 2004.

23. For instance, one need not accept Keller's formulation: "The problem is to say how there can be past- and future-tensed truths even though the past and future do not exist" (Keller 2004: p. 85) because the presentist need not deny that the past exists. Nor need the presentist accept a formulation according to which the problem is to account for truths about moments other than this one, now given that only present things exist—for the reason that he need not deny that only present things exist.

24. One can see examples of this view in Broad 1938: p. 316; Prior 1968: p. 11; Salmon 1987: p. 91; Cargile 1989: p. 162; and Craig 1997: p. 29.

25. This view is often formalized by means of primitive sentential operators, such as *P*, (PAST), or *WAS* for the past tense and *F*, (FUT), or *WILL* for the future tense. The semantics of these operators is simple:

WAS(*p*) is true if and only if *p* was true

WILL(*p*) is true if and only if *p* will be true,

where *p* represents any tenseless (or present tense) sentence.

The tense operators *P* and *F* are employed throughout the work of Arthur Prior; a semantics for these operators is presented at Prior 1957: p. 9. (PAST) and (FUT) are employed in Ludlow 1999 with the same semantics adopted by Prior (the operators are introduced on p. 100). *WAS* and *WILL* are presented and discussed at Sider 2001: p. 15. Whereas Sider is critical of the use of such operators insofar as one's objective is to provide insight into the metaphysics of time and the nature of temporal reality, Prior and Ludlow appear to believe that their use is instrumental in such a project.

26. The actualist presentism discussed by Matthew Davidson in Davidson 2003 is plausibly construed as positing three modes of existence *simpliciter*.

27. See Smith 2002.

28. For other, more baroque, responses to the truth-maker objection (responses which, it seems, are justifiably criticized) see Keller 2004.

29. In part III.

30. The constraints that *an entity must exist in order to bear any property whatsoever* and that *no obvious, uncontroversial truth regarding the structure in temporal reality need be denied* are operative here.

31. For the reasons discussed in "The Problem of Fixity."

32. See, for example, Davidson 2003. One version of the problem is discussed in Sider 2001: p. 25ff.

33. The two cases are not unrelated, for the features of the world necessary to account for the truth of a claim clearly depend on its meaning.

34. The claim *I am of the same political party as FDR* does not seem true; insofar as FDR does not exist, he is not a member of any political party—nor does he bear any properties or stand in any relations whatsoever. Nonetheless, the claim *I am of the same political party that FDR was* does seem to be true, made so, in part, by the atemporal existence of the simple fact that FDR was a member of the Democratic Party.

35. See, for example, Tooley 1997 and Bigelow 1996.

36. See Davidson 2003: p. 84ff.

37. Davidson 2003: p. 84.

38. Davidson 2003: p. 88.

39. Use of the present tense in stating a proposition does, it seems, license the further inference that the subject(s) of the proposition exist, whereas use of the past tense permits only the inference that certain atemporal, simple facts exist.

40. See note 34 above.

41. Davidson 2003: p. 89.

42. Singular propositions are also known as *Russellian propositions*.

43. One should note that the philosophical problems here—those associated with speaking meaningfully about what does not exist—are of a kind independent of one's account of temporal reality. Even if one rejected presentism, and believed that there is an infinite number of moments, each having the same ontological status as every other, one would be confronted with the "problem of singular propositions" with respect to fictional characters, myths, legends, and prevarication. So it is inappropriate to regard this as a problem for presentism per se.

44. Markosian 2004: p. 49. According to Markosian's solution to this problem, a sentence such as "Socrates was a philosopher," which contains a proper name of an entity which no longer exists, has "no propositional content, because it is 'trying' to express a singular proposition about the referent of 'Socrates,' and there is no such thing" (Markosian 2004: p. 67). Despite lacking propositional content, the sentence is not meaningless because it has a "linguistic meaning," associated with "the truth and falsity conditions for the

sentence." For present purposes, the details of Markosian's proposed solution are less pertinent than its upshot, to wit, that all sentences containing proper names of non-existent entities are false. Although the tenor of this putative solution is somewhat plausible, one should reject it, for the denial of the obvious truth that Socrates was a philosopher is inconsistent with the fourth constraint on the optimal version of presentism introduced above.

45. Sider 1999: p. 328. Sider cites the well-known work of Kripke, Donnellan, and Kaplan in support of his claim. Not only do these philosophers deny that proper names abbreviate descriptions, they also deny that other singular terms, such as indexicals and demonstratives, do.

46. The situation is unclear, for it is not obvious how the anti-descriptivist arguments of Kripke et al. would fare in explicitly temporal contexts and what conclusions, exactly, these arguments would support when examined in such contexts.

47. Sainsbury 2005.

48. See Sainsbury 2005: pp. 45–46 for a list of the main theses of the theory.

49. Sainsbury 2005: p. 62.

50. It is likely that the semantics of the disputed sentences considered in the preceding section—those seeming to attribute cross-temporal relations—will be particularly difficult to capture.

51. Saunders 2002: p. 279.

52. Sider 2001: p. 42.

53. Sklar 1981: p. 140.

54. For example, Hinchliff 1996, 2000; Stein 1968, 1991; Tooley 1997: chap. 11. Savitt 2000 challenges the views developed in Hinchliff 1996, 2000.

55. For an example of a neo-Lorentzian interpretation of the Special Theory of Relativity, see Craig 2000c.

56. See note 4.

57. Or an account of transient presentism in neo-Lorentzian terms.

58. I would like to thank Neil Tennant and two anonymous referees for *The American Philosophical Quarterly* for helpful criticisms and comments that improved this paper.

References

Adams, R. M. "Time and Thisness." Pp. 23-42 in *Themes from Kaplan*, edited by J. Almog, J. Perry, and H. Wettstein. Oxford: Oxford University Press, 1989.

Bergmann, M. "(Serious) Actualism and (Serious) Presentism," *Noûs* 33 (1999): 118–32.

Bigelow, John. "Presentism and Properties," *Philosophical Perspectives* 10 (1996): 35-52. Chapter 8 in this volume.

Broad, C. D. *Examination of McTaggart's Philosophy*. Cambridge: Cambridge University Press, 1938.

Callender, C., ed. *Time, Reality & Experience*, supplement to *Philosophy*, vol. 50. Cambridge: Cambridge University Press, 2002.

Cargile, J. "Tense and Existence." Pp. 193-200 in *Cause, Mind, and Reality*, edited by J. Heil. Dordrecht: Kluwer Academic Publisher, 1989.

Craig, W. L. "Is Presentness a Property?" *American Philosophical Quarterly* 34 (1997): 27–40. Chapter 4 in this volume.

———. "McTaggart's Paradox and the Problem of Temporary Intrinsics," *Analysis* 58 (1998): 122–27.

———. *The Tensed Theory of Time*. Dordrecht: Kluwer Academic Publisher, 2000a.

———. *The Tenseless Theory of Time*. Dordrecht: Kluwer Academic Publisher, 2000b.

———. *Time and the Metaphysics of Relativity*. Dordrecht: Kluwer Academic Publisher, 2000c.

———. "In Defense of Presentism." Pp. 391-408 in *Time, Tense, and Reference*, edited by A. Jokić and Q. Smith. Cambridge, Mass.: MIT Press, 2003.

Davidson, M. "Presentism and the Non-Present," *Philosophical Studies* 113 (2003): 77–92. Chapter 11 in this volume.

Healey, R., ed. *Reduction, Time and Reality*. Cambridge: Cambridge University Press, 1981.

Hinchliff, M. "The Puzzle of Change," *Philosophical Perspectives* 10 (1996) ("Metaphysics"): 119–36.

———. "A Defense of Presentism in a Relativistic Setting," *Philosophy of Science* 67 (2000) (Supplement): S563–S574. Chapter 14 in this volume.

Keller, S. "Presentism and Truthmaking." Pp. 83-104 in *Oxford Studies in Metaphysics*, edited by D. Zimmerman. Oxford: Oxford University Press, 2004. Chapter 15 in this volume.

Keller, S., and M. Nelson. "Presentists Should Believe in Time-Travel," *Australasian Journal of Philosophy* 79 (2001): 333–45.

Le Poidevin, R., ed. *Questions of Time and Tense*. Oxford: Clarendon Press, 1998.

Lewis, D. *On the Plurality of Worlds*. Oxford: Blackwell, 1986.

———. "Armstrong on Combinatorial Possibility." Pp. 196-214 in *Papers in Metaphysics and Epistemology*, by David Lewis. Cambridge: Cambridge University Press, 1999.

Lillehammer, H., and G. Rodriguez-Pereyra, eds. *Real Metaphysics: Essays in Honour of D. H. Mellor*. London and New York: Routledge, 2003.

Ludlow, P. *Semantics, Tense, and Time*. Cambridge, Mass.: MIT Press, 1999.

Markosian, N. "A Defense of Presentism." Pp. 47-82 in *Oxford Studies in Metaphysics*, edited by D. Zimmerman. Oxford: Oxford University Press, 2004.

McTaggart, J. M. E. "The Unreality of Time," *Mind* 18 (1908): 457–84.

Mellor, D. H. *Real Time*. Cambridge: Cambridge University Press, 1981.

———. *Real Time II*. London and New York: Routledge, 1998.

Merricks, T. "On the Incompatibility of Enduring and Perduring Entities," *Mind* 104 (1995): 523–31. Chapter 18 in this volume.

Oaklander, L. N. "Craig on McTaggart's Paradox and the Problem of Temporary Intrinsics," *Analysis* 59 (1999): 314–18.

———. "Presentism, Ontology and Temporal Experience." Pp. 73-90 in *Time, Reality & Experience*, edited by C. Callender. Cambridge: Cambridge University Press, 2002.

———. "Presentism: a Critique." Pp. 196-211 in *Real Metaphysics: Essays in Honour of D. H. Mellor*, edited by H. Lillehammer and G. Rodriguez-Pereyra. London and New York: Routledge, 2003. Chapter 5 in this volume.

Plantinga, A. "Actualism and Possible Worlds," *Theoria* 42 (1976): 139–60.

Prior, A. N. *Time and Modality*. Oxford: Oxford University Press, 1957.

———. "Changes in Events and Changes in Things." Pp. 1-14 in *Papers on Time and Tense*, by Arthur Prior. Oxford: Clarendon Press, 1968. Chapter 3 in this volume.

Sainsbury, R. M. *Reference without Referents*. Oxford: Oxford University Press, 2005.

Salmon, N. "Existence," *Philosophical Perspectives* 1 (1987): 49–108.

Saunders, S. "How Relativity Contradicts Presentism." Pp. 277-92 in *Time, Reality & Experience*, edited by C. Callender. Cambridge: Cambridge University Press, 2002.

Savitt, S. "There's No Time Like the Present (In Minkowski Spacetime)," *Philosophy of Science* 67 (supplement) (2000): S563–S574.

Sider, T. "Presentism and Ontological Commitment," *Journal of Philosophy* 96 (1999): 325–47.

———. *Four-Dimensionalism: An Ontology of Persistence and Time*. Oxford: Oxford University Press, 2001.

Sklar, L. "Time, Reality and Relativity." Pp. 129-42 in *Reduction, Time and Reality*, edited by R. Healey. Cambridge: Cambridge University Press, 1981.

Smith, Q. *Language and Time*. Oxford: Oxford University Press, 1993.

———. "Absolute Simultaneity and the Infinity of Time." Pp. 135-83 in *Questions of Time and Tense*, edited by R. Le Poidevin. Oxford: Clarendon Press, 1998.

———. "Time and Degrees of Existence: A Theory of 'Degree Presentism.'" Pp. 119-36 in *Time, Reality & Experience*, edited by C. Callender. Cambridge: Cambridge University Press, 2002.

Stein, H. "On Einstein-Minkowski Space-Time," *Journal of Philosophy* 65 (1968): 5-23.

———. "On Relativity Theory and Openness of the Future," *Philosophy of Science* 58 (1991): 147–67.

Tooley, M. *Time, Tense, and Causation*. Oxford: Oxford University Press, 1997.

Zimmerman, D., ed. *Oxford Studies in Metaphysics*, vol. 1. Oxford: Oxford University Press, 2004.

Part IV

The Case against Presentism

Section 1

Presentism and Inter-Temporal Relations

Introduction: Presentism and Inter-Temporal Relations

Ernâni Magalhães

Since presentism claims that everything is present, a leading problem for the view is to account for those phenomena that seem to imply the existence of non-present things. The truthmaking objection (part IV) is an example. True statements seem to require something in reality to make them true. The truth of some statements—"Socrates was a philosopher," for example—seems not to be captured by present reality. Thus the truth of those statements must make reference to non-present things. The argument from relations is another example. Present things seem to stand in relations to non-present things. Present things are after non-present (i.e., past) things; they are caused by non-present things; they resemble non-present things; and so on.[1] But one thing cannot be related to another unless both things exist. Then these non-present things that precede, cause, and resemble present things must exist. So, contrary to presentism, there are some non-present things.

The argument has two premises: one a principle about relations; the other, a claim about relations that actually obtain between present and non-present things. Presentist responses have largely conceded the principle of relations—that two things cannot be related if one does not exist.[2] Their focus has been on the factual claim that there are relations between present and non-present things. Matthew Davidson (chapter 11) examines and rejects what he takes to be the most promising strategies for denying or mitigating the factual claim. One desperate strategy is to deny that present things are caused by, resemble, and succeed non-present ones (see Sider 1999). Even if nearby "paraphrases" can be found for strictly untrue inter-temporal claims, as Davidson points out, it seems too high a price to pay for a view to deny such obvious truths as that my current writing was caused by some previous intentions (p. 200).

Most presentist responses concede—or rather emphasize—that present things are connected to non-present ones. In contrast to the paraphrase approach, they claim that the truth of assertions about such relations does not involve or require the obtaining of relations between existing and non-existing things. Two of Davidson's criticisms about this approach are worth emphasizing. First, Davidson worries that any attempt to ground such relations without conceding there are non-present things will require commitment to an array of "questionable entities" like political parties or other non-concrete objects (p. 200). In this, Davidson confirms John Bigelow's contention that presentism can have its spare concrete world without past or future objects, only at the cost of postulating an extravagant realm of abstract entities (see chapter 8). And, as we will see, Rafael De Clercq's (chapter 12) riposte offers further confirmation.

Another of Davidson's concerns is this. Many of the apparent relations that cause trouble for presentism can connect both present to non-present things and present to other present things. *Belonging to the same party* and resemblance are examples of the latter. Davidson rightly points to a tension for presentist accounts of the latter relations. According to Davidson even if in some cases sentences about belonging to the same party can be understood in terms of something other than an inter-temporal relation, at least in some cases such sentences must affirm the existence of genuine relations. Thus the presentist is forced to say that the proposition which asserts that I belong to the same party as FDR implies one sort of fact, while one that describes my belonging to the same party as Barack Obama involves a dramatically different one. But "surely the logical form of the proposition I express doesn't change simply because one of the objects doesn't exist because it's not present" (p. 204).

Rafael De Clercq takes up this very challenge. Although he speaks of "paraphrases" (p. 209), it seems clear that De Clercq's response on behalf of presentism aims to identify the content expressed by the problematic apparently inter-temporal sentences. De Clercq's strategy involves "a leap to the a-temporal ... realm of abstract entities (i.e., of types, properties, and propositions) where the original cross-time relation can be recast in terms that are unobjectionable by presentist standards" (p. 208). The strategy involves unapologetic commitment of a sort Davidson claims is problematic: my being taller than FDR is understood in terms of a relation between abstract items, my height and FDR's height. In general, present items are related to non-present items in virtue of being related to abstract entities which are, or were, in turn related to the non-present things.

De Clercq expends most of his energy trying to account for semantic relations to non-present things. How can "Napoleon" refer to Napoleon in light of the fact that Napoleon is not present? The word token "Napoleon" succeeds in referring to Napoleon because it expresses the type name **Napoleon**, which (timelessly) names the man (pp. 210-11). This descriptivist strategy[3] claims that the name "N" is equivalent to the description "the thing called 'N.'" Since type names always exist, **Napoleon** names Napoleon just as much today as it did in the eighteenth century.

But what about Davidson's objection: shouldn't the logical form of "Napoleon was defeated" be the same as that of "Petreaus was defeated"? The person uttering the latter may seemingly intend to express the same belief on two occasions which differ in that in one Petreaus is alive and in the other he is dead. De Clercq believes sentences involving names express propositions that include the named item as a constituent when the item named exists (p. 213). When the item named does not exist, the sentence expresses a general proposition involving the type name.

"Napoleon was defeated" means that "It was the case that something is defeated which is named **Napoleon**." In sum, De Clercq believes a presentist should accept that "our language changes when objects go out of existence" (p. 212). As the propositional form of a sentence can differ from its surface form, so two apparently similar non-simultaneous sentences can differ in their form.

Abstract names are also used to accommodate intentional relations to past entities. De Clercq claims the sentence, "There was an x who is (now) believed to have been a recursion theorist," can be interpreted as:

> For some name **n** there was an object named by **n** and there is a subject s such that **n** is believed by s to have named a recursion theorist.

What if x is (literally) anonymous? Or if s does not associate x with any name? De Clercq concedes that for the account to work it is necessary that the "object is given a name at some point" (p. 215). However, it is not necessary that the object have what is ordinarily understood to be a name: "One's intentional relation to x could be mediated by" any "currently existing object standing in a unique, non-intentional relation to x" (p. 220, n. 12). S's belief may be linked to x in virtue of x's toothbrush or dog.

But why should it be supposed that beliefs about past entities are invariably beliefs about names? De Clercq believes there is independent support for this in the fact that, typically, a subject thinks there was an x who did such-and-such when, and only when, she believes there is a name that picks out the x who did such-and-such. But in order for this account to "amount to the same thing" (p. 216) as the original, the two must not only be typically associated, it must be impossible to have beliefs about past objects that are not beliefs about names.

Accounting for the full array of apparently inter-temporal relations in a presentist-friendly way is a tall order, in part because these relations come with significant philosophical baggage. It seems fair to say presentists have only scratched the surface. A promising strategy for presentists may well be to save some effort and try to show that their view can piggy-back on one or more of the available accounts without violating the tenets of either presentism or the received analysis.

Notes

1. In these chapters, Rafael De Clercq and Matthew Davidson, following the pattern in the literature, focus on relations to past, rather than future, things. The focus on the past is usually chalked up to the sense that, intuitively, relations to the past have more basis in reality than purported relations to the future.

2. The one exception is Berit Brogaard (2006), who claims that the principle applies only to a proper subset of all relations. "Tenseless relations," in which one thing is

tenselessly R to another, require the existence of both relata. "Tensed relations," in which one thing is tensedly R to another, do not require that both relata exist.

3. Descriptivism is the view that a name is synonymous with a certain description. "George Washington" means, roughly, "America's foremost general and first president."

References and Further Readings

Brogaard, B. "Tensed Relations," *Analysis* 66 (2006): 194-202.

Chisholm, R. M. "Referring to Things That No Longer Exist," *Philosophical Perspectives* 4 (1990): 546–56.

Crisp, T. "Presentism and 'Cross-Time' Relations," *American Philosophical Quarterly* 42 (2005): 5-17.

Davidson, M. "Presentism and the Non-Present," *Philosophical Studies* 113 (2002): 77-92. Chapter 11 in this volume.

De Clercq, R. "Presentism and the Problem of Cross-Time Relations," *Philosophy and Phenomenological Research* 72 (2008): 386-402. Chapter 12 in this volume.

Markosian, N. "A Defense of Presentism." Pp. 47-82 in *Oxford Studies in Metaphysics*, volume 1, edited by D. Zimmerman. Oxford: Oxford University Press, 2004.

Oaklander, L. N. "Presentism, Ontology, and Temporal Experience." Pp. 83-99 in *The Ontology of Time*, by L. N. Oaklander. Amherst, N. Y.: Prometheus Books, 2004.

Sider, T. "Presentism and Ontological Commitment," *The Journal of Philosophy* 96 (1999): 325-47.

11
Presentism and the Non-Present

Matthew Davidson

Abstract

In this paper I argue that presentism has a problem accounting for the truth of statements whose truth conditions seem to require there be relations that hold between present and non-present objects. I motivate the problem and then examine several strategies for dealing with the problem. I argue that no solution is forthcoming, and this presents a *prima facie* problem for presentism.

Introduction

Presentism is the view that only the present is real.[1] Contrast this doctrine with *eternalism*, the view that past and future times are just as real as the present time. Or, past or future individuals are just as real as present individuals. They just happen to exist prior to the present, or after the present. One might compare presentism with actualism in the metaphysics of modality. Actualism is the view that the actual world enjoys a special ontological status over other worlds, if other worlds there be. Only the actual world is instantiated, or in some sense real. The presentist wants to say the same thing about times: Only the present time is real, and the present time enjoys a special ontological status over other times, if other times there be. On the other hand, one might compare eternalism with a Lewisian concretist understanding of possible worlds. On Lewis' view, each world is ontologically on a par. The actual world is no more real than any other possible world. Similarly, the eternalist wants to claim that no time enjoys any special ontological status with respect to other times. The present time is no more or less real than any other time.

It is important to note what the presentist isn't (or isn't necessarily) saying. First, the presentist need not say that there are no past or future times, if she takes times to be abstract objects.[2] But if she does say that there are abstract past and future times, these times will have a different status than does the present time.

Present, past, and future times all would *exist* on this view, yet only the present time would be *exemplified* or *instantiated*. Second, she can in some sense accept that there are past or future entities in that there *now* are individuals who did or will exist. What she cannot accept is that these entities have any reality beyond the reality they have now.

There appears to be a *prima facie* problem with presentism. If only the present is real, how can present entities stand in relations to past and future things which don't exist? It certainly seems to be the case that present objects do stand in relations to past and future things. Suppose that I'm a Democrat. I stand in the *being of the same political party* relation to FDR. The event consisting of my drinking coffee 20 minutes ago stands in a causal relation with my currently being alert. Suppose I have a crystal ball and I am able to view the life of an individual who is alive 100 years in the future. He and I might stand in the *being taller than* relation, or we might stand in the *being of the same ethnic background* relation. But how can any of this be if at least one relatum of the relation in question doesn't exist?

One consequence of these relations not holding is that sentences that appear to be true would seem to come out false. Consider "I am of the same political party as FDR." It appears to have roughly the logical form aRb, and its truth conditions appear to be that it is true if a stands in R to b; that is, if I stand in the *being of the same political party* relation to FDR. But if FDR doesn't exist, and this entails that I don't stand in this relation to FDR, then "I am of the same political party as FDR" would seem to be false. Note that this problem isn't solved by claiming that what is true here is that I am of the same political party FDR *was*. For the truth of this sentence too appears to depend on the *being of the same political party* relation holding between FDR and me. So we have many instances of propositions that certainly *prima facie* seem to be true, like *I am of the same political party as FDR*. Yet the truth of these propositions appears to require relations holding between presently existing and non-present entities, and therefore these propositions would seem to have to be false. Call this *the problem of relations*.

This problem rests on an assumption about relations. Call this proposition (P). We may state it thus:

> (P) Necessarily, if an individual a stands in a relation R to an individual b, then a and b exist.

Prima facie (P) appears to be true. If a relation holds between two individuals, then surely both individuals must exist.

There are several solutions one might try to give to the problem of relations. In this paper I wish to examine in detail what I take to be the four most plausible solutions. I conclude that none of these strategies work, and given this fact we have a *prima facie* case against the truth of presentism.

I. The Overlap Strategy

This solution to the problem of relations is suggested by Chisholm (1990). To solve the problem, Chisholm will attempt to show how present individuals can stand in relations to non-present individuals even though the latter don't exist. In his 1990 paper he is concerned about present individuals standing in relations to past individuals on a presentist picture, so I will present this solution to the problem of relations in these terms. Chisholm claims (I rephrase his proposal, but the general idea still is the same) that a present entity E may stand in a relation R to a past entity E' if there is something O which bore some relation R' to E, and E stands in some relation R'' to O. There always will be such an O if one believes in necessarily existing abstracta (or even abstracta that always exist). For instance, O might be the property *being blue*. So I am able to stand in a relation R to FDR in virtue of the fact that I stand in a relation R' to an object (say *being blue*) now that, was such that it stood in some relation R'' to FDR. There is an overlap in relations on this picture; two present objects overlap relationally, and one of these present objects is such that it did stand in a relation with the past entity in question.

I can't see how this works as a direct answer to the problem of relations. First of all, if this picture is to work there must be some specificity of the relations involved. Surely it's not sufficient for me to stand in the *being of the same political party* relation to FDR that I stand in any relation or other with an object that did stand in any relation or other to FDR. Can it really be that the fact that some relation or other holds between me and an object and some relation or other held between that object and another suffices for the fact that a *particular* relation—*being of the same political party*—holds between me and the other object?

Perhaps the account can be modified to avoid this objection. What if we required that the relation that is supposed to hold between the present entity and the past entity be such that it also is the relation that holds between the two present entities? For example, suppose that Frank is a Democrat, and he was alive when FDR was alive. So I now stand in the *being of the same political party* relation to Frank, and he did stand in this relation to FDR.

This makes the proposal seem more plausible, but it won't always work because it won't always be the case that the same relation holds between the two present objects and did hold between a present object and the past object. Suppose there was some obscure figure F in the fifteenth century in Rome who invented his own religion. He had a few followers, but the religion died out a long time ago. Suppose I discover this person's writings and convert to his religion, and I am its only convert. Then intuitively we shall want to say that I stand in the *being of the same religion* relation to F. But there is no present individual with whom I stand in the *being of the same religion* relation who did stand in the *being of the same religion* relation to F.

There is a more fundamental problem with this solution to the problem of relations. How does it make any difference as to whether I *now* stand in a relation to a past individual S that I now stand in some relation to some other individual S' who *did* stand in a relation to S? The existence of this other set of relations has nothing to do with whether or not I now stand in a relation to some past individual; Chisholm's truth conditions appear to be a complete *non sequitur* with respect to the truth of the proposition consisting of my now standing in a relation to a past individual. Hence, I conclude Chisholm's solution fails.

II. The Negative Paraphrase Strategy

The second strategy we will examine is what I shall term the "negative paraphrase strategy." On this approach it is conceded that (many, at least) propositions that appear to depend on relations holding between present and non-present entities are false.[3] So the proposition *I am of the same political party as FDR* would be false, since it seems to be true if I stand in the *being of the same political party* relation to FDR; again the sentence that expresses it certainly appears to have the logical form aRb. That these propositions are false clearly is a problem for the presentist, for they certainly seem to be true. But the presentist need not simply admit defeat and go home. She is free to present true propositions "in the neighborhood" of the false relational proposition whose truth will help to make the denial of the relational proposition more plausible. It should be noted that the negative paraphrase theorist doesn't claim to be providing a proposition or propositions that preserve the meaning of the relational sentence in question. The only work the "paraphrasing" is doing is to ease the sting of the claim that apparently true sentences are false.

The strategy will be to attempt to distinguish between two sorts of false relational claims: those that are "clearly" false and those that are false, but "closer to being true." To illustrate, consider again the sentence "I am of the same political party as FDR." Though on a standard reading of the logical form of this sentence it is false, because its truth depends on a relation between FDR and me holding, there is a proposition that is true that is very similar to the false proposition this sentence expresses. *There is a political party such that I am a member of it and FDR was a member of it*, (or in a form which is more presentist-friendly, *there is a political party such that I am a member of it and it was the case that FDR is a member of it*). Consider on the other hand the sentence "I am of a different political party than FDR." The first relational sentence is, loosely speaking, "nearly true," whereas, the second is "clearly false."

Above, the relation *being of the same political party* holds between a and b if it holds between b and a. What about non-symmetrical relations? Suppose I utter, "I am dumber than Einstein." Anyone who knows the work of Einstein and knows me will claim readily that this sentence is true. Yet on the negative paraphrase strategy it is false. To what proposition or propositions can we point to make our denial of this sentence more palatable? I think that we can take some guidance from our FDR example. There we found something, a political party, such that, FDR did bear a relation to it and I do bear a relation to it. Perhaps we can do similarly in this case. We might try "There is a level of intelligence such that my level of intelligence fails to meet that level of intelligence and Einstein's level of intelligence met or exceeded that level of intelligence."[4] The same sort of paraphrase could be used in the case of the sentence "I am taller than Goliath." This relational claim we might paraphrase as "There is a height such that my height is greater than or equal to that height and Goliath's height is smaller than that height."

So there emerges a general strategy for paraphrasing problematic relational claims, in our case assertions of relations that hold between present and non-present individuals. We quantify over an entity (e.g., a political party or a height), and we assert that a particular relation did hold between an entity (e.g., FDR or Goliath's height) and that entity, and a particular relation does hold between an entity (e.g., me, or my height) and that entity. If it turns out that the entity that is first quantified over in the paraphrase (e.g., a political party or a height) is objectionable (perhaps one winds up quantifying over a club, and one doesn't think that there are such things), further paraphrase to eliminate reference to the unpalatable entity may be undertaken. So suppose I think that there aren't any clubs, but there are sets of individuals whose members have certain intentions and goals and engage in social relations with each other in order to achieve these goals. So whereas I might appear to quantify over clubs and reify their existence, actually all I am quantifying over are the entities in the above paraphrase. (This may fail as a paraphrase of what a club is, but what is important for present purposes isn't correctly paraphrasing away objectionable entities here, but seeing how such paraphrases would fit into the general paraphrasing scheme of the negative paraphrase theorist.) So, for the negative paraphrase theorist, the paraphrase of "I belong to the same club as FDR" might begin as "There is a set of individuals whose members have certain intentions and goals and engage in social activities to achieve these goals . . ." The rest of the paraphrase will be tricky, though, because it won't be the case that FDR and I are members of the same set of individuals; rather we are members of distinct sets whose members overlap, or which are such that there is some sort of ancestral of an overlap relation. Anyway, again the point here isn't to get the paraphrase exactly correct, but simply to point out that sometimes the entity that is first quantified over in an attempted paraphrase of a relational statement may be objectionable to some, and paraphrase of the existence of that entity may itself be needed.

So, will this general strategy work for all instances of sentences that assert that relations hold between a present entity and a non-present entity? I think that for intentional relational statements, a similar yet different strategy will need to be adopted. Consider "I adore Eleanor Roosevelt." This seems to resist paraphrase in the above fashion. However, there is a similar sort of paraphrase available to the negative paraphrase theorist: "There is a set of properties P such that Eleanor Roosevelt exemplified the members of P and anyone who exemplifies the members of P I adore." No paraphrase of this form will be an entirely happy one for the presentist, though, for the universal quantifier in the second part of the paraphrase ranges only over presently existing objects, and hence not over ER.[5] But it still goes some way in accounting for why the past/present relational sentence appears to be true.

No doubt the negative paraphrase theorist needs to say more about how she will provide paraphrases that take the sting out the denial of apparently true statements that assert that relations hold between present and non-present objects. But we've seen that she can go some way toward providing these sorts of paraphrases. There remain problems, however. First, the negative paraphrase theorist, in at least many of her paraphrases, winds up quantifying over what might appear to be strange sorts of entities (e.g., hairstyles, levels of strength, senses of smell). If many of these entities require further paraphrase, and there is no unified account of how one might give paraphrases for all of these entities (as seems to be the case) the negative paraphrase account becomes piecemeal and inelegant. Second, despite the fact that she may be able to ease our intuitions somewhat by providing paraphrases, the negative paraphrase theorist still is committed to the fact that apparently true relational statements are false. This counts strongly against the negative paraphrase view. Just how strongly will depend on how well the paraphrases the negative paraphrase theorist produces ease our minds regarding the denial of what appear to be true relational propositions concerning present and non-present objects. But it certainly would seem preferable to have a solution to the problem of relations that didn't commit us to the falsity of a myriad of propositions which certainly seem to be true.[6]

III. Hidden Logical Form Solutions

We will consider two solutions to the problem of relations that center on the fact that "The proponent of a hidden logical form solution is able to deny that relations hold between present and non-present individuals, and yet claim that the propositions expressed by relational sentences such as "I am of the same political party as FDR" are true. We will look at two variations of the hidden logical form solution: what I shall call the "existential quantification solution" and the "relational properties solution."

a) The Existential Quantification Solution

On this solution to the problem of relations, it is claimed that the problematic relational propositions that appear to have the logical form aRb actually are existential quantifications of the sort that the negative paraphrase theorist employed in her paraphrases. In particular, the logical form of a sentence that appears to have the logical form aRb, where either a or b is non-present, typically will be $\exists x(aRx \;\&\; W(bRx))$.[7] So, as we saw before, we might claim that "I am of the same political party as FDR" expresses the proposition *There is a political party such that I am a member of it and FDR was a member of it.*

There are several problems with this solution. One of them already has been noted with the negative paraphrase theory: This sort of paraphrase appears to be committed to quantifying over all sorts of questionable entities, and it's not clear that there is any systematic, or even piecemeal method of eliminating these entities from one's ontology. Second, it's not clear that one always can come up with an appropriate x to fill in the schema $\exists x(aRx \ \& \ W(bRx))$. Take one of the most important relations, that of causation. How is one going to give an existentially quantified paraphrase of the claim *a caused b*, where a is a past event? I don't see any way of doing this. One might take events to be abstract entities or repeatable entities—though there still will be many instances where mere repeatability won't suffice for the truth of aRb where a and b are events and R is a causal relation. It won't always be the case that a repeats such that it occurs at the same time as b. I will raise more objections to the hidden logical form solution and hence to the existential quantification solution below.

b) The Relational Properties Solution

On the relational properties solution, the logical form of a sentence like "I am of the same political party as FDR" turns out to be Fa, where F is a *relational property*. A relational property is a property like *being taller than Tom* or *being meaner than Leroy*, a property that "makes reference" to (a) particular individual or individuals. Relational properties aren't relations. They're possessed by a single individual in the same sort of way as non-relational properties like *being blue* are possessed. They don't "hold between" individuals in the way that relations do; rather, they're exemplified by a single individual—they're monadic. Normally one might think that for each relational property that is exemplified by an individual S, there is a corresponding relation that S stands in to some other entity (or entities) S. For instance, if I have the property *being taller than Tom*, then a relation holds between myself and Tom, namely the *being taller than* relation. However, on the relational properties solution this correspondence won't hold. It can be the case that an individual S have a relational property that involves another individual S' in the same sort of way that *being taller than Tom* involves Tom, and it not be the case that there be the corresponding relation (analogous to *being taller than*) that holds between S and S'.

It should be noted that relational properties are structured sorts of entities, in the sort of way that many think that propositions are structured. They are complex entities, built up of more primitive entities that stand in certain relations to one another. In particular, a relational property will be composed of the relation that is "mentioned" in the expression of the property, and an individual essence of the individual that the relational property is "about." An *individual essence* is a property p such that if it is possibly exemplified by some entity o, then necessarily, if o exists, o has p, and necessarily, for any o', if o' exemplifies p, then o=o'. So, the relational property *being taller than Frank* is a complex entity, composed of the relation *being taller than* and an individual essence of Frank. That the relational property is composed of these two elements is of great importance. The fact that the relevant relation is included in the structure of the property gives us an explanation of why certain logical inferences will hold in virtue of individuals having certain relational properties. For instance, if Joe has the relational property, *being taller than Frank* and Frank has the relational property *being taller than Fred*, then Joe should have the relational property *being taller than Fred*. How can we account for this? It is because the relation *being taller than* that is a constituent of the relational property in question is a transitive relation. So we have some sort of metaphysical explanation of why certain inferences that ought to hold among propositions that predicate relational properties do hold.

The fact that these properties have as a constituent element an individual essence is crucial as well. Insofar as certain terms, such as names and indexicals are rigid, we need some sort of way of picking out a particular individual across worlds when we have a relational property that is expressed using a name or indexical. Were we the analogue of direct reference theorists about properties, we would think that the way to do this is to have as a constituent of the property the individual who is involved in the relational property. So the relational property *being taller than Frank* would contain Frank as a constituent. But this obviously won't work for the presentist in addressing the problem of relations; non-present individuals don't exist to be parts of relational properties. However, their individual essences, which exist necessarily, are available to go "proxy" for the individual in the structure of the property. So this will allow the relational property that is expressed by a gerundial phrase that uses a proper name or indexical to be about the same individual from world to world.

On the relational properties solution as we first will consider it, non-present and present individuals will have relational properties that undergird and will provide for the truth of relational propositions like *I am of the same political party as FDR*. In this case, FDR would have the property (now) of *being of the same political party as me* and I would have the property *being of the same political party as FDR*. Or if I say "He is more ruthless than Stalin," this sentence may express a true proposition not in virtue of a relation holding between a present individual and Stalin, but in virtue of Stalin's having a relational property, *being less ruthless than him*, and his having a property *being more ruthless than Stalin*.

However, I don't think that this first-blush strategy will work. Consider the view *serious presentism*—objects can exemplify properties or stand in relations only at times when they exist.[8] I think that there is a strong argument that presentism entails serious presentism. Consider the following argument.

(1) Necessarily, for any time t, whatever there is (in as temporally-neutral sense as one likes) exists at t. (presentism)
(2) Necessarily, for any property F or relation R and time t, if F or R is exemplified at t, there is/are (in as temporally-neutral a sense as one likes) something or things that exemplifies or exemplify F or R at t.
(3) Therefore, necessarily for any property F or relation R and time t, if F or R is exemplified at t, then F or R is exemplified by something or some things that exists or exist at t.
(4) Therefore, necessarily, objects exemplify properties or stand in relations at a time only if they exist at that time.

This argument looks to me to be sound. (4), serious presentism, entails that objects that don't exist now can't exemplify any properties, and hence can't exemplify relational properties. So neither Stalin nor FDR can have the relational properties that we ascribed to them. Note also one other consequence of serious presentism—it entails (P). So any attempt to solve the problem of relations by denying (P) is undercut by the truth of serious presentism.

So if we're going to use relational properties to deal with the problem of relations we will have to use those that are instantiated presently, such as the one I have, *being of the same political party as FDR*. Can we use presently instantiated relational properties to respond to the problem of relations? There are problems with this strategy. First, as noted it cuts the connection between a relational property and its corresponding relation. Intuitively, I exemplify the property *being of the same political party as FDR* if I bear the *being of the same political party* relation to FDR. But this isn't so on the relational properties solution.

Furthermore, there are problems that plague both versions of the hidden logical form solution. First, we have no independent syntactic evidence for thinking that the logical forms of sentences like "I am of the same political party as FDR" are anything but aRb. The posit of the modification of logical form is theory-driven; it is motivated by an attempt to avoid a seemingly significant problem for presentism. Since our other attempts (the overlap strategy and the negative paraphrase strategy) failed to solve the problem of relations, assuming that I have covered the options for solutions to the problem of relations (and I think I have), this posit of a hidden logical form is only as plausible as presentism is. That presentism is taken to commit us to the claim that there are significant divergences from the apparent logical forms of relational propositions expressed by relational sentences can be taken as a strike against presentism. It should be noted, though, that it seems that the relational properties solution fares better than does the existential quantification solution in this regard. *I am of the same political party as FDR* certainly seems to have the logical form aRb, and it doesn't seem to be an existential quantification, or a proposition of the form Fa. But at least on the relational properties strategy something *relational* is left in the paraphrase. As noted before, the relational property will contain as a constituent the relation which *prima facie* seems to hold between a and b. The existential quantification solution eliminates the relational element entirely. Thus, though purely syntactically both strategies posit significantly different logical forms for propositions which seem to have the form aRb, the structure in the "F" in the relational properties' rendering of the relational sentence allows it to mimic the relational element in the relational sentence were it taken straightforwardly to be a sentence of the form aRb. For this reason and reasons mentioned in the discussion of the existential quantification solution, if one is going to adopt either of the hidden logical form strategies, the relational properties solution seems preferable.

However, there is a significant difficulty with any hidden logical form solution. This sort of solution states that a proposition a sentence expresses will vary in logical form simply in virtue of the fact that one of the apparent relata doesn't exist. It can't be that all relational sentences express propositions of the form Fa or ∃

x(aRx &W(bRx)). Both of these modifications of logical form themselves involve relations. It is clear how the existential quantification solution makes use of relations. The relational properties solution uses some sort of relation to serve as the "glue" for the constituents of the complex relational property. On pain of regress, in neither of these cases can the relational claims involved in both of these strategies be taken to be either existential quantifications or instances of relational properties. So not all relational sentences will have a hidden logical form; some will be of the form aRb. Furthermore, the only reasonable bifurcation as to whether a relational statement has a hidden logical form or has its apparent logical form of aRb would be when we have a relational statement involving a present and a non-present object. Note that both of the uses of relations in the two hidden logical form solutions make use of relations that hold between presently existing objects. So once we allow that we have some relational statements that have the form aRb that relate present objects, it seems arbitrary not to say that all of them do. But this means that I'm saying something of a very different form when I say "I am of the same political party as FDR" and "I am of the same political party as Russ Feingold." The former is either of the form $\exists x(aRx \,\&\, W(bRx))$ or Fa, and the latter has the logical form it appears to have, aRb. Surely this can't be right; surely the logical form of the proposition I express doesn't change simply because one of the objects doesn't exist because it's not present. The truth of the relational statement may change, but surely the logical form of the proposition can't vary only with the existence or non-existence of the "related" objects.

I conclude that the relational properties solution is preferable to the existential quantification solution. But both of them run into serious difficulty with a seemingly arbitrary bifurcation of the sorts of propositions expressed by relational statements.

In this paper I have investigated what I take to be the contenders for a solution to the problem of relations. All contenders have been found wanting. Insofar as we think it's true that, e.g., I am of the same political party as FDR (and it certainly seems to be true), I conclude that this is a *prima facie* consideration against presentism. Certainly my attempt to show problems with all the possible solutions to the problem of relations doesn't amount to a conclusive argument that presentism is false. There are many other arguments for and against eternalism and presentism that must be considered. But I contend that the inability for the presentist to solve the problem of relations gives us at least some reason to think that presentism is false.[9]

Notes

1. There are some who think that presentism is either trivially true or obviously false. So, if we take presentism to be the view that necessarily, whatever there is, is present, one might ask what the sense of "there is" is in the universal quantification in the definition. Does "there is" range over presently existing objects only? Then presentism is trivially true. Or, is the "there is" "tenseless"; does it mean something like "there was, is, or will be?" If so,

then presentism is obviously false. It is contended that these are the only two ways of understanding the presentist thesis, so there is no real metaphysical dispute between the presentist and eternalist. I think that this is misguided. Presentism is to be understood in a manner analogous to the manner in which actualism is understood, where actualism is the view that necessarily, whatever there is exists actually. The universal quantifier in the statement of actualism is "loosed" so that it may range over possibilia. Similarly, with presentism, the universal quantifier in the statement of the view is "loosed" so that it may range over past and future objects. Both presentism and actualism employ unrestricted quantification in their definitions to avoid the trivially true/obviously false objection. Unfortunately, when this is pointed out to those who think that presentism is either trivially true or obviously false, they tend not to understand the notion of unrestricted quantification. Thus, many such people also don't think that actualism is a substantive philosophical thesis. It is difficult to know what to say to such people. I think I understand the sort of unrestricted quantification employed by the presentist, and thus I can make sense of the presentist's thesis. But I have no way of convincing those who are skeptical of the coherence of an unrestricted quantifier that presentism is a robust philosophical claim. See Plantinga (1985, pp. 314–16) for discussion of similar concerns with respect to actualism. I am indebted to Tom Crisp for discussion of this issue.

2. For instance, she might take times to be propositions or states of affairs that for some time t, entail only every true proposition or state of affairs that is true or obtains at t.

3. Or, if they're not false, they're at least not true. See Sider (1999). Sider adopts the notion of "quasi-truth" where a proposition is "quasi-true" informally if it is true "philosophical niceties aside" (p. 332). More formally, in the case of relations to the non-present, a proposition is quasi-true iff if eternalism were true there would be a true proposition that would entail it (p. 339). It's not clear to me exactly what quasi-truth amounts to; so far as I can tell he's not proposing it as a third truth value. I think that the best way to view his strategy is as a sort of negative paraphrase strategy. He is trying to provide true propositions that ease the sting of the denial of the relational statement, which is what the negative paraphrase strategist is doing. Sider has just given a specific way of generating the propositions which the negative paraphrase theorist will use to make more plausible the denial of the truth of the relational statements.

4. Again, in all of these examples we could turn the past tense existential quantifier into a sentence operator with present tense quantification ("There was an x" becomes "It was the case that there is an x").

5. An anonymous referee suggested that I use individual essences so that I may avoid the problem of the "anyone" only ranging over present individuals, and hence not over ER. So, rather than saying "anyone who exemplifies P, I adore," I should say "if there were an instantiation of an individual essence E such that that instantiation exemplified P, then I would adore the instantiation of E." The presentist, like the actualist, will want to make use of individual essences to "go proxy" for nonexistent individuals. This much is clear. But I can't see how the above paraphrase is any better than the one I give with respect to quantifying over only present objects, and hence not over ER. Any instance of E that exemplifies P will be a present object, and hence E will not be any individual essence of ER. Thus, the presentist still is left with the fact that the entity quantified over in the second part of the paraphrase isn't going to be ER, whether or not one employs individual essences.

6. Again, on Sider's view, these propositions won't be false; they'll be "quasitrue." But then the objection can be recast: There are serious problems with a view that doesn't allow to be true a myriad of propositions which certainly seem to be true. Also, a presentist may well say, "You're right, there are some costs to this solution. But the costs outweigh the costs of adopting any other view in logical space. So, I'm justified in holding to my presentism." I

leave it up to the presentist to judge for herself whether these costs outweigh the benefits of being a presentist. I present in this paper simply a *prima facie* case against presentism.

7. I leave aside the paraphrase of intentional claims (such as "I adore Eleanor Roosevelt"). The analysis I provide here suffices to capture at least an essential part of the existential quantification solution's picture. If I can show that there are problems paraphrasing sentences into this form, which must be done for many relational sentences, I will have shown that there are problems with the existential quantification solution. The "W" in the existentially quantified sentence is a past-tense temporal operator. It should be read as "It was the case that."

8. The name is an analogue of Plantinga's "serious actualism," see Plantinga (1985). See Salmon (1998), and Bergmann (1999).

9. I am indebted to Gordon Barnes, Tom Crisp, Dave VanderLaan, Mike Byrd, and Alan Sidelle for helpful discussion of these issues. Thanks to an anonymous referee from *Philosophical Studies* for helpful comments on the paper.

References

Bergmann, M. "(Serious) Actualism and (Serious) Presentism," *Noüs* 33, no. 1 (1999): 118–32.

Chisholm, R. M. "Referring to Things That No Longer Exist," *Philosophical Perspectives* 4 (1990): 546–56.

Plantinga, A. "Replies." Pp. 313-98 in *Alvin Plantinga*, edited by J. Tomberlin & P. van Inwagen. Dordrecht: D. Reidel, 1985.

Salmon, N. "Nonexistence," *Noûs* 32 (1998): 277–319.

Sider, T. "Presentism and Ontological Commitment," *Journal of Philosophy* 96 (1999): 325–47.

12
Presentism and the Problem of Cross-Time Relations

Rafael de Clercq

Presentism is the view that only present entities exist. Recently, several authors have asked the question whether presentism is able to account for cross-time relations, i.e., roughly, relations between entities existing at different times. In this paper I claim that this question is to be answered in the affirmative. To make this claim plausible, I consider four types of cross-time relation and show how each can be accommodated without difficulty within the metaphysical framework of presentism.

1. Introduction

According to presentists, only what is present exists: there are no past or future entities such as Napoleon and the clouds forming in tomorrow's sky. This view can be called common sense with some justification. After all, past entities seem to be entities that have ceased to exist, and future entities seem to be entities that have yet to come into existence. Nonetheless, the view is far from universally accepted, and few renowned twentieth-century philosophers have held it. Arthur Prior was probably its most famous advocate. (See, for example, Prior 1968.)

Presentism has been attacked from different angles. Historically, its apparent conflict with relativity theory was probably the most important problem.[1] However, philosophers have also discerned logical, semantic, and metaphysical difficulties.[2] Some of these are unique to presentism but others have been held to undermine any so-called "tensed" view of time, i.e., roughly, any view that wishes to regard as objective the distinction between what is present and what is non-present

In this paper, my focus will be on a metaphysical problem that is unique to presentism. In the recent literature, it is discussed by Adams 1986, Bigelow 1996 [chapter 8 in this volume—*Editors*], Zimmerman 1998 [chapter 17], Sider 2001, Crisp 2003, and Markosian 2004. The problem, roughly, is how presentism can account for relations holding between things existing at different times.[3] Here is how the problem arises. A relation can only obtain between existing entities. However, if presentists are right, the only entities that exist are presently existing entities. So there can only be relations between entities existing at the present time. As a result, there cannot be any cross-time relations, i.e., relations between entities existing at different times. Nonetheless, both ordinary and scientific talk is ridden with reference to cross-time relations. For instance, we often say that events in the past are responsible for current events *(causal relations)*, that past entities are currently being thought of in a certain way *(intentional relations)*, and that past and present entities are alike or different in certain respects *(comparative relations)*. Moreover, cross-time relations are also involved in our talk in less explicit ways. For instance, in calling a past entity by its name ("Julius Caesar," "The French Revolution") we assume that a *semantic relation* can link a present linguistic token with some object or event in the past.

My aim in this paper is to set the stage for a general solution to the problem of cross-time relations (as this problem arises in the context of presentism). By a "general" solution I mean a solution that works for cross-time relations of all kinds: semantic, causal, intentional, comparative, and what have you. Very roughly, the solution is expected to consist in a leap to the a-temporal (or perhaps just "sempiternal") realm of abstract entities (i.e., of types, properties, and propositions) where the original cross-time relation can be recast in terms that are unobjectionable by presentist standards.[4] A. N. Prior (1967, pp. 170-71) had already hinted at such a solution but he was in doubt about whether it could work for all cases.

One last preliminary remark. In what follows, logical notation has been kept to an absolute minimum. This was done in the interest of readability, and not to hide the fact that implicit use has been made of a relatively rich logical language.

2. Some Assumptions

In what follows, my concern will be with cross-time relations between past and present entities. Moreover, the entities in question will be assumed to be different. In principle, they could be identical (as Prior noted in 1967, p. 170), but no special difficulties arise in such a case and so it will be left out of consideration.

Perhaps cross-time relations can obtain equally well between present and future entities. However, it is controversial whether objects in the future *can* be named and whether all statements about the future have determinate truth-values.

Therefore, it seems more interesting to start with relations connecting past and present entities and to work out a solution for them. If the aforementioned controversies get resolved, then one might have to generalize the solution.

Finally, it will be assumed that presentism can allow at least some truths about the past, for example, that somebody's army was defeated in Waterloo. If this were not assumed, it would be impossible to bring out the problem of cross-time relations as a *special* problem. After all, if there are no truths about the past, then there are no truths about entities in the past, and if there are no truths about entities in the past, then *a fortiori* there are no truths about how past entities relate to present entities.

3. Comparative and Causal Relations

My main claim is (i) that cross-time relations "hold" in virtue of relations that are not cross-time and (ii) that relations of the latter sort can be invoked to paraphrase everything we may wish to say about (or in terms of) relations of the former sort. In what follows, no explicit attempt will be made to provide a principled argument for this claim. By and large it will have to be made plausible by considering the different types of cross-time relation that were distinguished in the beginning of this paper and by checking whether the proposed solution works in each case.

Consider, first, comparative relations. Clearly, when entities are compared, they are always compared in a given respect, e.g., in respect of size, weight, color or personality. Such a respect can be regarded as a determinable of which the entities in question exemplify a determinate, e.g., a particular size, weight, or color. Moreover, it is precisely because of the relation between their respective determinates that entities can be said to stand in a given comparative relation. For instance, Leon is taller than Napoleon (was) because Leon's *length* is greater than Napoleon's. (Problems arising from the use of "Napoleon" will be discussed in due course.) More precisely, for some lengths L and L', with L greater than L', it is the case that Leon has length L and it was the case that Napoleon had length L' (see Prior 1967, p. 170). In this way, the original cross-time relation between Leon and Napoleon has been recast as an a-temporal relation between Leon's length and Napoleon's length. (See figure 1.)[5] Such a relation does not present any problems for presentism.

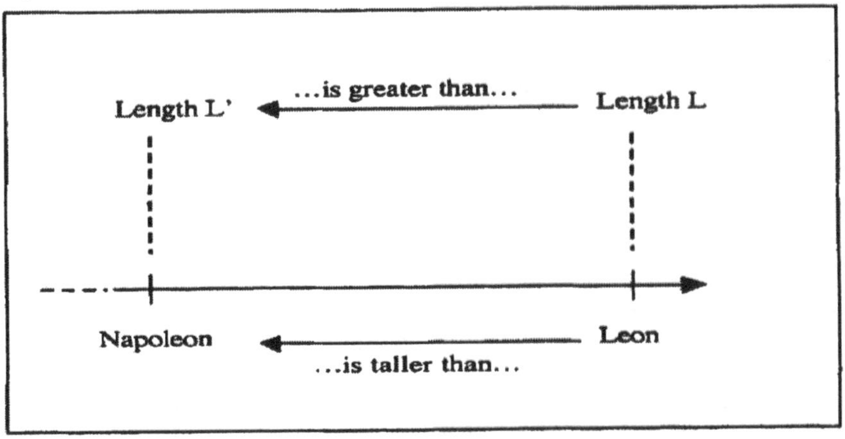

Figure 1

Next, consider causal (or explanatory) relations. If presentism is correct, then events cannot be the primary relata of the causal relation, since the events in question would have to exist at different times. However, it is also common to consider facts as the primary relata of the causal relation (among these facts there may of course be facts about the occurrence or non-occurrence of certain events). If this view is correct, then there is then no temporal distance to be bridged by means of a cross-time relation: the fact that the world *was* a certain way is a current fact about the world, and one which may be invoked to explain why the world now *is* a certain way (cf. Bigelow 1996). In other words, facts about the past are not entities in the past, and so causal relations between such facts and facts about the present do not involve bridging a temporal distance. As a result, it is possible to conceive of the causal relation as a relation that is not cross-time in the relevant sense. Moreover, this possibility presents itself as soon as one allows facts to play the role of causal relata. Apart from that, one does not have to assume anything specific concerning the nature of causation. (Thus, presentism is neutral in the realism-reductionism debate.)

4. Semantic Relations

Cross-time semantic relations can be accommodated in the manner of comparative relations, except that in their case the a-temporal relation does not obtain between two abstract entities but between an abstract entity and two concrete entities.

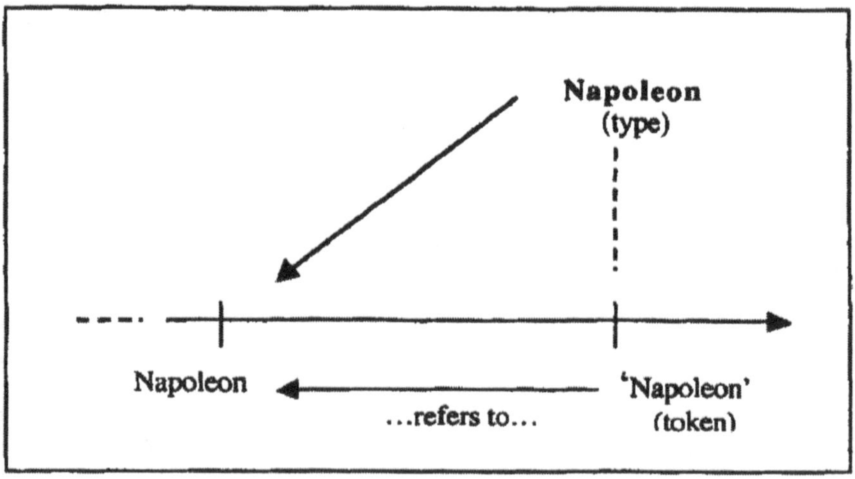

Figure 2

As Figure 2 shows, the abstract entity in question is a type-name which is supposed to have a present token and to have referred to some entity in the past. In other words, a statement such as "Napoleon was defeated" is to be read as saying:

PAST (There is an x [at most one] such that x is named by **Napoleon** and x is defeated).[6]

(Where "PAST" could be an undefined sentential operator predicating that "It has been the case that. . . ." or that "It was the case that. . . ."[7])

Of course, "Napoleon" may still be a referring term when someone is identical to Napoleon. Moreover, the above reconstruction of a cross-time naming relation is neutral on the question of whether token-names that are spelled the same are always to be regarded as tokens of the same type-name. In other words, it does not matter here whether we regard David Wiggins' first name and David Kaplan's first name as two homonyms or as one and the same, ambiguous name. In any case, i.e., whether or not we prefer homonyms over ambiguous names, contextual elements will have to intervene at some point—either, in the case of homonyms, to link a token to its corresponding type, or, in the case of ambiguous names, to link a type to the concrete entity named by it.[8] (Of course, if the concrete entity no longer exists, then the type cannot be linked to it. However, the type may still be linked to a context of use in which the type's reference is unambiguous.)[9]

Note also that this way of analyzing singular statements about the past is not original. In "The Philosophy of Logical Atomism," Bertrand Russell had already suggested that the name "Romulus" can be regarded as "a sort of truncated description" standing for "the person who was called 'Romulus.'"[10] In a similar vein, Arthur Prior (1968) paraphrased "Queen Ann has died" as "Once there was a person named 'Anne,' who reigned over England, etc., but there is not now any such person" (pp. 17-19). More recently, the idea has been considered in connection with presentism by Theodore Sider (1999) and Ned Markosian (2004).

Sider's objection against the proposal is that it is a version of the descriptivist theory of proper names, which we all know to be incorrect since Kripke's *Naming and Necessity* appeared (Sider 1999, pp. 327-28). However, this objection seems to overlook the possibility of regarding the naming function appealed to as *constant,* i.e., as assigning to all possible worlds the same value (for a given name). Thus, the proposal seems to be immune to Kripke's modal objection against descriptivist theories of names. Kripke's other objections seem even less threatening. Consider, for example, the objection that speakers may associate descriptions with a name that are not applicable to the bearer of the name. This problem cannot occur if, as the above proposal suggests, the description associated with a name **n** is simply "the bearer of **n**." Every competent speaker knows that when she is using the name **n**, she is referring to *someone by the name of* ***n***.

According to Markosian, the problem with the paraphrase ("PAST. . . .") is that its meaning differs from the meaning of the original sentence. Here is what he offers in support of this claim (p. 10):

- [The original sentence] seems to be about a man, while [the paraphrase] seems to be about a name.

- [The original sentence] has the form of a sentence that expresses a singular proposition, while [the paraphrase] has the form of a sentence that expresses a general proposition.

- [I]t seems pretty clear that [the original sentence] did not have the same meaning as [the paraphrase] back when Socrates [Napoleon in my example] was still present, and it would be strange to say that the two sentences differed in meaning at one time and then had exactly the same meaning at a later time, even though (we can assume) there were no changes in the interpretation of the relevant language between those two times.

None of these reasons is really convincing, I think. The problem with the first reason is that it relies on an unarticulated view of aboutness. Moreover, it is not intuitively clear why the paraphrase is to be read as being "about" a name instead of a man. After all, the paraphrase seems to say, *concerning* some (person) *x* that existed in the past, that it was defeated and named a certain way.

The second argument is also problematic, since it assumes that the underlying logical form of a sentence should be transparent. However, many philosophers—in particular, those sympathetic to Russell's theory of descriptions—will not accept that the logical form of a sentence can always be read off from its surface grammatical form. Moreover, even if sentences wear their logical form on their sleeves, it may still be true that the proposition which a speaker asserts, or intends to assert, by means of a particular sentence is more transparently revealed by another sentence with a different logical form.

Finally, a presentist is likely to accept that our language changes when objects go out of existence. For instance, a presentist is likely to accept that names lose their referents as time goes by. Maybe there are problems with this view, but the third argument does not show this. In assuming that all semantic change comes about through "changes in the interpretation of the relevant language," it pictures meaning as dependent only on use.

Now consider the following objection. Suppose that someone unaware of Napoleon's death utters "Napoleon was defeated." Surely (the objection goes) such a person must take herself to be referring to Napoleon at the time of her utterance. However, on my account of past-tensed statements, she would be mistaken about the meaning of her utterance or even about its meaningfulness.

One problem with this objection is that it is not clear why someone unaware of Napoleon's death *must* take herself to be referring to Napoleon when she utters "Napoleon was defeated." Why couldn't such a person think of herself as combining the following two beliefs:

> In the past, someone named "Napoleon" was defeated.
>
> There is someone named "Napoleon."

(where "Napoleon" is supposed to be an unambiguous or disambiguated name). What would be lacking if we assumed that she is ascribing this pair of beliefs to herself rather than the belief that "Napoleon" is a referring term in "Napoleon was defeated"? I would say "nothing" unless the person in question is an analytic philosopher possessing the conceptual resources to reason from

There is someone named "Napoleon"

to

"Napoleon" is a referring term

to

"Napoleon was defeated" expresses a singular proposition.

Admittedly, such a person could be mistaken about the meaning of her past-tensed utterance on my account. However, no ordinary person has semantic beliefs of such specificity, making such subtle distinctions of logical form.

This is not to say that ascribing a precise semantic belief to an ordinary person would have to count as a mistake. Rather, it means that a principle of benevolence may come into play here, favoring a true precisification over a false one. Such a principle governs of course also the attribution of non-semantic beliefs. This becomes clear when one considers the following version of "the objection from unnoticed deaths."

Suppose Mary believes that John was born in 1950, and that John is still alive. As a result, Mary may believe a singular proposition with John as a constituent (expressed by "John was born in 1950"). However, as soon as John dies, it seems that Mary's belief will have to change into a general (non-singular) one, if it is to remain true. This seems to be strange. It does not seem that Mary's belief about John's year of *birth* has to change when John dies. Moreover, Mary may be unaware of John's death, and it would be counterintuitive to suppose that her belief becomes empty unless she actively revises it.

Two things can be said in reply. First, when John dies, the content of Mary's belief does not have to change, only its logical form. This is, as it were, a purely structural change. Secondly, this change of logical form can go unnoticed and does not require any conscious effort on Mary's part. The new logical form is simply the one that a benevolent and well-informed interpreter would attribute to her belief after John's death.

Finally, consider the following objection. Suppose that Napoleon was not called "Napoleon" in his time and that we have forgotten how he was then called. It seems that we may still use the name "Napoleon" to refer to him. Yet in such a situation, it would not be possible to paraphrase "Napoleon was defeated" as "PAST (There is an x, at most one, such that x is named by **Napoleon** and x is defeated)." After all, the x in question was not named "Napoleon" in the past. Moreover, it is not possible to paraphrase the statement by using Napoleon's real name, since, by assumption, no one remembers it. Lacking an appropriate paraphrase, it seems that "Napoleon was defeated" is to be rejected in the circumstances envisaged, contrary to what common sense seems to suggest.

The objection assumes that someone is named by _ at *t* only if she is called _ at *t*. But that assumption is questionable. It is just as questionable as the assumption (concerning ordinary predicates) that something belongs to the extension of _ at *t* only if it is something to which people apply _ at *t*. And this latter assumption is plainly false. For although we no longer apply certain predicates (e.g., predicates belonging to forgotten languages), it is still true that currently existing objects belong to their extension. Similarly, although the word "carbon" may not be used by extraterrestrial beings, it is possible that, where the extraterrestrial beings live, there are things belonging to the extension of the English word "carbon." In other words, linguistic items may apply even when, and where, they are not *being* applied. So, returning to the example of Napoleon, it is possible that, in the past, the type-name **Napoleon** named Napoleon even though he was not (yet) called that way.[11] This would be an example of a case where a term applies before it is being applied.

Note that all this is evident if truth is defined somehow in terms of satisfaction. For regardless of what one's view is regarding the existence of non-present entities, a satisfaction-account of truth implies that formulas get satisfied at times when they are not in use. For example, the sentence "Diogenes was schizophrenic" is true on a very simple satisfaction-account if and only if, at some time in the past, Diogenes satisfies "is schizophrenic"—a predicate not yet in use in Diogenes' time, I presume.

In a recent book, Scott Soames has proposed a different way of dealing with singular statements about past entities (2002, pp. 90-95). On his approach, a singular statement about, say, Napoleon is to be interpreted as expressing a proposition that no longer exists (because one of its constituents, Napoleon, has gone out of existence). Clearly, this will not solve our problem but merely replace it. For instead of a semantic "naming" relation to some no longer existing object (e.g., Napoleon) we now get a semantic "expressing" relation to some no longer existing proposition (e.g., the proposition that Napoleon was defeated in Waterloo).

One extra advantage of the proposal under consideration is that it is readily applicable to sentences containing names of fictional characters. For example, the proposition expressed by "Holmes is a detective" could be represented as

> FICTIONAL (There is an x, at most one, such that x is named by **Holmes** and x is a detective.)

(Where "FICTIONAL" could be an undefined sentential operator comparable to "PAST.") Such an approach to fiction allows one to regard fictional statements as expressing propositions. At the same time it does not commit one to the existence of fictional entities such as Holmes. Moreover, this commitment can be avoided without having to conceive of fictional entities as abstract entities (as in Soames 2002, pp. 93-95).

5. Intentional Relations

Intentional relations appear to be a mixed bag. The distinction of central importance here is, presumably, the distinction between intentional relations to propositions (so-called "propositional attitudes") and intentional relations to objects.

First, consider intentional relations to *propositions* about past entities. Such relations do not pose any problems as long as the intentional relation—believing, desiring, hoping, fearing—is exclusively one between a subject and a proposition, e.g., a proposit ion believed to be true by that subject. (This is the situation we get if propositional attitudes are interpreted *de dicto*.) After all, in such a case the intentional relation is not itself cross-time. However, some intentional relations to propositions are more complex: rather than being dyadic relations between subjects and propositions, they seem to be triadic relations between a subject, a proposition, and an object. (This is the situation we get if propositional attitudes are interpreted *de re.)* Consider, for example, the sentence "There is an x whom Leon believes to be a recursion theorist." In this example, believing could plausibly be construed as a triadic relation between the believing subject (Leon), the proposition believed ("There is an x. . . .") and the object believed to be a certain way. Evidently, if the latter is an object of the past, then the relation of believing becomes cross-time and so problematic if presentism is correct.

Again, the solution is to be found at the level of abstract entities. More precisely, a sentence such as "There was an x who is (now) believed to have been a recursion theorist" could be read as saying:

> For some name **n** there *was* an object named by **n** and there *is* a subject s such that **n** is believed by s to have named a recursion theorist.

Two observations are in order here. First, **n** is a variable ranging over name-types, i.e., abstract entities. Second, for reasons mentioned earlier (in section 4), it is unimportant whether the name-types were in use when the object in question existed. Nonetheless, it is important that the object is given a name at some point, and here, admittedly, a piece of artifice—theoretical idealization—slips into the discussion. Now it becomes evident why, in the beginning, it was said that the paper would "set the stage" for a general solution.[12]

It might also be objected that, contrary to what the reformulation suggests, someone (allegedly) believing something of an *object* need not believe anything about the object's *name.* After all, the subject may not know what the name of the object is. However, if that is what is supposed to make the objection plausible, then it is not a very plausible objection. For it often happens that we are not able to recognize or identify the object of our belief. (Consider, for example, cases where the object of one's belief is a burglar or a germ.) Sometimes we are even ignorant about the *kind* of thing our belief is about. For instance, I may believe of the red spot in my face that it is a spider bite while it is just ordinary acne.

Nonetheless, the objection is on to something. For even if there is the possibility of being mistaken about the object of one's belief, it should not be accepted without reason that some beliefs about past objects are really beliefs about names. It is legitimate to ask what can be said in support of this bold claim and of the associated interpretation of *de re* beliefs.

What can be said in support of the claim is, of course, that it allows us, presentists, to make sense of *de re* beliefs without committing ourselves to the existence of past entities. Moreover, the associated interpretation of *de re* beliefs carries some independent plausibility because (it explains why) a subject is likely to assent to

For some name **n** there was a recursion theorist named by **n**

whenever she declares (or is correctly declared) to believe of someone in the past that he or she was a recursion theorist. Likewise, a subject is likely to dissent from that statement whenever she declares (or is correctly declared) *not* to believe of someone in the past that he or she was a recursion theorist.[13] Therefore, the most natural way of rendering *de re* beliefs ("There is an object x. . . .") and my own, somewhat contrived way ("There is a name **n**. . . .") may be said to amount to the same thing.

It may be objected that this treatment of *de re* beliefs severs the characteristic connection between the subject and the object of belief. For if the above account is correct, we need not be "acquainted" with the object of a belief. There is not even the need for a causal relation to it, since names, conceived as abstract types, do not stand in causal relations. Thus, anything whatsoever seems available as an object of belief.

In response to this objection, it may be pointed out that common sense does, as a matter of fact, ascribe *de re* beliefs in a rather libertarian way. Acquaintance, for instance, does not seem to be a precondition: biologists are said to believe things of species that have disappeared long ago, and astronomers are said to believe things of parts of the universe that are too remote to be observed. Causal interaction does not seem to be a precondition either: mathematicians are said to believe things of numbers, and linguists are said to believe things of syntactic structures. (Of course, the fact that practically anything is *available* as an object of belief does not mean that everything is *in fact* an object of belief.)

In addition to intentional relations to propositions, there seem to be (dyadic) intentional relations to *objects*. At least, this is suggested by such examples as "a admires b," "b sees c," "c thinks of d," and "d is intrigued by e." On closer inspection, a substantial number of these relations may turn out to be disguised (dyadic or triadic) intentional relations to propositions. For instance, it is arguable that "d is intrigued by e" just means "d thinks e is intriguing."[14] However, as a general strategy this will not work. For instance, *perceptual* relations to objects cannot be treated in a similar manner: seeing an object does not amount to taking an object to be visible.[15] However, on the face of it, perceptual relations to objects cannot be of objects in the past: we do not normally accept that people see, feel, hear, or smell things in the past. (Of course, one can smell a burnt toast even when there is nothing left of the toast, but arguably, what one smells in such a case is the smoke resulting from the burning process, or that a toast has been burnt, and not the toast itself.) So here the problem of cross-time relations does not seem to arise. Admittedly, there is a complication arising from the fact that there is an inevitable time-lapse between the event causing the perception of an object and the actual perception. This time-lapse is unproblematic as long as the object in question remains available as an object of perception. However, in cases where the object disappears before it is actually being perceived (think of distant, dying stars), the appropriate thing to say would be that the object is not being perceived. Of course, one may still see how the object looked, what its appearance was like, where it was located, etc. The object itself, however, is no longer there to be perceived. (Of course, one need not be aware of this fact.)

Intentional relations based on affective states such as liking and hating likewise resist the reductive strategy (i.e., the strategy that tries to reduce intentional relations to objects to propositional attitudes). However, here too it is hard to find any genuinely cross-time instances. For example, we do not normally say that we like or hate a deceased person; we say that we liked or hated them. (Of course, people do say that they love Mozart but then "Mozart" stands metonymically for his oeuvre. Moreover, it does probably happen that one still hates a person who is now dead, but then one's state of mind may be rendered as a stable, affect-laden belief that the person in question deserved one's hatred.)

Finally, it may be noteworthy that the foregoing does not presuppose that there exist *de re* relations or even intentional relations. The thesis defended was that if such relations exist, then they do not pose any serious problems for presentism.

6. A Different Kind of Presentism

The view put forward by Timothy Williamson in "Necessary Existents" (2002, pp. 244ff.) resembles presentism and at the same time avoids the need for paraphrases. On this view, past entities such as Julius Caesar have not ceased to exist: they have become "ex-concrete" (p. 246). Thus, instead of accepting as existent only what is present, this view accepts as concrete only what is present. Perhaps this is close enough to presentism to merit attention as a different type of solution.

However, Williamson's proposal faces several difficulties, even if one disregards the extravagant assumption that all currently existing entities (you, me, . . .) exist at all future times. First of all, the ghostly existence accorded to entities that were once present is characterized in a purely negative manner—such entities are said to be neither abstract nor concrete—which makes the proposed solution look very much ad hoc (in this context) and, also, difficult to evaluate. Secondly, the original problem seems to be replaced: instead of the problem of how we can refer to non-existents, we now get the problem of how we can refer to entities that are neither concrete nor abstract. Moreover, we get an additional problem: how is the referential connection preserved in the transformation from concrete to non-concrete? (For instance, how can the causal-informational links that underlie reference to concrete entities be maintained when an entity becomes non-concrete?) Or is the original connection replaced in the process? If so, how? Thirdly, questions arise concerning the further applicability of Williamson's idea. For instance, can it also be applied to processes and events? After all, these are also often given names (e.g., "The French Revolution") and we keep referring to them. Does that mean that these events and processes are now going on outside space and time? Or do they exist without actually going on? But then are we still referring to the same entity? Isn't the French Revolution necessarily an event, and aren't events necessarily things that happen?

It would be consistent to say in reply that, although the French Revolution is not necessarily an event, it is an event whenever it is concrete (i.e., in space and time). However, this reply leaves us with a puzzle: if the French Revolution can literally exist as a non-event—as something that does not happen and does not involve any action—as long as it is outside space and time, then why can't it exist as a non-event when it is within space and time—especially, given that some things in space in time are not events? Some explanation seems to be needed here.

7. Concluding Remarks

To recapitulate: the type of solution proposed (or adumbrated) in this paper consists in re-describing the original cross-time relation at an abstract level, with at least one abstract entity to serve as an argument of the substitute, a- temporal relation(s). No doubt more relations admit of cross-time instances than the ones that have been considered in this paper (causal, semantic, intentional, and so-called "comparative" relations). However, what is crucial is not that the list of considered relations is exhaustive but that it exhibits sufficient variety. For to the extent that it does, it is more likely that the type of solution applied to it will work also in the other, unconsidered cases.

One might wonder why it is so important to solve this problem: why bother about the problems and prospects of presentism? There are two reasons. First, presentism is taken seriously by philosophers not because recent scientific discoveries point in its direction—rather to the contrary—but because presentism is believed to be in harmony with (considerable parts of) common sense. For instance, most people will agree that the Berlin Wall no longer exists and that my children are still to come into existence. Moreover, the anti-presentist's notion of *existing at a time* (which decomposes into existing *simpliciter* and being located at a time) will not mean much to ordinary people. Finally, there is a striking asymmetry between the way we think of "now" and the way we think of "here." As Jeremy Butterfield notes: "We more readily take as real the presently existing objects, wherever they are, than the objects that are at some time located here. We are more apt to give sentences time-variable truth-values than space-variable ones. And we think of ourselves as sharing a common, albeit ever-changing, now, while we each have a different here" (1998, p. 61).

However, presentism is not just attractive because it is supported by common sense. It is also a theoretically appealing position because it offers an elegant solution to the problem of "temporary intrinsics," i.e., the problem of how a persisting object can have different intrinsic properties at different times.[16] More specifically, within a presentist framework the problem of intrinsic change simply does not arise because objects do not exist at more than one time.

Finally, one might wonder how a presentist can claim to defend common-sense ideas and at the same time attempt to paraphrase away all kinds of attributions (e.g., of cross-time relations) which look perfectly natural. In other words, if presentism is a common-sense view, then why does it require us to impose a ban on cross-time relations? After all, common sense does not seem to have any problems with such relations.

In response to this objection, I would say that the ontological commitments of common-sense attributions do not always lie at the (grammatical) surface. For instance, it is perfectly ordinary to say that Pegasus is a horse and that Holmes is a detective and yet it is also perfectly ordinary to say that neither Pegasus nor Holmes exists. There is no inconsistency here, as long as it is clear that some propositions about Pegasus and Holmes are meant to be fictionally true or true-in-a-fiction.[17]

Notes

1. Recent work on the (in)compatibility of relativity and presentism includes Crisp 2003, Saunders 2002, and Sider 2001.

2. Among these other problems the following are noteworthy: the existence of singular propositions about the past (Crisp 2003), the existence of cross-time relations (references in main text), the definition of temporal priority (cf. Tooley 1997), McTaggart's paradox (cf. Mellor 1998), the apparent commitment to instants in ordinary language (Ludlow 1999, Kuhn & Portner 2002), and the problem of nestling temporal operators (Ibid.).

3. In principle, it does not make sense within a presentist framework to speak in terms of "existing at ***a*** time," since things can only exist at the present time. However, for certain purposes presentism may well avail itself of representations of time picturing the present as just one instant among others. It is perfectly possible to interpret such representations in metaphysically less objectionable ways, e.g., by interpreting instants as (sets of) propositions (as in Prior 1968, pp. 122-23).

4. There is a worry about the status of abstract objects in presentist frameworks (cf. Meyer 2002). It seems that a presentist has to say either that abstract entities exist at all times (which would render them everlasting or "sempiternal") or that presentism is really a thesis about concrete entities. Fortunately, for the purpose of this paper I do not have to choose between these two possible views.

5. If one prefers to think of exemplification as a relation, then there are three relations involved: one between Leon and his length, one between Napoleon and his length, and one between Leon's length and Napoleon's length. By the way, by an "a-temporal" relation I mean a relation that does not hold between entities existing at different times.

6. Henceforth, bold text refers to types.

7. For objections to the view that "PAST" is primitive, see Tooley 1997, pp. 175-85. Tooley's objections do not strike me as very strong, however. For example, he argues that the property of being past is never the object of direct (non-causal or causal) awareness. In particular, he argues that direct (causal or non-causal) awareness of something's being past is not part of memory knowledge (pp. 182-83). As a result, he claims that "PAST" cannot be primitive, since "a descriptive concept cannot be analytically basic for a given person unless that concept picks out a property M a relation that, for the individual in question, is, or has been, either an object of direct, non-causal awareness, or an object of immediate perception" (p. 179). I do not see why this poses any special problems for operators such as "PAST." However one interprets statements about the past, primitive temporal notions (e.g., "being true at a time," "being earlier than") will enter the semantics at some point and it is unlikely that they will satisfy Tooley's acquaintance condition any better than "PAST" does. In addition, it may be asked whether Tooley's reasoning should not also cast doubt on the primitiveness of the modal operators and the primitiveness of the concept of identity. Finally, it should be noted that presentists are unlikely to attribute the property of being past to anything.

8. In other words, I am assuming that we can refer to unambiguous or disambiguated name-types. This does not seem to be a highly controversial assumption (cf. how Kripke formulated his thesis about proper names).

9. Thus, "Napoleon was defeated" could be read as "PAST (There is an x, at most one, such that x is named by **Napoleon** in context C and x is defeated)."

10. See Russell 1971, p. 243-46.

11. Of course, the type-name would then refer in virtue of how its tokens are used at a later time (and similarly for type-predicates). The clever reader may notice that, at this point, the paper is probably not neutral on the question of whether sentences regarding the future have determinate truth-values.

12. However, going beyond the idealization or the stage-setting does not seem too difficult. One only needs to realize that the type-name **n** could be any currently existing object standing in a unique, non-intentional relation to x. For example, it could be x's toothbrush, x's dog, or x's spatial position at a given time in the past. One's intentional relation to x could be mediated by any of these other objects instead of being mediated by **n**.

13. Of course, a subject will not assent to the quoted statement unless certain background conditions are met. For instance, the subject has to believe that the object in question has a name.

14. Perhaps it is possible to be intrigued by someone or something without realizing that one is so intrigued. On my interpretation, that would simply mean that one does not always realize what, or how, one thinks of something or someone.

15. Perhaps seeing an object amounts to seeing that an object has a certain appearance. However, seeing that may in turn decompose into seeing an object and having a certain belief about that object (and, probably, some other condition as well).

16. David Lewis, who coined the term "temporary intrinsics" and who is not himself a presentist, acknowledges this theoretical virtue in Lewis 2002, p. 2.

17. Versions of this paper were presented at conferences in Leuven and Belfast. I remember receiving very helpful comments on these occasions from, among others, Leon Horsten and Timothy Williamson. Thomas Crisp and an anonymous referee for this journal have also provided highly valuable comments.

References

Adams, R. M. "Time and Thisness," *Midwest Studies in Philosophy XI* 3,(1986): 15-29.

Bigelow, J. "Presentism and Properties,"*Philosophical Perspectives* 10 (1996): 35-52.Chapter8 in this volume.

Butterfield, J. "Seeing the Present." Pp. 61-75 in *Questions of Time and Tense*, edited by R. LePoidevin.Oxford: Clarendon Press, 1998.

Crisp, T. "Presentism." Pp. 211-45 in *The Oxford Handbook of Metaphysics*, edited by M.D. Loux & D. Zimmerman. Oxford: Oxford University Press, 2003.

Kripke, S. *Naming and Necessity*. Cambridge, Mass.: Harvard University Press, 1972.

Kuhn, S. T., andP. Portner. "Tense and Time." Pp. 277-346 in *Handbook of Philosophical Logic*, Vol. 7, edited by D. Gabbay & F. Guenther. Dordrecht: Kluwer, 2002.

Lewis, D. "Causation." Pp. 193-204 in *Causation*, edited by E. Sosa & M. Tooley. Oxford: Oxford University Press, 1993.

———. "Tensing the Copula," in *Mind* 111 (2002): 1-13.

Ludlow, P. *Semantics, Tense and Time: An Essay in the Metaphysics of Natural Language*. Cambridge: MIT Press, 1999.

Markosian, N. "A Defense of Presentism." Pp. 47-82 in *Oxford Studies in Metaphysics*, volume 1edited by D. Zimmerman. Oxford: Oxford University Press, 2004.

Mellor, D. H. *Real Time II*. London and New York: Routledge, 1998.

Meyer, U. "Prior and the Platonist," *Analysis* 62 (2002): 1-16.

Prior, A. N. *Past, Present and Future*. Oxford: Clarendon Press, 1967.

———. *Papers on Time and Tense*. Oxford: Clarendon Press, 1968.

———. "The Notion of the Present." Pp. 80-82 in *Metaphysics: The Big Questions*, edited by P. van Inwagen and D. Zimmerman. Oxford: Blackwell, 1998.

Russell, B. "The Philosophy of Logical Atomism." Pp. 177-281 in *Logic and Knowledge*, edited byR.C. Marsh. London: George Allen & Unwin, 1971.

Saunders, S. "How Relativity Contradicts Presentism." Pp. 277-92 in *Time, Reality, and Experience*, edited by C. Callender. Cambridge: Cambridge University Press, 2002.

Sider, T. "Presentism and Ontological Commitment," *The Journal of Philosophy* 96 (1999): 325-47.

———. *Four-dimensionalism: An Ontology of Persistence and Time*. Oxford: Clarendon Press, 2001.

Soames, S. *Beyond Rigidity: The Unfinished Semantic Agenda of Naming and Necessity*.Oxford: Oxford University Press, 2002.

Tooley, M. *Time, Tense &Causation*. Oxford: Ciarendon Press, 1997.

Williamson, T. "Necessary Existents." Pp. 233-51 in *Logic, Thought and Language*, edited by A. O'Hear. Cambridge: Cambridge University Press, 2002.

Zimmerman, D. "Temporary Intrinsics and Presentism." Pp. 206-19 in *Metaphysics: The Big Questions*, edited by P. van Inwagen and D. Zimmerman.Oxford: Blackwell, 1998. Chapter 17 in this volume.

Section 2

Presentism and Relativity

Introduction: Presentism and Relativity

Ernâni Magalhães

The objections to presentism considered thus far appeal to facts as accessible to philosophers a thousand years ago as they are today. Human knowledge has not stood still in the intervening time, however, and some of what has been discovered appears to many to be incompatible with presentism. Specifically, presentism appears to be incompatible with the aspects of time uncovered by the special theory of relativity (STR) and general relativity (GR).

Although STR does not say anything about what is or is not *present*, it has rather significant things to say about the nature of simultaneity and succession. Presentism requires that in order to exist something must be present. The crucial question about the compatibility of presentism and STR concerns how something's being present (or not) connects with its standing in the various relations of succession and simultaneity. It seems that given that X is present, something Y is present only if Y is simultaneous with X. How could X and Y both be present if X is either earlier or later than Y? But, according to a fundamental tenet of STR, whether X and Y are simultaneous depends on the frame of reference—the perspective—one adopts. X and Y may be simultaneous from the perspective of a moving train, but not from a station at rest. This frame variance of simultaneity appears to wreak havoc with presentism. There are a number of different ways to spell out this havoc. Whether X and Y are simultaneous depends on the frame of reference adopted. So whether Y is present depends on the frame of reference adopted. So whether Y exists depends on the frame of reference adopted. But whether something exists cannot depend on whether one—some conscious being—decides to consider Y from this or that point of view. So presentism has the absurd consequence, when combined with STR, that whether something exists depends on what framework conscious beings adopt to consider the thing.

Hilary Putnam (chapter 13) spells out the problem in another way. Suppose two collocated observers, I-now and you-now, are in motion relative to each other. And assume the principle he calls "There Are No Privileged Observers": if X is real if and only if (iff) it stands in R to me-now, and you-now are real, then Y is real iff it stands in R to you-now (p. 225). But then "there are events which lie in 'the future' according to *my* coordinate system and which lie in the 'present' of you-now according to *your* coordinate system. Since these things stand in the relation *R* to you-now, and you-now are real [There Are No Privileged Observers] . . . requires that I call these future things and events real!" (p. 231)

The fundamental problem, therefore, is that whereas presentism seems to require frame invariant temporal relations to accommodate its view of the significance of the present, neither STR nor GR makes room for a privileged simultaneity or succession. In response, presentists may either, to paraphrase Mark Hinchliff (chapter 14, p. 245), relativize presentism by defining it in terms of STR or make relativity safe for presentism by introducing privileged notions of temporal relations. The relativizing maneuver concedes that being present is a frame variant feature and tries to identify the property with some theoretical element in relativity theory. One of Hinchliff's suggestions will illustrate the approach. The "past cone" of an event, E1, is the set of events that can reach it by a light signal. Its future cone is the set of events which can be reached by a signal sent from E1. Hinchliff suggests identifying E1's present with the set of events in E1's past cone.

By far the more popular presentist response involves revising relativity to accommodate presentism. Neo-Lorentzians[1] like William Lane Craig (2000), Quentin Smith (1998), and others argue that while relativity's frame variant temporal relations do justice to sensible temporal relations, there is in addition a relation of absolute simultaneity that obtains among objects. The advocates of real or absolute simultaneity propose that there is a real foliation of the Minkowski spatio-temporal manifold determined by what is really simultaneous with what, and not only the arbitrary divisions allowed by relativity. Some neo-Lorentzians even claim that only their view, not orthodox relativity, is consistent with all the experimental phenomena (Craig and Smith 2008: 5-6).

The whole debate about the compatibility of STR and presentism seems less significant when one considers the fact that the latest evidence suggests that STR is a false theory. Why should it bother a presentist if presentism is not compatible with STR if STR does not reflect time's true character? Yet these more recent empirical developments may provide cold comfort to the presentist since the picture of time emerging makes it perhaps even less likely that the desired privileged temporal characteristics and relations exist.

Notes

1. H. A. Lorentz proposed that the length-contraction and time-dilation phenomena uncovered in the late nineteenth and early twentieth centuries could be accommodated in terms of "local times" which are functions of a frame-invariant universal time.

References and Further Readings

Balashov, Y., and M. Janssen. "Critical Notice of Presentism and Relativity," *British Journal of Philosophy of Science* 54 (2003): 327-46.

Barbour, J. "The Emergence of Time and Its Arrow from Timelessness." Pp. 405-14 in *Physical Origins of Time Asymmetry*, edited by J. Halliwell, J. Perez-Mercader, and W. H. Zurek. Cambridge: Cambridge University Press, 1994.

———. *The End of Time. The Next Revolution in our Understanding of the Universe.* London: Weidenfeld and Nicolson, 1999.

Craig, W. L. *Time and the Metaphysics of Relativity*. Dordrecht: Kluwer Academic Publishers, 2000.

Craig, W. L., and Q. Smith, eds. *Einstein, Relativity, and Absolute Simultaneity*. London: Routledge Press, 2008.

Crisp, T. "Presentism, Eternalism, and Relativity Physics." Pp. 262-78 in *Einstein, Relativity, and Absolute Simultaneity*, edited by W. L. Craig and Q. Smith. London: Routledge Press, 2008.

Einstein, A. "On the Electrodynamics of Moving Bodies." Pp. 35-65 in *The Principle of Relativity*, edited by A. Sommerfeld. Toronto: Methuen and Company, Ltd., 1923. Originally published as "Zur Elektrodynamik bewegter Körper," *Annalen der Physik* 17 (1905): 891-921.

Godfrey-Smith, W. "Special Relativity and the Present," *Philosophical Studies* 36 (1979): 233-44.

Hinchliff, M. "A Defense of Presentism in a Relativistic Setting," *Philosophy of Science* 67 (2000) (Supplement): S563–S574. Chapter 14 in this volume.

Minkowski, H. "Space and Time." Pp. 75-91 in *The Principle of Relativity*, edited by A. Sommerfeld. London: Metheun and Company Ltd., 1923. Originally presented as an address, Eightieth Assembly of German Natural Scientists and Physicians. Cologne, Germany, September 21, 1908.

Prior, A. "Some Free Thinking about Time." Pp. 47-51 in *Logic and Reality: Essays on the Legacy of Arthur Prior*, edited by B. J. Copeland. Oxford: Oxford University Press, 1996.

Putnam, H. "Time and Physical Geometry," *Journal of Philosophy* 64 (1967): 240-47. Chapter 13 in this volume.

Saunders, S. "How Relativity Contradicts Presentism." Pp. 277-92 in *Time, Reality & Experience*, edited by C. Callender. Cambridge: Cambridge University Press, 2002.

Savitt, S. "There's No Time Like the Present (In Minkowski Spacetime)." *Philosophy of Science* 67 (supplement) (2000): S563–S574.

Smith, Q. "Absolute Simultaneity and the Infinity of Time." Pp. 135-83 in *Questions of Time and Tense*, edited by R. Le Poidevin. Oxford: Oxford University Press, 1998.

Stein, H. "On Einstein-Minkowski Space-Time," *Journal of Philosophy* 65 (1968): 5-23.

———. "On Relativity Theory and Openness of the Future," *Philosophy of Science* 58 (1991): 147–67.

Wheeler, J. A. "The 'Past' and the 'Delayed Choice' Double-slit Experiment." Pp. 9-48 in *Mathematical Foundations of Quantum Theory*, edited by A. Marlow. New York: Academic Press, 1978.

13
Time and Physical Geometry*

Hilary Putnam

I think that if we attempted to set out the "man on the street's" view of the nature of time, we would find that the main principle underlying his convictions on this subject might be stated somewhat as follows:

(1) All (and only) things that exist *now* are real.

Future things (which do not already exist) are not real (on this view); although, of course they *will* be real when the appropriate time has come to be the present time. Similarly, past things (which have ceased to exist) are not real, although they *were* real in the past.

Obviously, we shall have to make some assumptions about the concept *real* if we are to discuss the "man on the street's" view at all. The assumptions I shall make are as follows:

I. I-now am real. (Of course, this assumption changes each time I announce that I am making it, since "I-now" refers to a different instantaneous "me.")

II. At least one other observer is real, and it is possible for this other observer to be in motion relative to me.

And, the most important assumption, which will be referred to (when properly understood) as the principle that There Are No Privileged Observers:

III. If it is the case that all and only the things that stand in a certain relation *R* to me-now are real, and you-now are also real, then it is also the case that all and only the things that stand in the relation *R* to you-now are real.

---

* This paper was read to a meeting of the American Physical Society in New York, January 27, 1966.

If the assumption III is not to be vacuous, it is necessary to understand the expression "the things" as referring not just to present things, but to all things past, present, and future, whether we regard all these things as "real" or not. (That is, III is to be understood as stated in "tenseless" language, except for the indexical terms "me-now" and "you-now.") Secondly, *R* must be restricted to physical relations that are supposed to be independent of the choice of a coordinate system (as simultaneity was in classical physics) and to be definable in a "tenseless" way in terms of the fundamental notions of physics. And, lastly, it must not depend on anything accidental (physically speaking) that all and only the things that stand in the relation *R* to me-now are real.

We note that, if we assume classical physics and take the relation *R* to be the relation of simultaneity, then, on the view (1), it is true that all and only the things that stand in the relation *R* to me-now are real, and the principle III is satisfied because the relation of simultaneity is transitive.

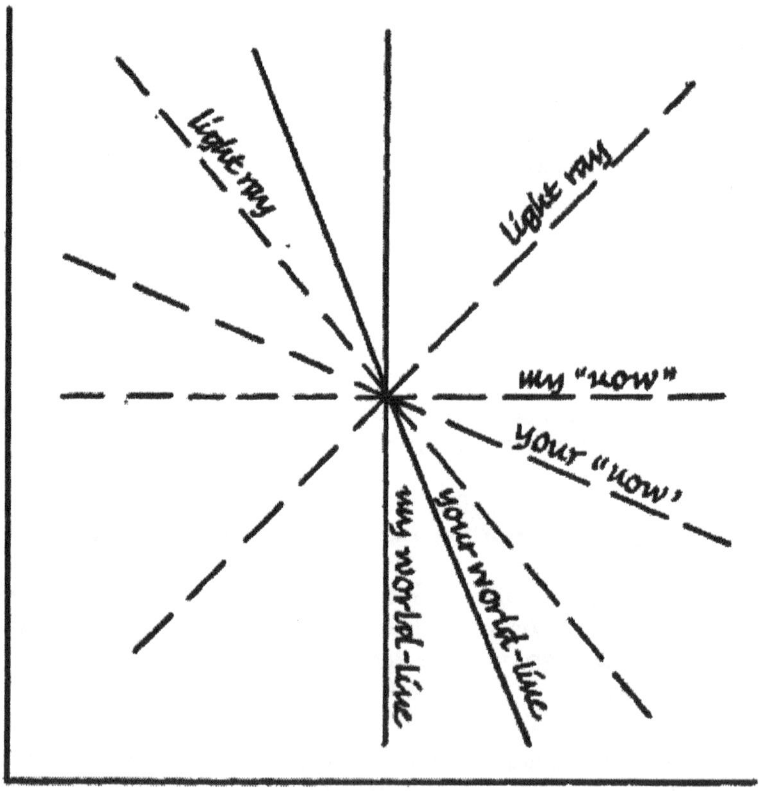

FIGURE 1

Finally, I shall assume Special Relativity.

We now discover something really remarkable. Namely, on every natural choice of the relation *R*, it turns out that *future* things (or events) are already real. For example, suppose I take the relation *R* to be the relation of simultaneity (as is suggested by (1)). For the remainder of this paper I shall assume the following (perfectly possible) physical situation to be actual: you-now and I-now are at the same place now, but moving with relative velocities which are very large (relative to the speed of light, which I take to be = 1). Thus, our world-lines look as shown in figure 1 (I have also drawn our "light-cone" for the purpose of the later discussion):

Now then, we cannot take the relation of simultaneity-in-my-coordinate-system to be *R* without violating the way in which the principle that There Are No Privileged Observers is intended to be understood. Rather, we have to take *R* to be the relation of simultaneity-in-the-*observer's*-coordinate-system. Then, if I assume that all and only those things which stand in *this* relation *R* to me-now are real, I find that you-now are also real (since I-now and you-now are simultaneous-at-a-point and, hence, simultaneous in *every* coordinate system). Hence, everything that stands in the relation *R* to you—everything that is simultaneous to you-now in *your* coordinate system—is also real. But it is well known that, as a consequence of Special Relativity, there are events which lie in "the future" according to *my* coordinate system and which lie in the "present" of you-now according to *your* coordinate system. Since these things stand in the relation *R* to you-now, and you-now are real, and it was assumed that all and only the things that stand in the relation *R* to me-now are real, the principle III requires that I call these future things and events real! (But, actually, I now have a contradiction: for these future things do not stand in the relation *R* to me-now, and so my assumption that all and only the things that stand in *this* relation *R* to me-now are real was already inconsistent with the principle that There Are No Privileged Observers.)

The difficulty is obvious: what the principle that There Are No Privileged Observers requires is simply that the relation *R* be *transitive*; i.e., that it have the property that from *xRy* and *yRz* it follow that *xRz*. Simultaneity-in-*my*-coordinate-system has this property, since if *x* is simultaneous with *y* in my coordinate system, and *y* is simultaneous with *z* in my coordinate system, then *x* is also simultaneous with *z* in my coordinate system; but simultaneity-in-*my*-coordinate system is not admissible as a choice of *R*, because it depends on the coordinate system. And the relation "*x* is simultaneous with *y* in the coordinate system of *x*" (which is essentially the relation we just considered), while admissible, is not transitive, since, if I-now am simultaneous with you-now in the coordinate system of me-now, and you-now are simultaneous with event *X* in the coordinate system of you-now, it does *not* follow that I-now am simultaneous with event *X* in the coordinate system of me-now.

Now then, if we combine the fact that the relation *R* is required by III to be transitive with our desire to preserve the following principle, which is one-half of (1):

(2) All things that exist now are real.

—then we quickly see that future things must be real.

For, if the relation *R* satisfies (2)—and I take (2) to mean (at least when *I* assert it) that all things that exist *now* according to *my* coordinate system are real—and you-now are as in figure 1, then you-now must stand in the relation *R* to me-now, since you exist both *now* and *here*. But, if the relation *R* always holds between all the events that are on some one "simultaneity line" in my coordinate system and me-at-the-appropriate-time, then (since the laws of nature are invariant under Lorentz transformation, by the principle of Special Relativity), the relation *R* must also hold between all the events on some one "simultaneity-line" in *any* observer's coordinate system and that-observer-at-the-appropriate-time. Hence, all the events that are simultaneous with you-now in *your* coordinate system must *also* bear the relation *R* to you-now. Let event *X* be one such event which is "in the future" according to *my* coordinate system (if our velocities are as shown in figure 1, then such an event *X* must always exist). Then, since the event *X* bears the relation *R* to you-now, and you-now bear the relation *R* to me-now, the event *X* bears the relation *R* to me-now. But we chose *R* to be such that all and only those events which bear *R* to me-now are real. So the event *X*, which is a *future* event according to my coordinate system, is already real!

The Space-fight Tomorrow. A different problem connected with time was studied by Aristotle. Aristotle was what we would today call an "indeterminist." He did not think that the outcome of certain future events—his example was, Who will win the sea-fight tomorrow?—was determined at the present time. Given this assumption, he found it difficult to believe that the statement "The *A*s will win the sea-fight" *already* has a truth value, that it is already true or already false. Rather he suggested that such "future contingent" statements have *no* truth value.

Some philosophers would say that this belief rested upon a simple *mistake* on Aristotle's part, that "is true" and "is false" are *tenseless* predicates, and that statements are "true" and "false" *simpliciter*, not "already true" or "now true" or "going to be true," etc. But this is begging the question. If any statement that is true at some time is true at every time, then indeed there is some point to dropping this way of speaking and making "true" and "false" tenseless (which they clearly are *not* in ordinary language, by the way). But the correctness of this view—whether you express it by saying that truth is eternal or by saying that truth is "tenseless" does not matter—was what Aristotle was challenging. It is unfair to assume a form of language which presupposes that Aristotle was wrong and then to use the assumed correctness of this linguistic formalism as an argument against Aristotle.

It would be easy to write down a list of predicates $P_1, \ldots, Pn$ such that, on Aristotle's view, the statement that any past or (completed) present event has one of the properties $P_1, \ldots, P_n$ always has a truth value, but the statement that a future event has one of the properties P_i does not have a truth value. Thus Aristotle's view is a sort of supplement to what we called the "man on the street's" view of time. The principle (1) asserts a fundamental difference between the present and both the past and the future; but it does not assert any difference between the past and the future; neither past things nor future things can now be called real. Aristotle would perhaps have accepted this, but he would have added that there is a fundamental difference, nonetheless, between the past and the future, viz., that past events are now determined, the relevant statements about them have now acquired truth values which will "stick" for all time; but future events are undetermined, and at least some statements about them are not yet either true or false. (The fact that you can change the future but not the past would have been explained by Aristotle in terms of this fundamental asymmetry between the two.)

Aristotle was wrong. At least he was wrong if Relativity is right; and there is today better reason to believe Relativity than to believe Aristotle, on this point at least. To see this, let us revert to the situation depicted in figure 1, and let there be a space-fight which is "in the future" in my coordinate system (i.e., above the "now" of "me-now") but "in the past" in your coordinate system (i.e., below the "now" of "you-now"). Then, since the space-fight is "in the future" (for me), I must say that the statement "the *A*s will win or have already won" has no truth value. But, if you say that this same statement *has* a truth value—as you must, if Aristotle is right, since it lies "in the past" for you—then you and I cannot both be right. Nor can just one of us be right, without becoming a Privileged Observer![1]

The point becomes even clearer when viewed a bit more abstractly. Let us assume that Aristotle was right. Then I shall show that we can define a relation of Absolute Simultaneity, contradicting Relativity. Namely, I define an event to be Absolutely Future if the statement that it has the property P_i (where P_i is one of the properties in the list alluded to before) has no truth value. I define an event to be Absolutely Present if it is *not* Absolutely Future and if every event that is in its proper future (i.e., occurs later on its world-line—this is relativistically invariant) *is* Absolutely Future. Then two events are Absolutely Simultaneous if and only if there is a time (in the coordinate system of *any* observer) at which they are both Absolutely Present. (Also, one can define *that* observer to be at Absolute Rest.)

It is no criticism of Aristotle's ability to say that he was mistaken, any more than it is a criticism of Newton that he failed to foresee Relativity. Had the world been different than it is, these sorts of objections could not be brought against Aristotle's view. But it is important to see that Aristotle's view depends upon an *absolute* "pastness" and "futurity" just as much as Newtonian physics does, and that it is obsolete for the same reason.

I might try to save the view that "future contingents" have no truth value by saying that even *present*-tense statements have no truth value if they refer to the outcome of events that are so far away that a causal signal informing me of the outcome could not have reached me-now without traveling faster than light. In other words, I might attempt saying that statements about events that are in neither the upper half nor the lower half of my light-cone have no truth value. In addition, statements about events in the upper half of my light-cone have no truth value, since they are in my future according to every coordinate system. So only statements about events in the lower half of my light-cone have a truth value; only events that are in "my past" according to *all* observers are determined.

This last move, however, flagrantly violates the idea that there are no Privileged Observers. Why should a statement's having or not having a truth value depend upon the relation of the events referred to in the statement to just one special human being, *me*? Moreover, the following highly undesirable consequence flows from this last view: let Oscar be a person whose whole world-line is outside of the light-cone of me-now. Let me-future be a future "stage" of me such that Oscar is in the lower half of the light-cone of me-future. Then, when that future becomes the present, it will become true that Oscar *existed*, although it will never have had a truth value to say in the present tense "Oscar exists now." Things could come to *have been*, without its ever having been true that they *are*!

Morals and Conclusions. What moral should we draw from all this? In the first part of this paper I showed that, if we couple the assumption I with the principle (2), we must say that future things are real, even if they do not exist yet. Strictly speaking, I showed only that *some* future things are real, and that only on the assumption that the situation shown in figure 1 obtains. But the argument can obviously be extended (bringing in certain further assumptions with which virtually no one would argue) to show that *all* future things are real ("things" here includes "events"), and likewise that all *past* things are real, even though they do not exist *now*. To spell this out: If I accept (2), then I must say that all those things are real (but not *only* those things, as we saw) which bear to me the following relation: the relation y bears to x just in case y is simultaneous with x in the coordinate system of x. I must describe the relation this way, and not as "the relation y bears to x just in case y is simultaneous with x *in the coordinate system of me-now*," because otherwise I would be committed to the view that the real things bear to me-now a relation which is not Lorentz-invariant and, hence, that they define an absolute simultaneity. But principle III then requires that I also count every thing and event which bears the *transitive closure* of R to me (i.e., which bears R to me, or which bears R to something that bears R to me, or which bears R to something that bears R to something that bears R to me, or. . . .) as real. But every thing and event in space-time bears the transitive closure of the above relation R to me, at least if there are enough observers. And, if we allow all physical systems (even electromagnetic fields, etc.) as "observers" (as why should we not?) and allow observers to use coordinate systems in which they are not at rest, then there are certainly "enough observers."

In the second part of this paper I showed that contingent statements about future events already have a truth value. In consequence, the "tenseless" notion of existence (i.e., the notion that amounts to "will exist, or has existed, or exists right now") is perfectly well-defined. This is fortunate, since the upshot of the first part of the paper could also have been stated by saying that the notion of being "real" turns out to be coextensive with the *tenseless* notion of existence.

I conclude that the problem of the reality and the determinateness of future events is now solved. Moreover, it is solved by physics and not by philosophy. We have learned that we live in a four-dimensional and not a three-dimensional world, and that space and time—or, better, space-like separations and time-like separations—are just two aspects of a single four-dimensional continuum with a peculiar metric which sometimes permits distance (y, x) = 0 even when x ≠ y. Indeed, I do not believe that there are any longer any *philosophical* problems about Time; there is only the physical problem of determining the exact physical geometry of the four-dimensional continuum that we inhabit.

In this paper I have talked only about the relativistic aspects of the problem of physical time: there is, of course, also the problem of thermodynamics, and whether the Second Law does or does not explain the existence of "irreversible" processes (the so-called "problem of the direction of time"), and the problem of the existence or nonexistence of true irreversibilities in quantum mechanics, which, I gather, is currently under hot discussion. I have not talked about these problems.

Notes

1. Nor can one avoid the difficulty by abandoning the objectivity of truth, and speaking of "truth-for-me" and "truth-for-you." For, if it is even true-for-me that the statement in question *has a truth-value-for-you*, then either it is true-for-me that the statement in question is true-for-you, or it is true-for-me that the statement in question is false-for-you (since "true-for-you" and "false-for-you" are the only two truth-values-for-you). But if it is true-for-me that "the *A*s have won or will win the space-fight" is true-for-you, then, to all intents and purposes, "the *A*s have won or will win the space-fight" is true-for-me too, since both facts—the outcome of the space-fight and the truth-value-for-you of this statement—have exactly the same status relative to me; and, likewise, if it is true-for-me that "the *A*s have won or will win the space-fight" is false-for-you, then, to all intents or purposes, "the *A*s have won or will win the space-fight" is false-for-me too. So I must not regard it as true that "the *A*s have won or will win the space-fight" is either true-for-you or false-for-you, if I wish to maintain that this future contingent has no truth value. And hence, if it is true that:

> For every contingent statement *S* and every observer *O*, (*S* is true-for-*O* or *S* is false-for-*O*) if and only if *S* refers to an event in the lower half of *O*'s light-cone.

—then this "truth" cannot itself be stated by any observer on pain of contradiction!

(Here I assume this much concerning the logic of "true-for-me": that, if *p* v *q* is true-for-me, then either *p* is true-for-me or else *q* is true-for-me. Of course, we might try still more radical alterations of logic. But it seems considerably preferable—to me at least—*not* to do this, and rather to concede the objectivity of future contingents.)

14
A Defense of Presentism in a Relativistic Setting

Mark Hinchliff

Abstract

Presentism is the view, roughly speaking, that only presently existing things exist. Though presentism offers many attractive solutions to problems in metaphysics, philosophy of language, and philosophy of mind, it faces threats from two main sources: McTaggart and the special theory of relativity. This paper explores the prospects for fitting presentism together with the special theory. Two models are proposed, one which fits presentism into a relativistic setting (the cone model) and one which fits the special theory into a presentistic setting (the surface model). It is concluded that there is no good reason arising from the special theory of relativity for rejecting presentism. Presentism is tenable here and preferable elsewhere.

1. Introduction

In "The Puzzle of Change" I argued that one could solve a traditional problem about the nature and possibility of change if and only if one adopted a view about the nature of time I called *presentism*. I believe presentism offers attractive solutions to philosophical problems not only in several areas of metaphysics but also in the philosophy of language and the philosophy of mind, such as the semantics of tensed discourse, our distinctive attitudes towards past, present, and future, the psychology of temporal experience, and the dynamics of continued belief. I think the presentist's rival, *the eternalist*, also has things to say in each of these areas, but I think the presentist's answers are better. Presentism, however, faces problems from two sources: McTaggart's argument against the A-theory, and the theory of relativity.

In unpublished work I have defended presentism against McTaggart's attack on the A-theory, and in the last section of "The Puzzle of Change," I took up in a cursory way the challenge posed by the special theory. In this paper I will take up the challenge again and hope to give a fuller answer on behalf of the presentist. Challenges arising from the general theory of relativity will have to wait for another day.

I will begin by saying what I take presentism to be and not to be. This will permit me to reply to Craig Callender's worry that the debate between presentists and eternalists is not a substantive one. I will then examine three models for reconciling presentism with the special theory—the point model, the cone model, and the surface model—discussing Steve Savitt's objections to these models, as well as raising some of my own. My overall conclusion will be that at least the cone and the surface models offer satisfactory presentist responses to the challenges posed by the special theory of relativity.

2. What Presentism Is and Is Not

Presentism is the view, roughly speaking, that only presently existing things exist, only presently red things are red, and so on. The presentist and his opposite, the eternalist, will agree that dinosaurs *existed* and do not presently exist; they will disagree over whether they *exist*. For the eternalist, time is like space; other times are like other places; there is nothing special about what exists *here*. For the presentist, time is like modality. Other times are like other possible worlds; they are not real. As there is something special about what is *actual*—it is all there is—so there is something special about the *present*—it is all there is. The way things are is the way things presently are.

It is possible to lose sight of the substantive issue that divides presentists from eternalists. The characteristic thesis of presentism—only present Fs are Fs—can be easily confused with closely related theses that are mere trivialities. Presentism can also be saddled with old baggage from previous battles over the nature of time, or expanded to include more controversial views on the nature of time. Successful attacks on the old baggage or the more controversial views can then be mistaken for successful attacks on presentism itself. It is therefore worth taking a few moments to say more about what presentism is and what it is not.

I am using the sentence "Only presently existing things exist" and its companions, such as "Only presently red things are red," to distinguish the presentists from the eternalists. To serve this purpose, these "test" sentences must be given a particular interpretation. The quantifiers should be read as unrestricted quantifiers; the tokens of "exists" should be understood as tokens of a nonindexical "exists" and as tokens of the same word; and "presently" should be read as an indexical tense operator. Working under these stipulations, we can see that the sentence "Newton exists" will express the same proposition at different times, and the sentence "Newton presently exists" will express different propositions at different times. This difference in the semantics enables us to express the substantive difference between presentism and eternalism.

Though more can be said on how to formulate the issue between presentism and eternalism, I have said enough about this matter to permit me to respond to Craig Callender's worry (1998, 2) that there is no substantive difference between presentism and eternalism. Callender uses the picture of a four-dimensional manifold of events, each carrying a light bulb that is on when the event exists and off when the event does not exist. Eternalism is then the view that all the light bulbs are on. Since a light bulb does not exist when it is not on, the only existing light bulbs, according to the presentist, are ones that are on—which is just what the eternalist says.

Right, everybody is going to say that every bulb that is on is on, that everything that exists, exists; and nobody is going to say that some bulb that is on is off. I agree with Callender that a substantive issue cannot be formulated in nonindexical terms. That is why I have formulated the issue with the aid of the indexical "presently." The eternalist thinks every bulb is on, but he does not think every bulb that is on is *presently* on; some bulbs are on that are off now. However, the presentist will think every bulb that is on is presently on, and no bulbs are on that are off now. Are any bulbs on that are not presently on? Contrary to what Callender contends, we have a substantive question and the debate is on.

An older, and perhaps more familiar, issue in the metaphysics of time is the dispute between the A and B theories of time. An A-theorist, or tenser, thinks events have tenses, the properties of presentness, pastness, and futurity, as well as degrees of pastness and futurity, such as being three days past, and nine months future. A B-theorist, or detenser, thinks there are no such properties, though there are the temporal relations of being-earlier-than and being-later-than, and relational properties composed of those, such as being nine months later than Monica's first visit.

Events do not change their relational properties. If an event is earlier than Clinton's testimony, it stays earlier. Events do, however, change their tenses. An event that is five days future, becomes present, then past. It was the continual gaining and losing of the tenses which was thought to account for the apparent flow or passage of time.

But when compared to masses or temperatures or charges, the tenses do seem to be a peculiar and suspicious class of properties. What differences could they make in the world? But the presentist is not committed to the tenses as irreducible properties. Sentences involving them can be paraphrased into relatively uncontroversial claims. For example, instead of saying "Clinton's testimony is past," we may say "Clinton *has given* his testimony." In short, sentences involving the tenses can be paraphrased into tensed sentences.

When the present is distinguished from other times, it becomes possible to draw further distinctions among the other times, to speak, for instance, of the past and the future, and to speak of differences between the past and the future. Familiar ideas about the purported difference are, for instance, that the past is fixed or closed or determined and the future is open or indeterminate. These ideas have played a role among some authors in the contemporary debates about the present in relativity theory. It has been thought sufficient to show that presentism is incompatible with the special theory to show that all events are "determined" or fixed. But presentism is not committed to any view about the difference between the past and the future or to any difference between the past and the future. Strictly speaking, it is not even committed to the past and the future; it only draws a distinction between the present and other times.

A second notion that has played a role in the recent history of the philosophy of time is a notion of *temporal becoming*. As a matter of grammar, the verb "become" is a linking verb. Like the "is" of predication, "become" takes a complement: particles become charged, students become rich, I become fat and bald. Temporal becoming differs from becoming by taking no complement. Events *become*. (The difference is like the difference between "is"s in "Gates is rich" and "Gates is.") Once again, the idea was to account for the apparent flow and passage of time by events becoming. Once again, the presentist is not committed to such an idea.

3. Open and Closed Avenues

I turn to the prospects for presentism in the special theory of relativity. In my paper on change, I discussed Hilary Putnam's argument from relativity against presentism (Putnam 1967) [chapter 13 in this volume—*Editors*]. Following Larry Sklar's 1981, I pointed out that Putnam's argument rests on two assumptions which the presentist is free to deny: a transitivity principle that A is real for C if A is real for B and B is real for C, and an assumption about what presentism is in a relativistic setting, that the real events are the events simultaneous with us now in our frame of reference. I said this latter assumption of relativized presentism was one proposal among many for fitting presentism into a relativistic setting, and I mentioned two others: identifying the present with the here-now and identifying the present with the surface of the past light cone. (In this paper I am going to call these two proposals the point model and the cone model.)

I then went on to suggest that we could even consider proposals that reverse the direction of fit between presentism and relativity, fitting the special theory, or at least a theory with its empirical consequences, into a presentist picture with a relation of absolute simultaneity—a type of proposal I'm going to call a surface model. Finally, I suggested in my paper that we approach this topic the way we would the problem of how to fit together our intuitive views of our mental lives and our best psychological theories. Proposals for understanding that relationship are varied and complex, and a "refutation" of one proposal is not taken to be a refutation of our intuitive views of our mental lives. No one believes the question about the mind to be easy to answer; the question about time seems just as hard.

Savitt 1998 has attempted to close each of the avenues I suggested were open for the presentist. About attempts to fit the present into a relativistic setting, Savitt concludes, "[Presentism] can not even be expressed in Minkowski spacetime." About attempts to fit relativity theory into a presentist setting, Savitt concludes that "taken at face value," the idea "really should not be respectable in the least." Savitt is thus led to advise that "the presentist should throw in the towel *with respect to STR*," but he offers me the hope that perhaps things will go better for me with the general theory of relativity. To which I say, "Beware Canucks bearing gifts."

4. The Point Model

What I am going to call "point presentism" is the view that in the special theory of relativity, the present is to be identified with the here-now. On this view, the way things are is the way things are here-now. Though the view has been discussed by several authors, I know of no one who actually holds this view. Howard Stein (1968) comes closest to being a point presentist. He does think the present is to be identified with the here-now. He writes, "In Einstein-Minkowski space-time *an event's present is constituted by itself alone*" (15). However, Stein is one of the philosophers I mentioned who wishes to define a notion of becoming definite and to use that notion to establish what is real as what has become definite. In Stein's terms, the absolute past has become real. So in my terms, Stein is not a point presentist.

The common objection to point presentism is that it is *lonely*. Only the here-now exists. Though I am moved by this objection, I have often thought I should not be. It is really just a re-statement of the view as an objection. It is like objecting to solipsism by saying the problem with solipsism is that there are no other people. Perhaps, though, in cases like these, an objection can be just a re-statement of the view.

The objection I prefer is one given by Putnam (1967, 246). If the present is the here-now, there are events that are past that were never present. These events were space-like separated from the earlier events on my world line. So they never were present. But they are time-like separated from me here-now; so they are here-now past. The objection is, in other words, that point presentism violates the "conceptual truth" that what *is* past *was* present.

Stein has an interesting reply to Putnam, interesting because (a) I find it unconvincing and (b) it is similar to some replies I am going to give myself. After saying, "in Einstein-Minkowski space-time an event's present is constituted by itself alone," Stein goes on to say,

> In this theory, therefore, the present tense can *never* be applied correctly to "foreign" objects. This is at bottom a consequence (and a fairly obvious one) of our adopting relativistically invariant language—since, as we know, there is no relativistically invariant notion of simultaneity. The appearance of paradox only confirms that the space-time of Einstein and Minkowski is quite different from pre-relativistic space-time (1968, 15).

Many of the objections and replies we will encounter have this general form. First, the presentist is urged to take relativistic space-time seriously. The presentist then makes a proposal about what his view comes to in a relativistic setting. An objection is then made to the presentist's proposal which is based on some principle from outside relativity theory, such as the principle that what is past was present, which fails to hold in the relativistic setting on the presentist's proposal. The presentist then replies that the principle's not holding is just what we should expect given the relativistic setting.

I do not know what to think about this familiar dialectic except (1) that not every objection like this will be fair, since some things will have to change for the presentist (and the eternalist) when we move to a relativistic setting, and (2) that not every reply like this will be acceptable, since we cannot call just anything presentism in a relativistic context.

I think it is a mistake to try to identify the "core" of presentism which must be preserved in a relativistic context. I think we should instead explore different proposals for fitting presentism in. We may conclude no proposal fits well enough to say that presentism is compatible with a relativistic spacetime. We may conclude that more than one proposal fits well enough. My own view is that point presentism does not fit well enough, but it may be that I just do not want to be alone.

5. The Cone Model

I am going to call "cone presentism" the view that in the special theory of relativity, the present for an event E is to be identified with the surface of E's past light cone. I believe this view was first suggested by William Godfrey-Smith (1979). One virtue of the view is that it captures the idea that what is present is what I am seeing now. A second virtue is that it identifies the present with an invariant feature of the special theory. A third virtue is that we are not alone. Nevertheless, cone presentism faces a number of objections, two of which Steve Savitt presses as "consequences that should make a red-blooded presentist squirm" (1998, 6).

Savitt presents the first of these squirm-inducing objections in the following passage:

> First of all, we have been observing for about four decades the Cosmic Microwave Background Radiation (CMBR), which is thought to have originated about 300,000 years after the Big Bang, or about 15 billion years ago (give or take a few billion years). On Godfrey-Smith's view, the origin of this radiation counts as part of the present of the contemporary astrophysicists who observe it! (1998, 6)

To the extent you find this objection compelling, you should be a presentist. The objection derives its force from the "fact" that CMBR originated 15 billion years ago. This "fact" comes from outside the special theory. The special theory is silent on this matter. Indeed, according to the special theory, there is no fact of the matter concerning how long ago this event happened. If we think it is a fact that this event happened 15 billion years ago, we must think there is a distinguished inertial frame which assigns events their "correct" dates. As a presentist, I say, "Great, the events that are simultaneous with me now in that special frame are the existing events." If we think there is no distinguished inertial frame, then we cannot appeal to alleged facts like the radiation's originating 15 billion years ago in objecting to cone presentism. If there is a distinguished frame, we can stay good old-fashioned presentists. If not, we can be cone presentists. Either way, presentism is unscathed by objections of this sort.

A point that emerges from this discussion is that the cone presentist also need not assign a unique temporal metric. It also is not a fact that the origination of the radiation is simultaneous with the physicist's observation. The cone presentist does not have to assign the same time to the two events, engendering worries about the infinite speed of light and a photon being two places at once. Presentism in a pre-relativistic setting was put in terms of which times were past, present, and future. In that setting, hyperplanes of simultaneity were invariant, hence real, features of the manifold. But in a relativistic setting, these planes are not invariant; simultaneity is relative. The invariant features are the spatiotemporal intervals. In that changed circumstance, the cone presentist should not say that the radiation's origination is simultaneous with the physicist's observation, but should say the origination is at spatiotemporal interval 0 from the physicist's observation.

Savitt's second squirm-inducing objection begins by making a similar point:

> [Cone presentism] seems to rest on the idea that events on the past light cone of E have a lightlike separation from E and hence the spacetime interval from E to (say) E' (on the past light cone of E) is 0. But then it seems arbitrary to exclude from the present events on the future light cone of E, which are also lightlike separated from E (1998, 6).

Let us call this idea that an event's present is to be identified with all the events lightlike separated from it "double-cone presentism." I find this position unappealing. It permits single events to be present twice; an event that is present because it is on the surface of my future light cone will be present again when it is on the surface of my past light cone. However, this objection poses no threat to cone presentism because the cone presentist is free to reject double-cone presentism. Savitt's reason for thinking otherwise is, to repeat, that "it seems arbitrary to exclude from the present events on the future light cone of E, which are also lightlike separated from E" (1998, 6). But the difference, as I understand it, between the past light cone and the future light cone is a non-arbitrary difference. The surface of E's past light cone is the set of events *from* which a light signal or ray could be sent *to* E. The surface of E's future light cone is the set of events *to* which a light signal or ray could be sent *from* E. The difference between the cones is due to the asymmetry built in to the nature of a light *ray* or *signal*. And that asymmetry arises from the asymmetric nature of causation itself, which is a non-arbitrary foundation on which to rest the distinction between cone and double-cone presentism.

Savitt's last squirm-inducing objection arises from a requirement of *achronality*, which says that if two events, X and Y, are present for E, then neither should be in the absolute past or future of the other. I agree with Steve that this requirement has some intuitive plausibility. How *can* an event X be in E's *present* when all observers agree that X is *earlier* than E? After all, it certainly seems that if X is earlier than E and E is present, then X is *past*. I also agree with Steve that cone presentism violates this principle. An event X on the past light cone of E is earlier than E in all inertial frames. The cone presentist thus says the event X is in E's present. The principle of achronality says it is in E's past. Something has to give.

I think it is the principle. (Here, I may go beyond Godfrey-Smith.) The principle of achronality derives its plausibility from classical settings, where I would not choose to deny it. But it loses that plausibility in a relativistic setting. In a classical spacetime where temporal intervals are invariant features of the manifold, presentism can be understood as a view about *time*. What the indexical "presently" picks out is, roughly speaking, the set of events at temporal interval 0 from the time of utterance. However, in Minkowski spacetime where temporal intervals are *not* invariant features of the manifold, presentism should be understood not as a view about time, but as a view about spacetime. The invariant features are the *spatiotemporal* intervals, and the present is not the events at temporal interval 0 from the time of utterance, but the events on the rearward cone at *spatiotemporal* interval 0 from the "spacetime" of utterance.

In a classical setting, to have an event be present that happened six million years ago would be a serious problem. An event would be present that was not at temporal interval 0 from us. But because the cone presentist does not define the present in terms of temporal interval 0, a present event that is not at temporal interval 0 poses no problem. So the objection, we see, rests on the account of the present the classical presentist gives and which the cone presentist believes must be supplanted by his own account in a Minkowski spacetime.

One might object that the cone presentist is no longer true to his school, that his view is no longer a view about *time*. This objector, however, seems not to be extending the same latitude and tolerance that we have come to expect elsewhere in the transition from classical to relativistic physics. Many of our ideas that we had thought were purely temporal or purely spatial turned out to be spatiotemporal in the setting of the special theory. If we are in sympathy with the idea famously expressed by Minkowski that "Henceforth space by itself, and time by itself, are doomed to fade away into mere shadows, and only a kind of union of the two will preserve an independent reality" (1923, 75), then we should also be in sympathy with the idea that henceforth presentism will no longer be a view about a mere shadow but a view about an independent reality, not about time but about spacetime.

There are two other objections to cone presentism which need to be addressed. The cone presentist must deny symmetry and transitivity for his notion of the present. It is not the case on the cone presentist's model that if X is present for Y, Y is present for X. An event can be in my present and I not in its. In fact, if a light signal can reach me now from E, I cannot reach E by a light signal. Furthermore, it is not the case on the cone presentist's model that if X is present for Y and Y is present for Z, then X is present for Z. An event E, timelike separated event from me in my past light cone, can send a signal to F, which is lightlike separated from me on the surface of my past light cone. E is present for F, and F is present for me, but E is not present for me. It is not lightlike separated from me but timelike separated; E is "past" for me.

We can certainly see that the symmetry and the transitivity of the "present-for" relation hold in a pre-relativistic setting. If you are present for me, I am for you. And if I am present for you and you are for Savitt, then I am for Savitt. It is equally clear these principles do not hold on the cone presentist's model. Are we to reject cone presentism for these reasons? It is not clear. I am inclined to think not.

The cone presentist ties in his notion of the present very closely with the nature of light. If light were an instantaneous signal, the present defined in terms of it would be symmetric and transitive. But because in the special theory of relativity light behaves in peculiar ways, the present defined in terms of it does too. The problem is not the presentism; it is the light![1]

I am inclined to think cone presentism gives us a better fit between presentism and the special theory than point presentism does. And I am inclined to think cone presentism gives us a good enough fit to say presentism is compatible with the special theory.

6. The Surface Model

The models we have discussed so far, point and cone presentism, try to fit presentism into a relativistic setting. The view I am going to call "surface presentism" reverses the direction of fit, and tries to accommodate the special theory or a theory with its empirical consequences into a presentist picture. There are a variety of ways of trying to do this, and I will not even try to cover them all. What they all have in common, however, is a commitment to an absolute relation of simultaneity.

In "The Puzzle of Change," I cited Arthur Prior 1970 in which Prior offers a hard-line version of this idea. Prior writes, "One possible reaction to this situation, which to my mind is perfectly respectable though it isn't very fashionable, is to insist that all that physics has shown to be true or likely is that in some cases we can never know, we can never find out, whether something is actually happening or merely has happened or will happen" (247).

In Prior 1968, Prior expands a bit on the idea expressed in the passage just quoted:

> I suspect that the infinity of "local proper times" which figure in relativistic physics amount simply to what appears from various points of view, or in various "frames of reference," to be the course of events. And given how the course of events appear from a certain point of view, your relativistic physicist will be able to calculate how it will appear from certain other points of view. He can also indicate what features of the course of events (what temporal orderings of those events) will be common to all points of view, and one can work out a "tense-logic" for that too. (It turns out to be slightly different in the special and the general theories of relativity.) What the relativistic physicists cannot calculate from how

> the course of events appears from certain points of view is how, in all its details, the course of events actually is. It is not clear to me that there is anything surprising or unacceptable in this conclusion, or that we should be driven by it to renounce the use of forms like "It appears from such-and-such a point of view that p," which assume that there is also a plain p which is or is not the case (133).

The sort of theory I think Prior has in mind is a fairly old and familiar one, going back to the period in physics preceding Einstein's 1905 paper. If we assume that we are in a Newtonian spacetime with absolute space and a relation of absolute simultaneity, and we assume that light travels at the same speed c in all directions in absolute space, then the speed of light in an inertial frame in absolute motion will not be c. But in such an inertial frame the speed is *measured* to be c. How can this be? An old answer is that clocks and rods in absolute motion undergo the familiar Lorentz-Fitzgerald contraction. The speed of light will then be *measured* to be the same in all inertial frames, even though it is not.

This "neo-Lorentzian" theory will, I believe, accommodate the same experimental results as the special theory, and it comes with a relation of absolute simultaneity. It is thus a way for us to fit the special theory—or a theory with the same experimental consequences—into a presentist setting. I think this is the sort of theory Prior had in mind, and I know it is the sort of theory other presentists have in mind. And though Michael Tooley is not a presentist (he thinks past *and* present are real), it is the sort of theory he defends (Tooley 1997) in the face of a similar challenge from relativity.

An obvious objection to this view is that it cannot be experimentally determined which inertial frames are at absolute rest. This point is often coupled with a verificationist principle of some kind to conclude that there is not as much spacetime structure as the neo-Lorentzian postulates.

I think the surface presentist is free to reject the verificationist principle and the conclusion following from it. On this matter, Prior writes,

> I'm sure there *are* questions which are perfectly genuine and intelligible questions but which seem to be incapable of being answered. For instance, I know perfectly well what it would be for you to see what I would call purple wherever I see red, and for you to see what I would call blue wherever I see purple, and so on round the clock; but I cannot imagine any procedure which would conclusively show that our respective visual experiences are, or that they are not, related in this way (1996, 50).

The surface presentist is still left, however, with the initial objection that no experiment can determine which frames are at absolute rest. Prior seems willing to live with this consequence. But Tooley (1997) and Timothy Maudlin (1997), who is also not a presentist, have each proposed that there might be experimentally determined facts which would reveal the frames at absolute rest. Both Tooley and Maudlin mention the apparently instantaneous collapse of the wave packet in quantum mechanics as experimental grounds for positing a relation of absolute simultaneity. Some presentists also mention facts about the cosmic background microwave radiation as providing possible grounds for determining which frames are at absolute rest. I cannot do full justice to these claims here. I only mention them to indicate that the surface presentist perhaps can determine which frames are at absolute rest.

7. Conclusion

In summary, then, I think the presentist can meet the challenges posed by the special theory of relativity along the lines of either the cone or the surface models. Though we have seen that these responses are not without their problems, I think we have also seen there are no good reasons arising from the special theory for rejecting presentism. Presentism is tenable here and preferable elsewhere.

Notes

1. I owe this phrase to John Bigelow, who has expressed a related thought in these terms.

References

Callender, C. "Shedding Light on Time," (paper presented at the Symposium on The Prospects for Presentism in Spacetime Theories, Philosophy of Science Association Meeting, Kansas City, Missouri, October 24, 1998).

Einstein, A. "On the Electrodynamics of Moving Bodies." Pp. 35-65 in *The Principle of Relativity*, edited by A. Sommerfeld. Toronto: Methuen and Company, Ltd., 1923. Originally published as "Zur Elektrodynamik bewegter Körper," *Annalen der Physik* 17 (1905): 891-921.

Godfrey-Smith, W. "Special Relativity and the Present," *Philosophical Studies* 36 (1979): 233-44.

Hinchliff, M. "The Puzzle of Change." Pp. 119-36 in *Philosophical Perspectives* 10, edited by J. Tomberlin. Cambridge, Mass.: Blackwell Publishers, 1996.

Maudlin, T. *Quantum Non-Locality and Relativity: Metaphysical Intimations of Modern Physics*. Oxford: Blackwell, 1997.

Minkowski, H. "Space and Time." Pp. 75-91 in *The Principle of Relativity*, edited by A. Sommerfeld. London: Metheun and Company Ltd., 1923. Originally presented as an address, Eightieth Assembly of German Natural Scientists and Physicians. Cologne, Germany, September 21, 1908.

Prior, A. "Tense Logic and the Logic of Earlier and Later." Pp. 116-34 in *Papers on Time and Tense*, by Arthur Prior. Oxford: Oxford University Press, 1968.

———. "The Notion of the Present," *Studium Generale* 33 (1970): 245-48.

———. "Some Free Thinking about Time." Pp. 47-51 in *Logic and Reality: Essays on the Legacy of Arthur Prior*, edited by B. J. Copeland. Oxford: Oxford University Press, 1996.

Putnam, H. "Time and Physical Geometry," *Journal of Philosophy* 64 (1967): 240-47. Chapter 13 in this volume.

Savitt, S. "There Is No Time Like the Present (in Minkowski Spacetime)" (paper presented at the Symposium on The Prospects for Presentism in Spacetime Theories, Philosophy of Science Association Meeting, Kansas City, Missouri, October 24, 1998).

Sklar, L. "Time, Reality and Relativity." Pp. 129-42 in *Reduction, Time and Reality*, edited by R. Healy. New York: Cambridge University Press, 1981.

Stein, H. "On Einstein-Minkowski Space-Time," *Journal of Philosophy* 65 (1968): 5-23.

Tooley, M. *Time, Tense, and Causation*. Oxford: Oxford University Press, 1997.

Section 3

Presentism and Truthmaking

Introduction: Presentism and Truthmaking

L. Nathan Oaklander

Throughout the selections there has been a recurring theme. It can be stated simply as follows: If only the present exists, then what makes propositions about the past and the future true? Alternatively, what is the ontological ground of past- and future-tensed truths such as the Japanese attacked Pearl Harbor on December 7, 1941, and Barack Obama's first term as President of the United States will end on January 20, 2013? Simon Keller (chapter 15) refers to this as the "truthmaking problem" and aims to show that the presentist can answer it only if she endorses some very controversial metaphysical views. Thomas Crisp (chapter 16) also addresses this problem for presentism which he calls "the grounding objection," but offers a rather different response to it. In this section I shall first explain what they take the truthmaking problem to be, why they reject certain responses to it, and how they attempt to solve it. I shall conclude with some critical remarks on their attempts.

Keller explains the truthmaker problem as the difficulty of reconciling three claims, one of which is presentism. According to the first claim (that he calls "Truthmaker"), "*What is true depends upon what exists*" (p. 260). This is not the stronger thesis that for every true proposition there is something that exists that makes it true, but the weaker thesis that can be expressed by saying that *truth supervenes on being*. In other words, if some true statements were false, or some false statements true, then there would have to be a difference in what things exist, or what fundamental properties and relations they instantiate.

Crisp expresses Truthmaker, or what he calls the "supervenience principle," in terms of possible worlds as follows:

> (SP) For any proposition p and worlds w and w^*, if p is true in w and not in w^*, then (a) according to w, something exists that doesn't exist in w^* (or vice versa), or (b) according to w, some objects instantiate a fundamental property or relation that they don't instantiate in w^* (or vice versa). (p. 278)

The second claim is that "*There are past- and future-tensed truths*" (p. 261). Such truths as, for example, "Anne Boleyn was executed," seem to imply the existence of Anne Boleyn, who, according to the presentist, does not exist.

The third claim is presentism: "*Only present things exist*" (p. 261). The problem for presentism is that it seems as though there is nothing that exists at present that could account for past- and future-tensed truths. That is, given presentism, things could have been just as they are at present and the past have been different. Thus, "it looks . . . as though presently existing things and their properties are not enough to ground past- and future-tensed truths" (p. 262).

Keller considers three responses to the truthmaking problem. The first is to deny that any past- or future-tensed statements are true, which he rejects out of hand. Keller argues that although it does make sense to deny that future-tense propositions are true or false, "The corresponding claim about past-tensed truths, however, is highly implausible" (p. 262).

A second response is to deny Truthmaker. There are two versions of this gambit and Keller rejects both of them. According to the first version, which Keller calls "Meinongian presentism," there are past and future things, but they do not exist. Hence, the Meinongian presentist denies that past- and future-tense truths are grounded in terms of what exists, but maintains, nevertheless, that past- and future-tense truth have an ontological ground. There are entities that exist and entities that do not exist and while present-tense truths supervene upon *existing* entities, past- and future-tense truths supervene upon *non-existent* ones. Keller rejects this view since it only nominally, but not substantially differentiates presentism from four-dimensionalism, which is his label for the view that past, present, and future things all exist. As Keller puts it,

> Either we restrict our use of "exists" to present things, adopt an ontology of non-existent objects, and call ourselves presentists, or we accept that all these things exist and call ourselves four-dimensionalists. [I]t's just a battle of linguistic convenience . . . and when it comes to metaphysics, it doesn't really matter. (p. 263-64)

The second way of denying Truthmaker is to claim that truth does not supervene upon anything. "Anne Boleyn was executed," on this view, is just plain true, and we don't need anything existing or non-existing to underlie its truth. Whatever initial plausibility this claim has regarding past- and future-tensed truths it is, Keller argues, "entirely implausible to say that present-tensed truth does not supervene upon something, and once this is conceded it becomes difficult to produce a respectable reason why past- and future-tensed truth should not supervene upon something as well" (p. 264).

Having argued for the plausibility of the first two claims of the truthmaker problem Keller sets his sights on the third, the thesis of presentism itself. He argues that what he calls Truthmaker-preserving presentism can be reconciled with Truthmaker by denying that we need non-present things to ground past and future tensed truths. On both versions of Truthmaker-preserving presentism, it is not true that the present could not be just as it is and the past (or future) be different from what it was (or will be).

Keller's versions of Truthmaker-preserving presentism takes its cue from a view put forth by John Bigelow in chapter 8. Recall that Bigelow attempts to ground the truths about past and future individuals by means of past and future tensed *properties* that *are presently exemplified*. Of course, these properties cannot be exemplified by past and future individuals since there are none, so there must be some *presently existing things* that exemplify them. Moreover, these properties must be fundamental or primitive in the sense that they are not analyzable in terms of the B-theoretic relations between individuals that exist at times earlier or later than the present time. Keller notes that Bigelow's theory would have to answer questions such as how would he explain the times in the past at which Anne Boleyn was executed and to which present individuals are these past- and future-tensed properties attached? Keller acknowledges the fantastical and lavish outlay of properties or entities that any Truthmaking-preserving presentism must countenance. Nevertheless, Keller develops two versions of Truthmaker-preserving presentism that are analogous to Bigelow's, that can answer the questions he raises, and at the end of his essay he suggests that the price of lavishness may be worth paying.

The first version he calls "haecceitist presentism" (p. 267). A *haecceity* is an individual's *uninstantiated thisness*; its property of being just that individual. This is a primitive uninstantiated property that the haecceitist presentist can use to account for the truth of *Anne Boleyn was executed in 1536*. Not only do persons have haecceities, but to make sure that the execution occurred in 1536 the presentist might treat times as individuals that have haecceities too. There are quasi-properties that mirror properties that are instantiated by present persons, only they are instantiated or exemplified by the haecceity of some individual. Thus, there is the haecceity of Anne Boleyn, the various quasi-properties and the property of being the time at around midday on May 19, 1536. These various abstract objects stand in certain relations to each other so as to provide a present ground for past- and future-tense truths.

In other words, when it is true that Anne Boleyn was executed there presently exists an individual's haecceity that stands in certain relations to the properties of being executed and the haecceity of the time at which she is executed, and that individual haecceity when it was instantiated was executed. It seems to me, however, that such an account is circular. The ground of truths about past and future individual is truths about haecceities that were or will be instantiated. What then is the basis of the past or future tense truths that a haecceity was or will be instantiated? Clearly, it cannot be that the haecceity of time involved is past or future, or earlier than the present time. (Note that at each moment there exist haecceities of every other moment.) Keller appeals to relations between these moments to ground their order, but these relations can hardly be said to mirror temporal relations since they can be none other than simultaneity relations.

Keller calls his second alternative "atomic presentism" (p. 269). Suppose the world is made up of a number of eternal (everlasting particles) and all the contingent truths about the world supervene upon the way in which those particles are presently arranged and the fundamental properties they presently instantiate. These particles carry their pasts and futures around with them, which is to say that each particle bears fundamental past and future tensed properties that give away the facts about where they have come from and where they are going. These are brute facts. Terms like "was" and "will be," then, are taken as primitively understood. Thus, Truthmaker is preserved since it is not possible that the world could be as it presently is, and the past and future be different since what did and will happen is encoded in the past and future properties of presently existing atoms.

These two views come with a price: A lavish ontology of *haecceities* on the one hand, and a lavish ontology of primitive tensed properties, on the other. Keller concludes his essay by noting that these unattractive ontological commitments may be worth the price.

Thomas Crisp, who is also attempting to respond to what he calls "the grounding problem," does not think that the presentist has to accept fundamental tensed properties like *being past* or *being future*. He is sensitive to the objection that they are not primitive enough or as Theodore Sider (2001) puts it, they are hypothetical and not categorical properties. For that reason, Crisp prefers not to countenance A-properties, and so proceeds to analyze them.

Recall, the problem for presentism is to reconcile the claim that *truth supervenes on being*, "where the idea here is that truth supervenes on what things there are and which fundamental properties and relations they instantiate" (p. 278) with the presentist view that past- and future-tensed truths are founded on what presently exists. Crisp, like Keller, is sympathetic with the version of presentism endorsed by Bigelow, which Crisp calls "Lucretianism," "the view that the world has this and many like past- and future-tensed properties: properties like *being a place where dinosaurs roamed,* and *having been inhabited by wooly mammoths*" (p. 280). Crisp proceeds to engage in a discussion regarding whether or not such properties are reputable. He concludes his discussion with the claim that objections to such exotic properties can be dismissed since "Properties like *being formerly roamed by dinosaurs* can be analyzed in terms of less exotic properties like *being a dinosaur* and *being past*" (p. 284).

Crisp is somewhat ambivalent regarding the tensed properties of *pastness* and *futurity*, since he questions whether or not they are really fundamental. Of course, even if they are fundamental one would think that a presentist would have no truck with them since there are at the root of McTaggart's paradox. Indeed previous presentists were at pains to reject them (see chapters 3 and 4). Thus, to overcome any problems with treating such properties as *being formerly roamed by dinosaurs* or *being past* as primitive or fundamental, Crisp argues that the work done by such properties can be done by treating temporal relations like the *earlier than* as fundamental and analyzing A-properties in terms of such relations.

We can begin our summary of the main elements of Crisp's response to the grounding objection by noting that he reduces monadic A-properties *pastness* and *futurity* to primitive B-relations or, more accurately, primitive *ersatz* B-relations between "the present time" and earlier and later times, and the property *being present* to the property *being true*. On the face of it, this appears to be more sympathetic with the B-theoretic view of time than presentism since B-theorists also maintain that tensed properties are ontologically reducible to B-relations. There are, however, fundamental differences as we shall see.

Temporal relations obtain between times although it is not the B-theoretic temporal relation, it is an analogue of a B-relation; it is an ersatz B-relation. Crisp eliminates B-relations between concrete objects, which relations according to his ontology do not exist, and replaces them with primitive, *ersatz* B-relations between times that do exist. An ersatz B-relation is "a primitive, unanalyzable relation on abstract times that is transitive, irreflexive, and asymmetrical and orders the times (better: some of the times) into an "ersatz" B-series representing the history of the world" (2005: 3).

Calling a relation between abstract propositions a "temporal" or *earlier than* relation is obfuscating. Admittedly, the relation Crisp defines purports to have the same logical properties as a B-theoretic temporal relation, but so does *greater than* and no one would claim that *greater than* is a temporal relation. Crisp claims that his ersatz B-relation is "the relation we learned at mother's knee when we learned that last Tuesday is earlier than today" (p. 288). I disagree. What we learned when we learned what "earlier than" represents is grounded in our experience of one event occurring before another, and not by contemplating a relation between abstract objects. In fact, his analysis of presentism even makes use of a B-theoretic notion of the *earlier than* relation for he says, "When the presentist says that everything is present, what she really means is that there is no *temporal distance* between any two things in our widest domain of quantification. (What is temporal distance? The general idea should be clear enough. *Suppose eternalism is true*. Then the temporal distance between Lincoln's assassination and now is a little over 139 years" (p. 289; some italics added). Since the notion of temporal distance incorporates a B-theoretic *earlier than* relation, I still do not see how he has made sense of a *temporal* relation between atemporal objects or "times."

Be that as it may, regarding "times" he says:

> A time, let us say, is any proposition that satisfies the following schema:
>
> > x is a time $=_{df.}$ For some class C of propositions such that C is maximal and consistent, x = [∀y (y∈C ⊃ y is true)],
>
> where (i) a class C of propositions is maximal iff, for every proposition p, either p or its denial is a member of C, (ii) a class C of propositions is consistent iff, possibly, every member of C is true, and (iii), "[∀y (y∈C ⊃ y is true)]," I assume, denotes a tenseless proposition. . . ." (p. 286)

Crisp later says about these abstract times that:

> I think the presentist should hold that there is one and only one ersatz B-series . . . but that it does not include *all* the abstract times among its members. It counts

among its members only *some* of the abstract times—those that *did, do*, or *will* represent the world. (p. 290).

Thus, Crisp rejects the A-properties *pastness* and *futurity*, and claims that they are eliminable in terms of (tenseless) ersatz B-relations which obtain between and among times construed as abstract objects, i.e., maximally consistent sets of (tenseless) propositions: "intuitively, an abstract representation of an instantaneous state of the world" (p. 286). And he rejects *presentness* and analyzes it in terms of the non-temporal property *being true*.

Crisp summarizes his analysis of the tenses in the following passage:

> We can now see how the presentist can do without primitive pastness, presentness, and futurity. She need simply take the foregoing *earlier than* relation as primitive, and say that a time is *past* iff df. it is earlier than the present time, that a time is *future* iff df. it is later than the present time, and that the present time is whatever time happens to be *true*. In general, she'll say that a proposition *p* is past (present, future) iff *p* is *included* in a past (present, future) time, where *p* is included in *t* iff *t* entails *p*. (p. 291)

On this picture, Bigelow's property *being an x such that dinosaurs roamed x* turns out to be something more like: *being an x such that the proposition that dinosaurs roam x is included in an earlier time*.

Thus, Crisp avoids the grounding objection by accepting Truthmaker or what he calls the supervenience principle (SB), but rejects the claim that there is nothing about the present time that grounds past- and future-tensed truths. The claim that *dinosaurs roamed the earth* expresses a past-tensed truth that is analyzed in terms of there being an ersatz B-series in which the present time represents what is true, that is, what exists, and the time (abstract representation of an instantaneous state of the world) at which the proposition that dinosaurs roam the earth is true is earlier than the present time. For that reason, the presentist can reject the claim that things could have been just as they are at present and the past have been different. Thus, only the present time is the true time and propositions about the past or the future are grounded in those times which are earlier or later than the present time. Whether Crisp's reductive analysis of the tenses and his reply to the grounding problem succeed are questions that we will leave for the reader to decide.

References and Further Readings

Cameron, R. "Truthmaking for Presentists." In *Oxford Studies in Metaphyiscs,* volume 6, edited by D. W. Zimmerman and K. Bennett. Oxford: Oxford University Press, forthcoming.

Crisp, T. "Review L. Nathan Oaklander, *The Ontology of Time*," *Notre Dame Philosophical Reviews*, http://ndpr.nd.edu/review.cfm?id=2201 (2005).

———. "Presentism and the Grounding Objection," *Noûs* 41 (2007): 90-109. Chapter 16 in this volume.

Keller, S. "Presentism and Truthmaking." Pp. 83-106 in *Oxford Studies in Metaphysics*, volume 1, edited by D. Zimmerman. Oxford: Oxford University Press, 2004. Chapter 15 in this volume.

Kierland, B., and B. Monton. "Presentism and the Objection from Being Supervenience," *Australasian Journal of Philosophy* 85 (2007): 485-97.

Lewis, D. "Truthmaking and Different-Making," *Noûs* 35 (2001): 602-15.

Mulligan, K., P. Simons, and B. Smith. "Truth-Makers," *Philosophy and Phenomenological Research* 44 (1984): 287-321.

Rhoda, A. "Presentism, Truthmakers, and God," *Pacific Philosophical Quarterly* 90 (2009): 41–62.

Sider, T. *Four-dimensionalism*. Oxford: Oxford University Press, 2001.

Tallant, J. "Ontological Cheats Might Just Prosper," *Analysis* 69 (2009a): 422-30.

———. "Truth-makers and Presentism," *Erkenntnis* 71 (2009b): 407-16.

15
Presentism and Truthmaking

Simon Keller

1. A problem for presentism

Saint Augustine's boyhood teachers told him that there are three types of time: past, present, and future. Are we then to say, Augustine asks in his *Confessions*, that the past and future exist? At first, he thinks that they do. It seems, he says, that "when the present emerges from the future, time comes out of some secret store, and then recedes into some secret place when the past comes out of the present." If the past and future are in some secret place then they must *exist* in that secret place. Further, Augustine says, we can remember the past and predict the future, and "to see that which has no existence is impossible."

Augustine soon changes his mind. Supposing, he says, that past and future things exist, they must, wherever they are, be present. For if they were past in some secret place, they would be there no longer, and if they were future in some secret place, they would not be there yet. The notion of a secret place for past and future things, Augustine concludes, is not very helpful. When we remember the past and predict the future, he now says, we are really making observations about the present. When we remember a past event, we are looking on its image in the present, and when we predict a future event, our prediction is based upon things that are already present and can be seen. The only things that really exist, he concludes, are present things.[1]

Once Augustine has dispensed with the past and future, it is difficult to see how a memory or a prediction could ever be *right*. What happens when a past event has differing images in the present—when people have different memories of the same event? Is it possible to have a false memory? What is there to make it false? Could there be past events that no one remembers? Augustine seems to think that past- and future-tensed statements can have truth-values that are independent of what is going on in the present; he thinks, for example, that the prophets have a special ability to make *correct* predictions. I think that Augustine never quite gets rid of the idea that the past and future are, in some sense, real, and I think that this problem can also be found among some of his philosophical descendants.

These days, those who agree with Augustine's conclusions about time call themselves *presentists*. Presentism is the belief that only present things exist. If something doesn't exist now, says the presentist, then it doesn't exist at all.[2] The most popular alternative to presentism says that past, present and future things all exist; the universe, on this view, is a four-dimensional space-time manifold, containing everything that has happened, everything that is happening, and everything that will happen. This is a view that I will call *four-dimensionalism*.[3] (Note that this label is sometimes applied to a different view in the philosophy of time, namely, the view that objects have temporal parts—that they persist through time by perduring, rather than enduring.[4] This is not the view that I have in mind.) Four-dimensionalists believe that there are places at which past and future things exist, but that those places are not so secret. Past things exist in the past, and future things exist in the future. Saint Augustine exists in fourth-century Italy.[5]

In this paper, I will describe an ancient and powerful problem for presentism, and consider the ways in which the presentist might respond.[6] The problem is to say how there can be past- and future-tensed truths even though the past and future do not exist. I aim to show that the presentist can avoid the problem only if she endorses some very controversial metaphysical views. While the truthmaking problem (as I will call it) does not stand as a refutation of presentism, it does show that presentism comes at a price. Whether or not presentism is plausible depends upon whether or not that price is worth paying.

The truthmaking problem is best presented as a set of three claims. Two of them seem to be true, and the third is presentism. The problem for presentism is that it is difficult to see how the three claims could all be true together. To solve the problem, the presentist must either deny at least one of the first two claims, or explain how their truth does not force us to deny the third.

First claim: *What is true depends upon what exists*. If some true statement were false, or if some false statement were true, then there would have to be a difference in what things exist, or in what the existing things are like. I will call this claim *Truthmaker*.[7]

It is important to distinguish Truthmaker from the stronger thesis that for every truth there exists something that makes it true.[8] This stronger thesis would serve just as well in an argument against presentism, but it is far more controversial. There may be some truths—negative existential truths, for example—for which it is not the case that there is some existing thing that makes the truth true.[9] But negative existentials do depend upon what exists in the way that the weaker Truthmaker thesis demands. There would have to be a difference in what exists in order for the proposition *Unicorns do not exist* to be false; unicorns would have to exist.

Other truths, however, cause trouble even for the weaker thesis. The obvious examples are truths about fictional entities—how would the existing world have to differ in order for *Sherlock Holmes lives on Baker Street* to be false?—and we could perhaps add mathematical, modal, and moral truths (although the truthmaking problem for necessary truths is not so easy to formulate). Nevertheless, Truthmaker is very plausible. It expresses an attractive philosophical picture according to which reality is made up of all and only the things that exist, and contingent propositions derive their truth or falsity from the relations in which they stand to the existing things.

(The nature of these relations, it must be admitted, is difficult to pin down, and it may be a little misleading to speak of "truth*making*"—as though the world is populated by a team of truthmakers working hard to cause propositions to change their truth-values. Really, the relations in question are explanatory; the truth of a proposition is somehow explained by a story about things that exist. One way to bring this out is to note that discharging the truthmaking burden with regard to some true proposition often involves pointing to an analytic connection between the proposition and some claim about the existing world. The following, for example, is surely analytic: if there is some existing thing that is identical to John and another that is identical to Mary, and if the first is such as to love the second, then the proposition *John loves Mary* is true. It would be nice to think that finding truthmakers for a proposition *always* involves finding such analytic connections, but there are reasons to be cautious here. First, negative existentials; *Unicorns do not exist* is not analytically implied by any claim about what *does* exist. Second, it might be wise to leave room for the possibility that the connection between a proposition and its truthmakers is sometimes synthetic, although still a priori. In any case, Truthmaker should be regarded as a thesis about a certain mode of explanation, a thesis about what is involved in a proposition's being true.)

For many of us, it is hard to imagine that Truthmaker could be incorrect, regardless of how many headaches it causes. Sometimes, it is difficult to explain how certain propositions that we normally regard as true could be made true by what we normally take to constitute the entirety of what exists. When metaphysicians notice such difficulties, however, they do not tend to reject Truthmaker. They are more likely to try to explain how the things that we normally take to exist really are enough to explain why the propositions are true, or to suggest that there are more existing things than we would otherwise have thought, or to argue that the apparent truths are not true after all.[10] Still, there are those who reject Truthmaker, and we will ask later whether this is an option that the presentist should take.

Second claim: *There are past- and future-tensed truths*. More precisely, there are past- and future-tensed truths that seem offhand to mention entities whose existence the presentist denies. It is true, for example, that Anne Boleyn was executed.

Third claim: *Only present things exist*.

I said that the first two claims, when taken together, pose a problem for presentism. But they certainly do not *imply* its falsity. After all, an exactly analogous set of claims could be constructed in order to pose a problem for the view that fictional characters do not exist—we just need Truthmaker, plus the truth that Sherlock Homes lives on Baker Street. The way to respond in that case would be to say that *Sherlock Holmes lives on Baker Street* is not strictly speaking a truth about Sherlock Holmes, who doesn't exist. Really, it's a truth about certain existing stories, or about existing thoughts in the mind of Arthur Conan Doyle, or something like that, and it's because of this that we can say that the truth of *Sherlock Holmes lives on Baker Street* supervenes upon what exists. It's not clear how the truthmaking story would go, but most of us feel sure that it can be accomplished without abandoning Truthmaker and without positing an existing Sherlock Holmes.[11]

Truths about fictional entities seem to depend upon what exists in the non-fictional world, but past- and future-tensed truths do not seem to depend upon existence in the present. It seems as though the present could be just as it is, without the past- and future-tensed truths being just as they are. Most would agree, for example, that there is nothing about the present world, considered in itself, that rules out the possibility that the world sprung into existence five minutes ago, or that someone combed his hair with his left rather than his right hand on a particular morning last century. So the reason why there is a problem for presentism is that it looks at first as though presently existing things and their properties are not enough to ground past- and future-tensed truths.

Three broad responses are available to the presentist. One is to deny that any past- or future-tensed statements are true. With regard to future-tensed statements, the claim in question might not be so crazy; Aristotle suggests that statements about the future are neither true nor false, because the events that would ground their truth or falsity are yet to come into existence.[12] The corresponding claim about past-tensed truths, however, is highly implausible—Anne Boleyn *was* executed[13]—and I cannot see how the presentist could make it attractive.[14]

A second option is to deny Truthmaker, either by denying that the things upon which past- and future-tensed truth supervenes *exist*, or by denying that past- and future-tensed truth supervenes upon anything. This is an initially tempting strategy, but I will explain why I don't think that either form of Truthmaker-denying presentism (as I shall call it) is acceptable.

Finally, the presentist can be a Truthmaker-preserving presentist, accepting Truthmaker while denying that we need non-present things to ground past- and future-tensed truths. This form of presentism is the most promising, and I will explore two ways in which it might be developed.

2. Truthmaker-denying presentism

2.1 "There are things that don't exist!"

There are, as I say, two ways in which the presentist might try to deny Truthmaker. She can deny Truthmaker either by saying that the things upon which past- and future-tensed truth supervene do not exist, or by saying that past- and future-tensed truth does not supervene upon anything. I'll begin by considering the first of these views, a view that I will call Meinongian presentism.[15] It is an odd view, and it probably represents the less obvious form of Truthmaker-denying presentism, but it is a good place to start.

The Meinongian presentist maintains her presentism by making a very simple move. She accepts that there are non-present things upon which truth supervenes, but denies that they exist. She has no trouble explaining how there can be past- and future-tensed truths; she has all the truthmaking resources of the four-dimensionalist. There are past things, says the Meinongian presentist, but they no longer exist, and there are future things, but they do not exist yet. *Anne Boleyn was executed* is true, says the Meinongian, because Anne Boleyn has the property of having been executed, even though she does not have the property of existing.

There are, however, good reasons to reject a view that relies upon non-existent entities. First, we might find it hard to make sense of the idea of objects that do not exist, but nevertheless *are*.[16] Second, and this is the objection that I'll take up first, we might worry that once we have admitted non-existent objects into our ontology, existence—and presentism—cease to be very interesting.

What do the Meinongian presentist and the four-dimensionalist disagree about? They agree that Anne Boleyn was executed, and that this is made true by Anne Boleyn, a sword, a swordsman especially brought over from France, and the relations in which those things stand. They agree that Anne stands in a certain relation to the Tower of London, and that this is the same Tower that exists now. Their theories, in fact, are exactly the same, except for the four-dimensionalist's insistence that all those past and future things *exist*.

This general objection can be put another way.[17] If the only difference between what exists and what does not is in a label, then why should we think that *we* are present and existing, while Anne Boleyn is non-present and non-existing? Anne, like us, has all sorts of properties and stands in all sorts of relations to the things around her, and she, like us, thinks that she exists. Who's to say that she is wrong and we are right? And why should we care? If Anne Boleyn were present instead of us, then I would be doing whatever I am doing, and you would be reading this paper. We wouldn't exist, but so what?

At this point, the Meinongian presentist might complain that I am willfully refusing to understand the significance of existence. I say that there is nothing to choose between two theories that agree on what there is while disagreeing about which of those things exist. The Meinongian presentist replies that we all know what existence means, and we all care about what exists and what doesn't, and that's enough to show that the two theories are interestingly different.

Do we have a robust pre-theoretical conception of what it means for something to exist? I think that we do. My own robust conception of existence, however, cannot cope with the notion that there is an X—there *really is* an X—and it has properties—it *really has* properties—but it doesn't exist. Once the presentist commits herself to such things, she loses her right to call upon our unanalyzed conception of existence. That is not to say that philosophers are not allowed to posit non-existent entities, but it is to say that when they do, they owe us an explanation of what, apart from a label, sets the existent apart from the non-existent.

There is no real difference between Meinongian presentism and four-dimensionalism. Either we restrict our use of "exists" to present things, adopt an ontology of non-existent objects, and call ourselves presentists, or we accept that all these things exist and call ourselves four-dimensionalists. If it's just a battle of linguistic convenience, then the one who posits non-existent objects is going to have a lot of trouble persuading us to adopt her way of speaking. And when it comes to metaphysics, it doesn't really matter.

2.2 "Truth does not supervene upon anything!"

A better way to deny Truthmaker, perhaps, is to say that the truths that are purported to create trouble for presentism are not made true by anything. *Anne Boleyn was executed*, on this view, is just plain true, and we don't need anything, existing or not, to underlie its truth. We might motivate such a move by pointing to all the other truths for which the truthmaking problem arises. If it's so difficult to say what underlies truths about fictional objects and the rest, we might say, then that's a reason to think that the truthmaking impulse should be resisted.[18] This refusal to worry about truthmaking is not unprincipled, but I do not think that it can be sustained. The problem, put briefly, is that it is entirely implausible to say that present-tensed truth does not supervene upon something, and once this is conceded it becomes difficult to produce a respectable reason why past- and future-tensed truth should not supervene upon something as well. That's what the following comments are intended to show.

Start with a present-tense truth: *The Tower of London is on the Thames*. This is a truth about the Tower of London and the Thames, two things that exist and are next to each other. If either the Tower or the Thames did not exist, or if they were not next to each other, then *The Tower is on the Thames* would be false. So there's one proposition whose truth depends upon certain things' existing and exhibiting certain characteristics. Now consider a past-tensed truth about those same things: *The Tower of London* was *on the Thames*. That too is a truth about the existing Tower and the existing Thames, and it's true because those very things *were* next to each other; if either the Tower or the Thames had never existed, or if they had never been next to each other, then *The Tower was on the Thames* would be false.

Next, *Anne Boleyn spent time at the Tower*. Here we have another truth about the existing Tower, and another truth that seems to report a relation that holds between the Tower and something else—not an existing Anne, perhaps, but definitely *something*. And if the Tower had never existed, or if the Anne-like thing was not such as to make it the case that Anne spent time at the Tower, then *Anne Boleyn spent time at the Tower* would not be true. But now we have an Anne-like thing, and surely that is the very thing about which *Anne Boleyn was executed* is a truth. If the Anne-like thing were different in certain respects, then it would not be true that Anne was executed. So the truth of *Anne Boleyn was executed* does, after all, depend upon a certain thing's being a certain way.

Can the presentist depart from this line of thought? There are two points at which it seems vulnerable. The first is in the claim that if *The Tower* is *on the*

Thames is made true by two existing things and the relation in which they stand, then the same must be the case for *The Tower* was *on the Thames*. The presentist might say that *The Tower was on the Thames* is true because there *were* two things that stood in a certain relation. It happens that both of those things still exist, but that is not at all to say that we need presently existing things to be a certain way in order for *The Tower was on the Thames* to be true.

This response, however, entails implausible claims about the nature of propositions. *The Tower was on the Thames*, the presentist agrees, is a relational truth, in that it posits a relation that either does or did exist. Normally, we would think that a proposition is a relational proposition because it has constituents that stand, or are claimed by the proposition to stand, in some relation. Our Truthmaker-denying presentist, however, cannot endorse this view. To do so would be to accept that *The Tower was on the Thames* has constituents and that a certain relation holds between those constituents, and that's to say that the proposition's truth does after all depend upon there being things that stand in a certain relation. (It's not enough to say that the proposition's constituents *did* exist and *did* stand in a certain relation, because that would only show that the proposition *did* exist. *The Tower was on the Thames* is true *now*.) And if the presentist rejects the idea that propositions have constituents, then how is she to account for the fact that *The Tower was on the Thames* is a relational truth but *The Tower was a tower* is not?[19]

Another way in which the presentist might resist the argument is to deny the claim that if *The Tower was on the Thames* reports a relation between two things then *Anne spent time at the Tower* must do so as well. The presentist, as someone who doesn't believe that Anne exists, perhaps has a reason for saying that any truth that seems to say something about Anne cannot really be saying something about anything, and hence that there is nothing that has to be a certain way in order for propositions that seem to mention Anne to be true. To make this claim, however, is to say that the question of how a sentence should be analyzed sometimes depends upon contingent questions about what exists. Suppose that we are unsure about whether or not the Tower was destroyed in last night's earthquake. If the view in question is correct, then we should also be unsure about whether or not *The Tower was on the Thames* is a relational truth (although we could be sure that the statement is true). Whether *The Tower was on the Thames* reports a relation between two things, on this view, depends upon whether the Tower still exists. And that just seems absurd.

Perhaps the presentist should respond to such arguments by positing presently existing surrogates for non-present entities. *The Tower was on the Thames* can be a proposition that has constituents, runs the thought, without those constituents having to be the Thames and the Tower. And the reason why the proposition is a relational proposition, we could add, is that it posits a relation between two present surrogates for the Tower and the Thames. This strategy is best discussed in the next section, on Truthmaker-preserving presentism.

3. Truthmaker-preserving presentism

3.1 Tensed properties of present things

Truthmaker-preserving presentism accepts Truthmaker while saying that past- and future-tensed truth does not supervene upon non-present being. Earlier, I said that the truthmaking problem is a special problem for presentism because truths about the past and future look at first glance to be underdetermined by truths about the present. The present world could be just as it is, it seems, even if the past- and future-tensed truths were very different. But perhaps the presentist can disagree. Perhaps she can say that the present world does the truthmaking for past- and future-tensed truths.

A version of Truthmaker-preserving presentism is sketched by John Bigelow. The present world, Bigelow says, has many, many past- and future-tensed properties, and these properties are what underlie past- and future-tensed truth. Where four-dimensionalists accept the existence of every past, present, and future individual, Bigelow writes, "presentism trades off this ontological expenditure against a lavish outlay on properties of present individuals."[20] A similar sort of proposal is presented by Roderick M. Chisholm, who speaks not of tensed world properties, but of the past- and future-oriented properties of eternal entities (where an example of an eternal entity is the property of redness).[21]

If the presentist is to claim that tensed properties of present things do the truthmaking for past- and future-tensed truths, then she must see present things as having *fundamental* tensed properties, by which I mean properties that are not to be explained by reference to anything else (like other individuals or other, more fundamental, properties). After all, the four-dimensionalist will agree that the present world has the property of being a world in which Anne Boleyn was executed, but will say that that property is grounded in the properties of the world as it is at another time, namely, the time at which Anne Boleyn is being executed. What is distinctive about the presentist view is the claim that the tensed properties of present things are not to be explained by way of the properties of past or future individuals. Only the presentist says that the present world could not be just as it is and yet have a different past or future.[22]

The commitment to past- and future-tensed properties, as Bigelow says, is lavish, and it looks especially lavish when we consider those fundamental properties, if there are any, that have been or will be instantiated but are not instantiated presently.[23] Suppose that there is some property P that is not instantiated by anything now, but will be instantiated by some future individual. Our presentist needs the future-tensed version of P, so she must say either that past- and future-tensed properties can exist even when their present-tensed correlates do not, or that properties like P can exist even when uninstantiated.

Further, the presentist must take care of the fact that non-present instantiations of properties are not just in the past or future, but are at particular moments in the past or future. It's not enough to say that this world is one in which Anne Boleyn was executed; we have to be able to say that it's a world in which Anne was executed at around midday on May 19, 1536. One way for the presentist to deal with such questions is to commit herself to an even greater abundance of properties, properties that encode facts about the times at which relevant events did or will occur; then, the property of having been Q at midday last Thursday is one fundamental property, and the property of having been Q at 1 pm last Thursday is another. Depending on her views about certain other matters, the presentist may have other options available. She might think that she can get the facts about the times at which properties are instantiated just by positing further facts about which properties are instantiated before and after which others.[24] Or, she might posit presently existing surrogates for non-present moments, so that a relation can hold between a present individual, a past- or future-tensed property, and a surrogate for a past or future time.

In any case, in positing her past- and future-tensed properties the presentist is taking on commitments that some will find unpalatable. But another important question remains: to which present individuals are these properties attached?

Past- and future-tensed truths, as we saw earlier, have structure, and it is hence not enough for the Truthmaker-preserving presentist to simply posit an abundance of past- and future-tensed world properties. The property of being a world in which Anne was executed, for example, cannot be just a bare property of the world, but must involve the attribution of a property-like thing to an Anne-like thing.[25] I want to consider two stories that the presentist might tell here. I am sure that they are not the only stories that could be told, but their telling will make clear the moral for Truthmaker-preserving presentism.

3.2 Haecceitist presentism

Let me explain how I will use some terms. An individual's *thisness* is its property of being just that individual. Your thisness is a property of you, and of nothing else. The nature of thisnesses is controversial, but one view is that thisnesses are primitive—that they are elements of reality that cannot be reduced to anything more fundamental.[26] Can an individual's thisness exist without the existence of the individual itself? To put it another way, are there uninstantiated thisnesses? If there are, then thisnesses are *haecceities*.[27]

If there are haecceities, then the presentist can suppose that although there exist no non-present individuals, there exist the haecceities of non-present individuals. And if thisnesses are primitive—if they are not constructed out of qualitative properties or "suchnesses"—then the presentist can use them to give the right sort of structure to past- and future-tensed properties. Here's how a haecceitist presentist might try to account for the truth of *Anne Boleyn was executed in 1536*.

Among the haecceities that presently exist, the presentist can say, are the thisnesses of Anne Boleyn, of the sword with which Anne was executed, and of the swordsman who was especially brought over from France. These properties themselves, says the presentist, instantiate a relation that somehow mirrors the relation that the four-dimensionalist claims to be instantiated by Anne, the sword, and the swordsman. When it comes to making sure that the execution occurred in 1536, rather than at some other time, the presentist might treat times as individuals that themselves have haecceities, and say that the pertinent relation actually holds between the haecceities of Anne, the sword, the swordsman *and* the property of being a time at around midday on May 19, 1536.[28]

Why do I say that the haecceitist presentist posits a relation between haecceities that "somehow mirrors" the relation that the four-dimensionalist believes to hold between individuals? Because the presentist and the four-dimensionalist cannot be positing exactly the same relation. Where the four-dimensionalist has a relation between individuals, the presentist has a relation between properties—you can execute a person, but you can't execute a property. To take a simpler example, the entities that underlie the truth of *Anne was sophisticated* cannot be just Anne's haecceity and the property of sophisticatedness. Rather, the presentist says that there exists the property of sophisticatedness—a property that can be instantiated by present persons—and there also exists a different property—call it sophisticatedness*—that can be instantiated by thisnesses. Sophicatedness* is of such a nature that if it is instantiated by the haecceity of some individual, then it is true that when that haecceity was or will be itself instantiated, the mentioned individual was or will be sophisticated. Anne's thisness is sophisticated*, and that's enough to make it true that Anne, when she existed, was sophisticated.[29]

There are two objections to haecceitist presentism worth mentioning here. The first is obvious; the view posits the existence of strange, unappealing entities that have strange, inelegant properties. I think that the only way for the haecceitist presentist to respond is to try to show that the positing of such entities is a sacrifice worth making—more on this later. The second objection, if successful, strikes more deeply. In committing herself to the existence of haecceities, runs the objection, the presentist is trying to pull much the same trick as was attempted by the Meinongian presentist. Like the Meinongian, the haecceitist is positing one entity for each of the four-dimensionalist's entities, and using them as truthmakers in much the same way. The Meinongian presentist turned out to be a four-dimensionalist with a funny way of speaking. Can the haecceitist presentist really do any better?

I think that she can. First, she can express her presentism in terms that we all can understand; she not only says that no non-present things exist, but that *there are* no non-present things. Second, the haecceitist presentist can say that the entities to which she is committed are of an importantly different type from the corresponding entities that appear in the four-dimensionalist's ontology. When the four-dimensionalist talks about Anne Boleyn, he is talking about a person—a concrete thing—but when the presentist talks about Anne Boleyn, she is talking about a property—an abstract thing.

As David Lewis emphasizes, the presentist still faces the question of what the difference between abstract and concrete entities really comes down to.[30] The correct answer, I think, is that the entities postulated by the presentist count as abstract, rather than concrete, because they are not spatially located; they exist now, but not in any particular place. But how can the presentist say that her entities are not located at non-present times, and are not spatially located at all, when she is committed to the truth of propositions like *Anne Boleyn was executed at the Tower of London in 1536*? Anne was in London in 1536—doesn't this give her location away?

No, the presentist can respond, because what underlies the truth of that statement is not Anne, but her haecceity. Anne's thisness is not spatially located and is not in the past. It does, however, stand in relations to certain other thisnesses, and these relations are such as to make it the case that on May 19, 1536, Anne did exist—her thisness was instantiated—and she was to be found at the Tower of London. This sounds very odd, and the objector may reject it as fantastical. Fair enough, but by now the objector is not saying that the presentist is just a four-dimensionalist in disguise, but that it is very difficult to believe that there could be the things that the haecceitist presentist postulates. And that has already been conceded.

3.3 Atomic presentism

I now want to outline a second strategy that the Truthmaker-preserving presentist might employ. Suppose that the world is made up of a number of particles, and that all the contingent truths about the present world supervene upon the way in which those particles are presently arranged and the fundamental properties that they presently instantiate.[31] *The Tower is on the Thames* is true, on this account, because there are certain particles that bear such properties and are arranged in such a way as to constitute the Tower, and there are a number of particles that bear such properties and are arranged in such a way as to constitute the Thames, and these two collections of particles stand in a relation that suffices for the Tower's being next to the Thames.

Next, suppose that each of the particles that make up the world has always existed and always will exist. Finally, suppose that these particles carry their pasts and futures around with them, which is to say that each particle bears fundamental past- and future-tensed properties that give away the facts about where they have come from and where they are going. No particle could be just like another, according to the theory, unless it had just the same history and just the same future. It is, on this view, a brute fact about any given particle that it was in a certain place and instantiated certain properties in 1536, and it is a brute fact about the particle that it will be in a certain place and will instantiate certain properties in 2036. These are said to be *brute* facts because they are not to be explained in terms of any more basic facts—facts about what is happening in 1536 or 2036, for example. Terms like "was" and "will be," then, are taken as primitively understood.[32]

How does all of this help the presentist to account for past- and future-tensed truth? Well, there is, according to atomic presentism, no need for non-present things to tell us what the world was like in 1536 or what it will be like in 2036, because the information is all encoded in the past- and future-tensed properties of presently existing atoms. The fundamental past-tensed properties of present particles are enough to make it the case that the particles had certain properties and were arranged in a certain way in 1536, and that is enough to make it the case that Anne Boleyn was executed at the Tower in 1536. Given the properties of the things that make up the present world, says the atomic presentist, there's no other way that things could have been, and that's why the world has the property of being a world in which Anne was executed.

Atomic presentism does not posit haecceities, but it does commit its own form of ontological extravagance. If atomic presentism is correct, then the things that we find around us have many more fundamental properties than we would normally suspect. Things that we would normally think to be duplicates of each other turn out not to be, because they do not have the same history and future.

Further, the atomic presentist is committed to the claim that there exist eternal elemental particles that constitute the world at every point in its history. As a claim about the actual world, this is ambitious, but not entirely implausible. Consider, however, the following claim. If X is presently P, then it will be the case at every future time that X was P. That's true, and seems indeed to be necessarily true; it does not seem possible, for example, that Elizabeth II is Queen now and yet there is a future time at which it will not be the case that Elizabeth II was Queen. But if atomic presentism is correct, then what underlie the truth of *X was P* at future times are the existence and tensed properties of the very same particles that presently underlie the truth of *X is P*. So, if it's necessary that if X is presently P then it will always be true that X was P, then it's necessary that each of the world's presently existing particles will exist at all future times. The atomic presentist must take it to be a necessary truth that the world is made up of eternally existing particles.

That is not a happy result. Normally, we would think it quite possible that matter could be destroyed, that some fundamental constituent of being could suddenly drop out of existence.[33] And we do not normally think that if a bundle of matter were to be utterly destroyed then a number of past-tensed truths would go along with it. The atomic presentist, however, must either deny the possibility of particles whose lives are not coextensive with the life of the world, or deny that the truths about what is happening now make it necessary that certain past-tensed truths will hold at all future times. If this is not a devastating dilemma, then it is certainly embarrassing.

3.4 Evaluating Truthmaker-preserving presentism

I've tried to show that Truthmaker-preserving presentism is the only viable form of presentism, and I've sketched two ways in which the view might be developed. Each version of the view involves troublesome commitments, but neither is utterly implausible. There are probably other ways in which the presentist might try to describe the present world in such a way as to ground past- and future-tensed truth, but I think that our examination of those two theories has been enough to show that any form of Truthmaker-preserving presentism will come at a high cost. If the Truthmaker-preserving presentist doesn't posit uninstantiated haecceities or fundamental tensed properties, and if she doesn't make surprising claims about what is and isn't possible, then she will surely take upon herself some other undesirable commitments.

The truthmaking problem does not refute presentism, but it does leave the presentist with the twin burdens of choosing an account of what underlies past- and future-tensed truths and of showing that it is worth making the unattractive commitments that such an account will inevitably involve. I will not try to address the question of whether presentism has benefits that outweigh the costs associated with responding to the truthmaking problem, but I do want to close by considering a possible advantage of presentism that arises naturally from the issues addressed in this paper.

I said earlier that the truthmaking problem is a little different when it comes to the question of what underlies necessary truths. The point is that it does not make sense to ask what things would have had to have been like in order for some necessary truth to fail to be true, so if there is to be a truthmaking problem here then we will have to find some other way to formulate Truthmaker. Still, truthmaking concerns do arise for necessary truths in one form or another, as is evidenced by the philosophical debates over what sorts of ontological commitments are required in order to account for moral, mathematical, and modal truths, and it is with regard to the truthmaking problem for modal truths that the presentist may be on stronger ground than her four-dimensionalist rival.

Suppose, as many do, that it is difficult to account for what is possible using only those things that are actual, and suppose, as many do, that realism about possible objects is too high a price to pay to secure the truths of modality. It may be that the only way to navigate between these two suppositions is to posit actually existing surrogates for merely possible objects—perhaps the uninstantiated haecceities of things that never actually exist, or perhaps the irreducibly modal properties of the things that do. If this is right, then the presentist's commitments may no longer seem so unattractive; placed next to the commitments that are required to solve the truthmaking problem for modality, they are relatively benign. Then, the truthmaking problem might not seem like such a problem. Whether the debate about modality ever gets to this point, however, is another question entirely.[34]

Notes

1. Augustine 1991 (Book XI, sections 22-24).

2. For some defenses of presentism, see Prior 1970, Geach 1972, Zimmerman 1998 [chapter 17 in this volume—*Editors*], Hinchliff 1996, and Bigelow 1996 [chapter 8].

3. For good descriptions of the four-dimensional view of time, see Smart 1966 and Williams 1968. I do not mean to suggest that presentism and four-dimensionalism are the only possible, nor the only plausible, views about the nature of time. For one well worked-out alternative, see Tooley 1997.

4. See, for example, Jackson 1998 (p. 138). For the reasons why "four-dimensionalism" is a poor label for perdurantism, see Parsons 2000, pp. 399-400.

5. A little more needs to be said about how presentism and four-dimensionalism come into conflict. It is natural to say that where the four-dimensionalist believes that the world is four-dimensional, the presentist believes that it is three-dimensional. If presentism is just the claim that only present things exist, however, then the presentist need not be a three-dimensionalist. She could say that only present things exist, but some of those things are four-dimensional. All the things that exist exist presently, the claim would go, but some of them are not instantaneous—they extend into the past or future. Most presentists are in fact committed to three-dimensionalism, but it will be enough for my purposes to assume that there are at least some past or future things in the four-dimensionalist's ontology whose existence is denied by the presentist. Anne Boleyn can serve as an example; the four-dimensionalist, but not the presentist, believes that Anne Boleyn exists. (Thanks to Josh Parsons for clearing my thoughts on this question.)

6. A version of the problem, the "argument from relations," is presented in Bigelow 1996 (p. 38). Bigelow finds traces of the argument among the Stoics and Epicureans. For other recent discussions that mention some version of the truthmaking problem, see Adams 1989 (pp. 32-33), Lewis 1999a (p. 207), and Lewis 1999b.

7. I take the name "Truthmaker" from the helpful discussion in Bigelow 1988 (pp. 121-34). Now Truthmaker, as it stands, may seem to be trivially true. "Propositions exist," we might say, "so of course it is true that if some proposition had a different truth-value then there would be a difference in what some existing thing is like." But Truthmaker is supposed to make a stronger claim than this. The idea is that there are no contingent truths that hang free of reality; the truth of any contingently true proposition is underwritten by certain existing things that are not themselves propositions. A better way of putting Truthmaker is as follows. *Truth is supervenient upon what things exist and which perfectly natural properties and relations they instantiate, where "things" do not include such things as propositions, and "perfectly natural properties and relations" do not include properties and relations that refer to such things as propositions.* (This is a modification of the formulation settled upon in Lewis 1999a (p. 207)—see Lewis's discussion for a more extensive explanation of why this is the best way to express the claim.) From now on, when I talk about the ways in which things would have to be different in order for some contingent proposition to have a different truth-value, I am not talking about differences that exist only at the level of propositions themselves. (Thanks to Mike Fara for help with this question.)

8. The stronger truthmaker thesis is advocated in Armstrong 1997. For discussion of truthmaker theses of various strengths see Lewis 1999a (pp. 196-214) and Lewis 1998.

9. See Lewis 1999a (pp. 204-5).

10. When it comes to talk of possible worlds, the first strategy is taken in Armstrong 1989 and in Rosen 1990, and the second strategy is taken in Lewis 1986. With regard to moral talk, the third strategy is taken in Mackie 1977.

11. As is pointed out in Bigelow 1996 (p. 39). Note that many philosophers *are* prepared to posit existing fictional entities; see, for example, Van Inwagen 1977 and Salmon 1998.

12. Aristotle, *De Interpretatione*, ch. 9. See also Tooley (1997).

13. See, for example, Hibbert 1992 (p. 18).

14. Well, there might be a more principled way of defending the claim that there are fewer past-tensed truths than we might previously have thought. In Prior 1960 and Adams 1989, it is argued that not-yet-existing individuals are not available to be the constituents of propositions, which is to say that there are no true singular propositions about not-yet-

existing individuals—it was not true in 1066 that Anne Boleyn would be executed in 1536. Perhaps, then, we could treat no-longer-existing individuals in just the same way, saying that because Anne Boleyn does not exist now, there are no singular propositions about her (and hence none that are true). This view, however, has the fatal feature of implying that it is not true that Anne Boleyn was executed. (Adams makes this point himself on p. 29, saying that what can plausibly be said about the future cannot plausibly be said about the past.) Now, it might be thought that this feature is not really so fatal, because we can get almost all that we want in the way of past-tensed truth from tensed propositions that involve definite descriptions, rather than past individuals themselves. While there are no singular propositions about Anne, we could say, it is true that someone who was married to the king, gave birth to a daughter, was called "Anne" . . . was executed. Supposing, however, that there are properties that were once instantiated but are instantiated no more, and that there are no uninstantiated properties, the presentist will run into difficulties in constructing the required definite descriptions—some of the properties that she needs do not, on her view, exist. For further discussion of this general line of thought, see part 3 below.

15. Named after Alexius Meinong, who is credited with founding the doctrine of non-existent objects. The best recent exposition of the doctrine is in Parsons 1980. For a defense of Meinongian *presentism*, see Hinchliff 1988.

16. If W. V. Quine's views on how to do ontology are correct, then the Meinongian presentist, as someone who quantifies over non-present entities, is not a presentist at all. See Quine 1953 (p. 13).

17. See Williams 1962 (p. 752).

18. I do not know of anyone who has defended this view in print, but it often comes up in conversation.

19. There are probably things that the presentist could say here. Perhaps there is just a brute difference between relational and non-relational past- and future-tensed truths, or perhaps the difference is merely syntactic. We would never make such claims about the difference between relational and non-relational present-tensed truths, however, so these responses appear to be somewhat desperate.

20. Bigelow 1996 (p. 47).

21. Chisholm 1990. As far as I can tell, the Bigelow and Chisholm strategies have much the same virtues, and run into much the same problems.

22. Bigelow does not speak of "fundamental" world properties, but they seem to be what he has in mind when he speaks of "the property of being burdened with a certain sort of past, or (as Leibniz put it) being pregnant with a certain sort of future" (Bigelow 1996, p. 47). The relevant burdens and pregnancies are not thought to be grounded in any non-present things.

23. As Lewis points out, there may be some fundamental properties that were had by individuals early in the history of the universe, but are had by no individuals now. See Lewis 1999b (p. 7).

24. See Chisholm 1990. Chisholm's proposal is complicated, but the thought, put roughly, is that events, plus the facts about which events are parts of other events, plus the relations in which events stand to eternally existing entities, are enough to make it the case that events occur at particular times.

25. It's not clear what Bigelow and Chisholm would say here. Bigelow speaks of the presentist's investment in the properties of present *individuals* (Bigelow 1996, p. 47), but at another point says that he prefers the language of *world* properties (Bigelow 1996, p. 46).

26. Another view, associated with Leibniz, is that to have a thisness is to have all the qualitative properties in some large, maximally consistent set, so that to be Anne Boleyn is to

give birth to a daughter, to marry the king, to be executed in 1536, to be called "Anne Boleyn," to not have a tail, and so on and on. This Leibnizian view is probably more compatible with what I call atomic presentism—the view that I discuss in section 3.3.

27. A clear description of the debate about the nature of thisness is given in Adams 1979 (pp. 5-10). I am following Adams's terminology. On the conditions for something's being a qualitative property (or a "suchness"), see Adams 1979 (pp. 7-9).

28. Or, if she prefers, she could adopt one of the other two strategies suggested a few paragraphs back.

29. Some further complications. First, the presentist will have to make sure that Anne's haecceity's being sophisticated* has some connection with certain times in the sixteenth century; depending on the time-indexing strategy that she favors, the presentist might want to say that Anne's haecceity is [sophisticated at certain times in the sixteenth century]*. Second, for the sake of consistency, the presentist could add that all instantiations of properties by individuals are mirrored by instantiations of properties by haecceities; if you are presently sophisticated, then your haecceity is sophisticated*. Furthermore, we might add, your haecceity exists*, but Anne Boleyn's doesn't (although it did). Obviously, there are further details to be worked out here.

30. Lewis 1999b (pp. 7-9).

31. It's easiest to imagine such particles as being very, very small, but it might be possible to pursue this strategy by postulating any sorts of eternally enduring particulars which together constitute the entire world. There could, conceivably, be just one such particular—space, perhaps, or God, or the world itself. For reasons that should become clear from my discussion, however, I think that the presentist would have to say not only that the world consists in one eternally enduring particular, but that all the parts of that particular are also eternally enduring—in which case we are back where we started. Note also that there is no need for the atomic presentist to claim that everything in the world just *is* a collection of particles. It's consistent with the view say that a person, for example, cannot be identical with a collection of particles, although a person can only exist if there exists a collection of particles bearing certain properties and arranged in a certain way. Finally, note that there is nothing in the view that claims that all of the particles in the world are *physical* particles.

32. We had to add extra details to the haecceitist's story in order to account for the fact that events do not just happen, but happen at times, and a similar move is required of the atomic presentist (as I will call her). She might say, for example, that the various past- and future-tensed instantiations of properties by particles stand in (perhaps quasi) temporal relations, so that it can be true of some particle that it was in place P and of nature X *before* it was in place Q and of nature Y. Then, it can be claimed that the facts about the times at which events happen supervene upon the facts about which events happen before and after which other events, and which events are simultaneous with which others.

33. An exactly analogous problem, of course, arises with regard to the possibility that particles could suddenly pop into existence. We'd normally think it to be a necessary truth that if X is presently P, then it was the case at all past times that X will be P. But if the particles that make for X's being P only came into existence last year, then it was not true until last year that X will be P.

34. I owe large debts to David Lewis and John Bigelow, who have greatly influenced my thinking about this material. Thanks also to Karen Bennett, Michael Fara, Josh Greene, John Hawthorne, Bradley Monton, Michael Nelson, Josh Parsons, Ed Zalta, and Dean Zimmerman.

References

Adams, Robert Merrihew. "Primitive Thisness and Primitive Identity," *The Journal of Philosophy* 76, no. 1 (1979): 5-26.

———. "Time and Thisness." Pp. 23-42 in *Themes from Kaplan*, edited by J. Almog, J. Perry, and H. Wettstein. Oxford: Oxford University Press, 1989.

Aristotle. *De Interpretatione*.

Armstrong, D. M. *A Combinatorial Theory of Possibility*. Cambridge: Cambridge University Press, 1989.

———. *A World of State of Affairs*. Cambridge: Cambridge University Press, 1997.

Augustine. *Confessions*. Translated by H. Chadwick. Oxford: Oxford University Press, 1991. Chapter 1 in this volume.

Bigelow, J. *The Reality of Numbers*. Oxford: Oxford University Press, 1988.

———. "Presentism and Properties." Pp. 35-52 in *Philosophical Perspectives* 10, edited by J. E. Tomberlin. Cambridge, Mass: Blackwell Publishers, 1996. Chapter 8 in this volume.

Chisholm, R. M. "Events without Times: An Essay On Ontology," *Noûs* 24 (1990): 413-28.

Geach, P. T. "Some Problems about Time." Pp. 302-18 in *Logic Matters*, by P.T. Geach. Berkeley: University of California Press, 1972.

Hibbert, C. *The Virgin Queen: The Personal History of Elizabeth I*. London: Penguin Books, 1992.

Hinchliff, M. *A Defense of Presentism*. Dissertation. Princeton University: University Microfilms, 1988.

———. "The Puzzle of Change." Pp. 119-36 in *Philosophical Perspectives* 10, edited by James E. Tomberlin. Cambridge, Mass: Blackwell Publishers, 1996.

Jackson, F. "Metaphysics by Possible Cases." Pp. 133-53 in *Mind, Method and Conditionals* by F. Jackson. London: Routledge, 1998.

Lewis, D. *On the Plurality of Worlds*. New York: Basil Blackwell, 1986.

———. "Truthmaking and Difference-making." Paper presented at the meeting of the Australasian Association of Philosophy, July 1998.

———. "Armstrong on Combinatorial Possibility." Pp. 196-214 in *Papers in Metaphysics and Epistemology*, by D. Lewis. Cambridge: Cambridge University Press, 1999a.

———. "Tensed Quantifiers." Paper presented at the meeting of the Australasian Association of Logic, July 1999b.

Mackie, J. L. *Ethics: Inventing Right and Wrong*. London: Penguin Books, 1977.

Parsons, J. "Must a Four-Dimensionalist Believe in Temporal Parts?" *The Monist* 83 (2000): 399-418.

Parsons, T. *Nonexistent Objects*. New Haven, Conn.: Yale University Press, 1980.

Prior, A. N. "Identifiable Individuals," *Review of Metaphysics* 13 (1960): 684-96.

———. "The Notion of the Present," *Studium Generale* 23 (1970): 245-48.

Quine, W. V. "On What There Is." Pp. 1-19 in *From a Logical Point of View*, by W.V. Quine. Cambridge, Mass: Harvard University Press, 1953.

Rosen, G. "Modal Fictionalism," *Mind* 99 (1990): 327-54.

Salmon, N. "Nonexistence," *Noûs* 32 (1998): 277-319.

Smart, J. J. C. "The River of Time." Pp. 213-27 in *Essays in Conceptual Analysis*, edited by A. Flew. London: MacMillan, 1966.

Tooley, M. *Time, Tense and Causation*. Oxford: Clarendon Press, 1997.

Van Inwagen, P. "Creatures of Fiction," *American Philosophical Quarterly* 24 (1977): 299-308.

Williams, D. C. "Dispensing with Existence," *The Journal of Philosophy* 59 (1962): 748-63.

———. "The Myth of Passage" Pp. 98-116 in *The Philosophy of Time*, edited by R. M. Gale. London: MacMillan, 1968.

Zimmerman, D. W. "Temporary Intrinsics and Presentism." Pp. 206-19 in *Metaphysics: The Big Questions*, edited by P. v. Inwagen and D. W. Zimmerman. Malden, Mass: Blackwell, 1998. Chapter 17 in this volume.

16

Presentism and the Grounding Objection

Thomas Crisp

Presentists think that only present things exist.[1] But according to a popular line of objection, their thesis has an unacceptable implication: if there were only present things, there would be nothing to *ground* past and future truths.[2] It is a plausible principle that contingent truths like *Socrates existed* are true only if there is some thing or things in the world that *ground* their truth, that *make* them true. But what among the present things—which according to the presentist, exhaust reality—grounds a truth like *Socrates existed*? Says the objector: nothing. So presentists face an uncomfortable dilemma: reject the principle that contingent truths need grounding or give up obvious truths about the past like *Socrates existed*. Neither option is attractive.

Such is the grounding objection to presentism, roughly stated. I argue in this paper that presentists have an adequate reply. After setting out the objection in proper detail, I expand on a reply once advocated by John Bigelow (1996) [chapter 8 in this volume—*Editors*] and defend my expanded reply against objections.

1. The Grounding Objection Stated

1.1 Grounding and Truthmakers

The grounding objector holds that contingent truths need grounding. But what does she mean? What does her talk of *grounding* come to? A standard answer takes its start from the *truthmaker axiom*, which is typically construed along these lines:

(TA) $\Box\forall p(p \text{ is contingently true} \supset \exists xs\Box(\text{the } xs \text{ exist} \supset p \text{ is true}))$.[3]

In English: necessarily, for any proposition *p*, if *p* is contingently true then there are some *xs* such that, necessarily, if the *xs* exist then *p* is true. A truth is *grounded*, we might say, when there are some things, the *xs*, such that the proposition that the *xs* exist entails the truth in question.

If (TA) is right, then, necessarily, all contingent truths are grounded. Suppose it *is* right. Then there's trouble for presentism. Consider the contingent truth that dinosaurs existed. What present things are such that the proposition that they exist entails the proposition that dinosaurs existed? Isn't it *possible* that things be just as they presently are and there have been no dinosaurs? (For example, isn't it possible that things be just as they presently are except that God created the world five minutes ago?) It would seem so. But if so, then no present things are such that the proposition that they exist entails the proposition that dinosaurs existed. Wherefore, given (TA) and the presentist's claim that the present things exhaust reality, it is not true that dinosaurs existed. But this looks bad for presentism: dinosaurs did, after all, exist.

Many presentists will reply by denying (TA). There are well known reasons for being skeptical about such principles.[4] So, for example, is there some thing or things such that the proposition that they exist entails the proposition *there are no unicorns*? Not obviously. For that matter, is there some thing or things such that the proposition that they exist entails *Bush is president*? Bush doesn't fit the bill (he could exist without being president), nor does the pair comprising Bush and the property *being president* (it could exist without Bush being president). Friends of (TA) will likely reply that the "state of affairs" *Bush's being president*—the non-mereological fusion of Bush and the property *being president*—does the job. But it's not just obvious that there are such things as these non-mereological fusions.

Perhaps these and like worries can be worked up into a good objection to (TA) and an adequate reply to the objection to presentism of paragraph before last. Perhaps. But even if so, the presentist has little cause for comfort since there's a widely discussed variation on (TA) that makes just as much trouble for presentism and is vastly more difficult to deny.

1.2 Grounding, Truth, and Supervenience

The variation I have in mind is formulated by David Lewis and John Bigelow as the claim that *truth supervenes on being*, where the idea here is that truth supervenes on what things there are and which fundamental properties and relations[5] they instantiate.[6] In terms of possible worlds:

> **(SP)** For any proposition *p* and worlds *w* and *w**, if *p* is true in *w* and not in *w**, then (a) according to *w*, something exists that doesn't exist in *w** (or vice versa), or (b) according to *w*, some objects instantiate a fundamental property or relation that they don't instantiate in *w** (or vice versa).[7]

(SP) is plausible. It fails only if there are worlds in which truths about what *non*-fundamental properties and relations are instantiated—truths about what grue-like properties and relations are instantiated—fail to supervene on truths about what exists and what fundamental properties and relations are instantiated. But it is plausible that, necessarily, truths of the first sort supervene on truths of the second. At any rate, so think many of us.

(SP) looks pretty good. But presentism would seem to contravene it. Dinosaurs existed. Since, one thinks, present things could have been just as they are and dinosaurs *not* have existed (here again, couldn't things have been just as they presently are except that God created the world five minutes ago?), the truth that dinosaurs existed would seem not to supervene on what present things there are and which fundamental properties and relations they instantiate. But presentism says that present things exhaust reality. So given presentism, the truth that dinosaurs existed would seem not to supervene on what things there are and which fundamental properties and relations they instantiate. So given presentism, (SP) would seem to be false. But (SP) surely isn't false; so presentism must be.

Let it be understood henceforth that the foregoing bit of reasoning expresses the grounding objection to presentism. In a word: presentism is incompatible with an obvious truth, (SP); so presentism is to be rejected.

2. The Grounding Objection Examined

The grounding objection as we've stated it depends on two principles, the supervenience principle (SP), and what we might call a *temporal recombination* principle:

> **(TR)** Things could have been just as they are at present (same things presently in existence, same fundamental properties and relations presently instantiated) and the past have been different.[8]

Sensible presentists will object to the latter. There are various ways of proceeding; I'll focus on a proposal once advocated by John Bigelow.

2.1 Lucretianism

Bigelow (1996) endorses a view we might call Lucretianism. The Epicurean philosopher Lucretius had an exceedingly sparse ontology. He seems to have thought that reality is exhausted by present atoms in the void and their properties. Responding to the objection that there must be something more to the world than this because, after all, there is the fact that the Trojans were conquered, and this fact isn't to be found among the present atoms in the void and their properties, Lucretius had this to say:

> Again, when men say it *is* a fact that Helen was ravished or the Trojans were conquered, we must not let anyone drive us to the admission that any such factual event *exists* independently of any object, on the ground that the generations of men of whom these events were accidents have been swept away by the irrevocable lapse of time. For we could put it that whatever has taken place is an accident of a particular tract of earth or of the space it occupied. (Lucretius 1994, 21)

According to Lucretius, then, to say that the Trojans were conquered is to predicate a certain accident or property of a bit of land or a region of space—as we might put it, the property of being the place where the Trojans were conquered. *Lucretianism*, let us say, is the view that the world has this and many like past- and future-tensed properties: properties like *being a place where dinosaurs roamed*, and *having been inhabited by wooly mammoths.*[9]

The Lucretian presentist will be unimpressed by (TR). She'll think it impossible that a world should have the same past-tensed properties as ours but a different past. (So she'll think it impossible that a world should have, say, the past-tensed property of having been roamed by dinosaurs, but not have been roamed by dinosaurs.) But given (SP), truths to the effect that our world has thus-and-such past-tensed properties supervene on what things exist and what fundamental properties and relations they instantiate. So given that she accepts (SP), and I see no reason why she shouldn't, the Lucretian presentist will think that things couldn't have been just as they are—same things presently in existence, same fundamental properties and relations presently instantiated—and the past different. So she'll think (TR) false.

Since without (TR), the grounding objection is a non-starter, we see that, for the Lucretian presentist anyway, the grounding objection gives little cause for concern. In the sequel, I consider what might be said against Lucretianism.

3. Lucretianism and Ontological "Cheating"

Theodore Sider (2001, 39-41; 2003) proposes that Lucretianism is of a piece with a variety of ontologies that "cheat" by invoking dubious irreducibly *hypothetical* properties. Some examples of irreducibly hypothetical properties (Sider 2001, 39-41): First, maybe the causal dispositions of a thing are brute—in no way explained by or reducible to its non-dispositional properties (or its non-dispositional properties together with the laws of nature). The glass has the disposition to shatter when dropped. On the "dispositions are brute" view, this property is a brute feature of the glass, not to be analyzed in terms of its structural properties, physical make-up, and the like. Such brute dispositional properties are in an important sense *hypothetical* because, to borrow Sider's phrase, they "point beyond" the way things actually are—they are "outward looking." The dispositional property of shattering if dropped points beyond the way things are to the way things would be were something dropped. Hypothetical properties contrast with *categorical* properties:

intuitively, those properties that do not point beyond their instances; they characterize a thing as it actually is, not how it could have been, or would have been, or was.

Second: brute counterfactual properties. Molinists propose that, from all eternity, the world had the property of being such that, were Quine to consider writing *Word & Object* under such-and-such circumstances, he would freely decide to do so. From the Molinist's perspective, this property is not analyzable in terms of non-counterfactual properties. So, for example, the world's having this property is not a matter of its having certain physically specifiable initial conditions and being governed by certain physical laws. Nor is it a matter of Quine's having certain propensities: the world had this property long before Quine existed. Molinists will think of this property as primitive, a part of the world's fundamental ontology. Such properties are paradigmatically hypothetical: they point beyond the way things are to how things would be under certain conditions.

Third: irreducible modal properties. Some actualists think that the world includes among its primitive properties modal properties like *being possibly true*, and *being essentially human*. Such properties point beyond their instances to how things could have been and are thus hypothetical properties.

And finally, the tensed properties of Bigelow's Lucretian. Properties like *being formerly roamed by dinosaurs* are hypothetical since they point beyond their instances to how things were or will be. If we think of them as primitive, unanalyzable into more basic materials, we get another example of an irreducibly hypothetical property.

Sider (and others too I expect) regards any ontology that invokes such irreducibly hypothetical properties as dubious. Says Sider: such ontologies "cheat" by postulating disreputable properties to solve their problems.

Sider's suggestion raises several interesting questions. First, what exactly does the hypothetical/categorical distinction come to? As it stands, it's a bit undercharacterized. Second, why are ontologies that postulate irremediably hypothetical properties ipso facto suspect? Why are these properties untoward? And third, is there any reason to think that the ontologies Sider mentions are committed to *irreducibly* hypothetical properties, or could it be that certain of their hypothetical properties reduce to categorical ones?

3.1 On the Hypothetical/Categorical Divide

I begin with the first question: What does the hypothetical/categorical distinction come to? This is a hard question. I *think* I understand the distinction between properties like *being round* and *being possibly round*, or *being short* and *being formerly short*well enough. It can be thought of as follows.

A couple points by way of setup. I shall follow George Bealer (1982) in abbreviating the singular term ⌜the proposition that S⌝ with ⌜$[S]$⌝, and in referring to the property of being an x such that x is F with the singular term "$[Fx]_x$," and the relation among $x_1, \ldots, x_n$ such that $Rx_1, \ldots, x_n$ with the singular term "$[Rx_1, \ldots, x_n]_{x1 \ldots xn}$." I shall join Bealer in assuming that these singular terms may contain externally quantifiable variables. For example, using Bealer's bracket notation, we may represent the proposition that, for some *x*, Joe believes the proposition that *x* is *F* as "$\exists x(jB[Fx])$."

Finally, I shall assume that expressions like "is possible," "is necessary," and "is past" are predicates, expressions that combine with a singular term *t* to yield a sentence that expresses a proposition about the referent of *t*. More, I shall assume that such predicates express properties. So: to say that $\Diamond[S]$ is to say that $[S]$ is possible—that is, that $[S]$ has the property *being possible*. To say that WAS$[S]$ is to say that $[S]$ is past—that $[S]$ has the property *being past*. I leave it an open question for now whether such properties are analyzable into more basic properties or relations.

I can now characterize the distinction between the property *being round* and the property *being possibly round*. The latter property, I propose, is identical with the property $[\Diamond[Rx]]_x$—that is, it is identical with the property of being an *x* such that the proposition that *x* is round is possible. Some related examples:

$$\textit{being necessarily divine} = [\Box[Dx]]_x$$

$$\textit{being formerly short} = [\text{WAS}[Sx]]_x$$

Properties like the foregoing have this in common. Each is such that, for some properties *G* and *H* (such that *H* is not equivalent to the property *being true*[10]), it is identical with the property of being an *x* such that the proposition that *x* has *G* has *H*. In terms of our bracket notation: each is such that, for some properties *G* and *H* (such that *H* is not equivalent to the property *being true*), it is identical with the property $[H[Gx]]_x$. (So, e.g., the property *being possibly round*, I say, is identical with the property $[\Diamond[Rx]]_x$—the property of being an *x* such that $[Rx]$, the proposition that x has *roundness*, has the property *being possible*, where the latter property, notice, is not equivalent to the property *being true*.) Not so for the property *being round*. For no properties *G* and *H* is it the case that *being round* is identical with the property $[H[Gx]]_x$ (though it certainly is *equivalent* to such properties).[11]

Here then we have a clean distinction between two sorts of property. Let us call properties like $[\Diamond[Rx]]_x$ *hypothetical* properties, and define the notion of a hypothetical property thus:

> *F* is a hypothetical property $=_{\text{df.}}$ for some properties *G* and *H* (such that *H* is not equivalent to the property *being true*), $F = [H[Gx]]_x$.

Categorical properties, say, are non-hypothetical properties.

If this is what the hypothetical/categorical distinction comes to, then I understand it well enough. But if this is what the distinction comes to, it's hard to see why the actualist, the Lucretian, and their ilk should be criticized for admitting hypothetical properties into their ontologies. To say that *x* is possibly *F* is just to say that the proposition that *x* is *F* has the property *being possible*; to say that *x* is formerly *F* is just to say that the proposition that *x* is *F* has the property *being past*. And what's to complain about with properties like *being possible* and *being past*? They aren't exotic or obviously untoward. Note too that if this is what the distinction comes to, the actualist, the Lucretian et al are not committed to *irreducibly* hypothetical properties. To say that *being possibly round* is identical with *being an x such that the proposition that x is round is possible* is to give a reductive analysis of *being possibly round* in terms of categorical properties: as the distinction is currently drawn, properties like *being round* and *being possible* are categorical: plausibly, for no properties *G* and *H* is it the case that *being round* or *being possible* is identical with $[H[Gx]]_x$.

In brief: if the hypothetical/categorical distinction is what I say it is, the properties invoked by the Lucretian—properties like *being an x such that dinosaurs formerly roamed x*—are neither irreducibly hypothetical nor obviously offensive. They are hypothetical, to be sure, but are analyzable into non-hypothetical properties like *being a dinosaur* and *being past*. Since the latter properties aren't dubious in any obvious sense, nor are the Lucretian's more exotic properties.

Sider will likely retort that (a) the hypothetical/categorical distinction is *not* what I say it is, and (b) even if I'm right that we can reductively analyze properties like *having been round* and *being possibly round* in terms of properties like *being past* and *being possible*, this is no real gain since the latter properties are just as dubious as the former. Take (a) first. Comments of Sider's suggest that, by his lights, properties like *being possible*, *being necessary*, *being past*, and *being future* are hypothetical properties.[12] Since my analysis of the distinction puts these properties on the categorical side of the divide, my analysis gets his distinction wrong.

Well, it would come as no great shock if my analysis gets the hypothetical/categorical distinction wrong: it's notoriously difficult to say just what this distinction comes to. But for present purposes, not much hangs on this. What matters, I say, is that properties like *being formerly tall* can be analyzed as above in terms of simpler properties like *being tall* and *being past*, and that the latter properties aren't obviously objectionable.

Sider will reply that the property *being past is* objectionable because it is irreducibly hypothetical. This takes us to (b). In reply, I'm not sure whether it's irreducibly hypothetical: if the hypothetical/categorical distinction isn't what I suggest above, then I'm not sure what the distinction comes to. (I doubt anyone else does either.) But even if *being past* were irreducibly hypothetical, why should this count as a mark against it? What exactly is wrong with these properties? Sider, I think, will grant that it's exceedingly hard to say just what the hypothetical/categorical distinction comes to, but will argue that, even if we can't say exactly, we can just *see* that there is something dubious about properties like *being possible*, *being past*, and *being future*.[13] Maybe it's difficult to say just what makes these properties hypothetical as opposed to categorical. No matter: we can just *see* that these properties, taken as primitives, are inappropriate ontological posits. Since it's very difficult to see how the Lucretian presentist could reduce tensed properties like *being past* or *being future* to less objectionable properties, the Lucretian who postulates their existence stands convicted of ontological cheating.

By way of reply, what one would like to see here is an *argument* for the claim that primitive properties like *being past* are somehow untoward. Surely we *can't* just see that such properties are inappropriate ontological posits. (How could one just see a thing like that?) Every ontology has its primitives; ontologies that postulate primitive tense are not so far any worse off than any other ontologies. Since, as best I can tell, Sider gives no argument for thinking there is something wrong with these ontologies, by my lights, his objection to Lucretianism doesn't cut very deep. The Lucretian need not postulate exotic primitive properties. Properties like *being formerly roamed by dinosaurs* can be analyzed in terms of less exotic properties like *being a dinosaur* and *being past*. And so far, we've no reason for thinking that tensed properties like *being past* are inappropriate ontological posits.

3.2 But Are These *Fundamental* Properties?

"Fair enough," you might reply, "we've no reason for thinking that such tensed properties are inappropriate posits in the sense that one can just *see* that there are no such things." But we do have reason for thinking that such properties are ill-suited for the work the Lucretian presentist needs them to do. Here's why. Given that our Lucretian presentist accepts (SP) (something she surely *should* do: (SP) looks pretty good), she would seem to be committed to the claim that at least some tensed properties or relations are fundamental. For suppose for *reductio* that (SP) holds but that no tensed properties or relations are fundamental. Suppose too, as seems plausible, that the world could've been just as it is at present—*modulo* the instantiation of tensed properties and relations—and the past have been different. Then it follows that the world could've been just as it is at present—same present things in existence, same fundamental properties and relations instantiated—and the past have been different. So it follows that truths about the past do not supervene on what things exist and what fundamental properties and relations are instantiated. So it follows that (SP) is false, which completes our *reductio*. Given that she accepts (SP), then, the Lucretian would seem to be committed to thinking that at least some of the tensed properties and relations are fundamental.

Suppose so. Then there is this problem. None of the tensed properties and relations postulated by the Lucretian seems well-suited to the role of fundamental property/relation. Properties like *being an x such that the proposition that x is roamed by dinosaurs is past* certainly don't. Fundamental properties 'carve nature at the joints'; sharing of them makes for sameness of intrinsic qualitative character.[14] But surely the property *being an x such that the proposition that x is roamed by dinosaurs is past* does not mark a joint in nature; and surely two things aren't similar with respect to intrinsic qualitative character by virtue of each having been roamed by dinosaurs.

We discussed earlier the primitive tensed property *being past*, but it seems no better suited to the role of fundamental property than the property *being an x such that the proposition that x is roamed by dinosaurs is past*. Surely sharing a property like *being past* doesn't make for intrinsic qualitative similarity. Past things don't resemble each other by virtue of being past any more than things that exist on Tuesdays resemble each other by virtue of their existence on a Tuesday.

So our Lucretian presentist has this problem. "She needs at least some tensed properties or relations to be fundamental, but it's hard to believe that any of the tensed properties and relations she postulates *are* fundamental."[15]

By way of reply, though, it's not just obvious (to me anyway) that *being past* is not well-suited to the role of fundamental property. Remember, as our Lucretian presentist thinks of things, *being past* is a property of propositions. Is it just *obvious* that propositions wouldn't resemble each other by virtue of sharing such a property? Not to me. I have very few intuitions about what would and wouldn't make for qualitative similarity among propositions and no clear intuition to the effect that propositions that shared the property *being past* wouldn't be similar by virtue of this fact.

That said, it's also not just obvious (to me anyway) that the property *being past* *is* well-suited to the fundamental property role. I suppose the Lucretian could dig in her heels and insist that it is, but it'd be nice if she didn't have to. It'd be nice if she could show that the property *being past* and like properties are reducible to more basic properties and relations, properties and relations that are themselves more obviously suited to the role of fundamental property/relation. I shall urge next that she can.

4. The Ersatz B-Series

I shall argue in this section that the work done in a presentistic metaphysic by primitive tensed properties like *being past*, *being present*, and *being future* can be done equally well by the B-theoretic *earlier than* relation. Since it is plausible that temporal relations like the *earlier than* relation are fundamental, if the Lucretian can show that properties like *being past* are analyzable in terms of such relations, she'll have in hand an attractive reply to the worry of the previous section.

4.1 Taking Tenselessness Seriously

We begin by taking *tenselessness* seriously.[16] Call the properties *being past*, *being present*, and *being future* the "A-properties." A tenseless proposition, let us say, is a proposition *p* such that it is possible to grasp or conceive of *p* without thereby grasping or conceiving of an A-property. To illustrate, I'll assume there is such a thing as the tenseless "is"—henceforth, "IS." A few words about the meaning of "IS." When we say that *x* IS *F*, we indicate thereby that the two-term *instantiation* relation connects *x* and the property *F*, but say nothing about whether *x*'s possession of *F* is past, present,or future. This as opposed to the present-tensed "is": when we say that *x* is (present-tensed) *F*, we indicate thereby that *x*'s possession of *F* is *present*—that *x* has *F at* the present time. Well, suppose so. Then [Lincoln IS president] is plausibly taken as an example of a tenseless proposition: one can grasp it, one thinks, without thereby grasping an A-property. Given above assumptions about the tense operators, [WAS[Lincoln IS president]] is an example of a proposition that is *not* tenseless.

One "takes tenselessness seriously," let us say, iff one thinks there are tenseless propositions. In what follows, I shall suppose there are. I shall suppose henceforth that expressions like $\ulcorner[\Phi x]\urcorner$, $\ulcorner[\forall x \Phi x]\urcorner$ (where Φ does not express an A-property) denote such propositions.

(It's an interesting question whether there are such things as tenseless propositions. I'm inclined to think so, but I don't have any knock-down arguments. For present purposes, though, I don't need any. My target here is the philosopher who doubts there are fundamental properties like *being past*, *being present*, and *being future*. Such philosophers are not likely to doubt the existence of tenseless propositions. It is to these thinkers that the arguments of this section are addressed. I want to convince them that we presentists who join them in taking tenselessness seriously needn't suppose that properties like *being past*, *being present*, and *being future* are fundamental. If you're one of those philosophers who doubts there are tenseless propositions, you're likely comfortable with an ontology that includes these and/or like properties as fundamental and can skip this section. I hope that enough has been said in previous sections to persuade you that the presentist has in Lucretianism an attractive reply to the grounding objection.)

4.2 On Times

I join Roderick Chisholm,[17] Edward Zalta,[18] and others[19] in thinking of a "time" as a certain sort of maximal abstract object: intuitively, an abstract representation of an instantaneous state of the world. Let us think of a time as any proposition that satisfies the following definition:

> x is a time $=_{\text{df.}}$ For some class C of propositions such that C is *maximal* and *consistent*, $x = [\forall y(y \in C \supset y \text{ is true})]$,[20]

where (i) a class C of propositions is maximal iff, for every proposition p, either p or its denial is a member of C, (ii) a class C of propositions is consistent iff, possibly, every member of C is true, and (iii), "[∀y(y∈C ⊃ y is true]," I assume, denotes a tenseless proposition (lest my attempt to give a reductive account of tensed properties fall into unhappy circularity).

Three comments. First: Patrick Grim's arguments to the effect that there is no set of all truths (Plantinga and Grim 1993) raise doubts about whether there is any maximal, consistent class of propositions. I am not sure what to think about Grim's arguments, but not much hangs on it. Nothing interesting changes in the discussion to follow if instead of thinking of times as above, we think of them thus:

> x is a time $=_{\text{df.}}$ For some propositions, the *ps*, such that the *ps* are maximal and consistent, $x = [\forall y(y \text{ is one of the } ps \supset y \text{ is true})]$,

where (i) "the *x*s are maximal" is a variably polyadic predicate that applies to some propositions, the *p*s, iff, for every proposition *q*, either *q* or its denial is one of the *p*s, and (ii) "the *x*s are consistent" is a variably polyadic predicate that applies to some propositions, the *p*s iff, possibly, every one of the *p*s is true. That said, I'll assume the former definition in the discussion to follow.

Second: my definition of times won't work on just any metaphysic. The B-theoretic eternalist will find it odd to think of my times as "abstract representations of an instantaneous state of the world." She'll think that my times look more like abstract representations of the whole of history. She'll think, that is, that my times look more like abstract possible worlds than like abstract times. But given an A-theoretic or a dynamic view of time—the view that, necessarily, what exists and what is true are constantly changing—my definition is fine.

And third: well-known complications arise for my definition if *existentialism* is true. (Existentialism: the widely held doctrine that singular propositions depend for their existence on the individuals they are about, in this sense: for any $x_1 \ldots x_n$ and singular proposition $[Fx_1 \ldots x_n]$, necessarily, $[Fx_1 \ldots x_n]$ exists only if $x_1 \ldots x_n$ exist.)[21] So if we read ⌜possibly *S*⌝ in the above characterization of consistency in the standard way (i.e., ⌜possibly *S*⌝ is true in a context *C* iff the proposition expressed by *S* in *C* is true in some possible world), and we read ⌜*p* is true in world *W*⌝ in the standard way (i.e., *p* is true in a world *W* iff, were *W* actual, *p* would be true), we get trouble given existentialism. For given existentialism and plausible assumptions, it follows that there is no maximal, consistent class *C* of propositions such that [~(Bush exists)] is a member of *C*. Here's why. Suppose there is such a class; call it *C*. Then there's some possible world *W* such that, were *W* actual, every member of *C* would be true. So were *W* actual, [~(Bush exists)] would be true. But were [~(Bush exists)] true, Bush wouldn't exist, and given existentialism, neither would [~(Bush exists)]. So were *W* actual, [~(Bush exists)] would be true but fail to exist. But, one wonders, how could a proposition be true but fail to exist? It's a plausible assumption that this is impossible. If so, then there's no maximal, consistent class of proposition such that [~(Bush exists)] is a member of it. But if there is no maximal, consistent class of proposition such that [~(Bush exists)] is a member of it, then *every* maximal, consistent class of propositions includes [Bush exists] among its members. (A class *C* of propositions is maximal, recall, iff, for any proposition *p*, either *p* or its denial is a member of *C*. So if no maximal, consistent class of propositions includes [~(Bush exists)] among its members, it follows that every maximal, consistent class of propositions includes [Bush exists] among its members.) But from this and the above definition of times, it follows that, for any time *t*, were *t* true, Bush would exist—i.e., that Bush *always* exists. Since this clearly won't do, we see that the combination of existentialism and the above definition of times is an unhappy one.

The upshot of this is that my definition needs fiddling with if existentialism is true. I am persuaded by Plantinga (1983) and Bealer (1993)—for reasons that are independent of the question whether presentism is true—that existentialism is *false*, so I'm happy to leave the definition as is. Friends of existentialism will need to adjust it. One well-explored option is to join Robert Adams, Kit Fine, and other existentialists in distinguishing between truth *in* a world and truth *at* a world, where truth in a world is characterized as above, and truth *at* a world is thought of along the following lines.[22] Consider once again the proposition [~(Bush exists)] and let $W_{\text{no-Bush}}$ be a world according to which Bush never exists. Given existentialism, [~(Bush exists)] is not true *in* $W_{\text{no-Bush}}$. Still, though, there's a sense in which [~(Bush exists)] accurately describes what goes on in $W_{\text{no-Bush}}$. The language of truth *at* a world is intended to express the relationship that holds between a proposition *p* and a world *W* when, whether or not *p* is true *in W*, it accurately describes what goes on in *W* in the way that [~(Bush exists)] accurately describes what goes on in $W_{\text{no-Bush}}$.

Of course, it'd be nice to see this filled in some. For just what is the relationship that holds between [~(Bush exists)] and non-Bush worlds if [~(Bush exists)] isn't true in those worlds but nevertheless manages to describe what goes on in them? This turns out to be a ticklish question. I'm not sure how to answer it. But many existentialists think *they* have an answer to it. Suppose you're one of them. Then you can adjust the above definition of times by reconstruing its notion of consistency in terms of your notion of truth *at* a world. As follows:

> a class *C* of propositions is consistent iff there is a possible world *W* such that every member of *C* is true *at W*.

The problem with our earlier definition of times arose when we interpreted its talk of consistency in terms of truth *in* a world. Since, given existentialism, [~(Bush exists)] is true *in* no possible world, we got that no maximal consistent set of propositions includes [~(Bush exists)], and thus that every maximal consistent set of propositions includes its denial. But this problem does not arise on the new account of consistency, for the existentialist will think that there are plenty of worlds *at* which [~(Bush exists)] is true, and thus plenty of maximal consistent sets of propositions including [~(Bush exists)] among their members.

I realize that much more could be said to fill this in, but I shall leave that to my existentialist friends. I only wish to point out here that (a) existentialists who wish to endorse the view of tensed properties I'll develop below will need to amend my definition of times slightly, and that (b) there are well-known suggestions on offer in the literature on how to do so.

Thus far, three elaborations on my definition of abstract times. A word now about how these abstract times are related to one another. Here is a prosaic truth: some times are earlier than the present time, others are later.[23] We presentistic believers in abstract times need not doubt this. Abstract times, we can suppose, form an ersatz B-series, where an ersatz B-series is a series of abstract times ordered by a primitive *earlier than* relation. This suggestion raises several questions. Let us look briefly into three of them.

First, what is this *earlier than* relation ordering the members of an ersatz B-series? I'm inclined to answer that it requires no further explanation as it's the relation we learned at mother's knee when we learned that last Tuesday is earlier than today. Maybe you reply: no, the relation learned at mother's knee is a relation that holds between concrete things (events, times, or maybe spacetime points), not *abstract* times. I reply that this isn't obvious; not much hangs on our dispute though. My purposes are served well enough by supposing that the relation linking abstract times is an analogue of the eternalist's *earlier than* relation. Some points of analogy: Both relations link times—concrete times in the one case, abstract times in the other. The eternalist's relation, as usually construed, is transitive, irreflexive, and asymmetrical; my analogical *earlier than* relation is also transitive, irreflexive, and asymmetrical. (We need not be dogmatic about the formal properties of either relation; for instance, if time is eternally cycling, neither will be irreflexive or asymmetrical.) The eternalist's relation is non-ending and dense; so too with my analogical surrogate. (Though here again, we need not be dogmatic. Maybe time isn't dense or has an end.)

Second, I propose that abstract times are related by the *earlier than* relation. But isn't this hypothesis at odds with presentism? According to the presentist, everything is present. But then nothing—not even an abstract time—can be earlier than the present time; else not *everything* is present.

We presentists say that everything is present, but what do we mean by *present*? Some will invoke a primitive property of presentness and say that everything has that property. I think there's a better option. When the presentist says that everything is present, what she really means is that there is no *temporal distance* between any two things in our widest domain of quantification. (What is temporal distance? The general idea should be clear enough. Suppose eternalism is true. Then the temporal distance between Lincoln's assassination and now is a little over 139 years. As I understand it, the notion of temporal distance has broad application: spacetime formulations of Newtonian physics (e.g., Friedman 1983, chap. 3) postulate a temporal distance relation linking spacetime points; relativistic spacetime theories postulate a different temporal distance relation linking spacetime points. Both theories, I say, postulate temporal distance relations.) When she says that everything is present, then, she has in mind presentness understood thus:

$$x \text{ is present}_1 =_{\text{df.}} \forall y(x \text{ has no temporal distance from } y).$$

Everything is present, says the presentist; that is, everything is present$_1$. But if she believes in abstract times, she'll think there's another sort of presentness—presentness$_2$ let us say—such that only one among the infinity of times—the present time—has this property. What is this presentness$_2$? I'll suggest below that a time is present$_2$ iff it is the *true* time, the time which has the property *being true* (abstract times, recall, are just propositions). Our analogical *earlier than* relation will need to work in such a way that all times, no matter how much earlier or later than the present$_2$ time, are present$_1$. All times are present$_1$, at no temporal distance from anything, though some are earlier and some are later. (So times are like the actualist's worlds: The actualist says that everything is actual. Nevertheless, she believes that some worlds are merely possible. What gives? She'll explain it thus: Everything is actual$_1$, where something x is actual$_1$ iff, according to the actual$_2$

world, x exists, and something y is $actual_2$ iff y *obtains* or is *instantiated* or is *true* (depending on your brand of actualism). Though all possible worlds are $actual_1$, most aren't $actual_2$—or as it's usually put, most are merely possible.)

So understood, the supposition that abstract times form ersatz B-series is not inconsistent with presentism, construed as the claim that everything is $present_1$.

And third, abstract times, I say, form an ersatz B-series ordered by the *earlier than* relation. Is it that *all* abstract times are organized into a single B-series, or that there are many ersatz B-series, or what? Presumably not the first option: Below I'll suggest that a proposition p has the property *being past* (*being future*) iff p is entailed by an abstract time t such that t is earlier (later) than the present (i.e., the true) abstract time. But this suggestion is plainly crazy if *all* abstract times are organized into one, big ersatz B-series. (The suggestion is crazy because it's plausible that there is a maximal, consistent set of propositions that includes among its members the proposition, say, that pigs fly. But then given my definition of times, it follows that there is an abstract time t_{pig} that entails the proposition that pigs fly. If all propositions are organized into one, big ersatz B-series, then t_{pig} is either earlier or later than the present time. Suppose it's earlier. Then given my suggestion that a proposition p has the property *being past* iff p is entailed by an abstract time t such that t is earlier than the present time, we get the unfortunate result that [pigs fly] has the property *being past*—i.e., that WAS[pigs fly]. Since this plainly won't do, I'm committed to rejecting the suggestion that all abstract times are organized into a single ersatz B-series.)

Presumably, then, the idea is that there are *many* ersatz B-series. But then there is this question: What differentiates the *actual* history of the world—the ersatz B-series representing the way history actually unfolds—from the merely possible ones, the ones representing how history could have unfolded? Is it that there is some kind of primitive *actuality* property that distinguishes the one actual history from the infinity of merely possible histories? One hopes not: thus supplemented, the theory is starting to seem a tad baroque. But if not a primitive actuality property distinguishing the one actual history from the infinity of merely possible ones, then what?[24]

There are various options here. I think the presentist should hold that there is one and only one ersatz B-series, all right, but that it does not include *all* the abstract times among its members. It counts among its members only *some* of the abstract times—those that did, do, or will represent the world. (So it does not count the above t_{pig} among its members.) The rest of the abstract times aren't members of it or any other ersatz B-series on account of the fact that they bear the *earlier than* relation to no times whatsoever. To be sure, many of these abstract times are such that they *could* have born the *earlier than* relation to other times, but as a matter of contingent fact, they don't. What abstract times get "hooked up" by the *earlier than* relation, on this picture, is a contingent matter that varies from possible world to possible world.

(So the suggestion here is that the *earlier than* relation connects certain abstract times and not others, though which abstract times it connects is something that could have been different. Well, one might wonder, why wasn't it different?

Why does the *earlier than* relation connect just the times it does? What explains the fact that it connects these times and not others?

Note that similar questions arise for the eternalist who believes in concrete times. Why do the concrete times come temporally ordered in the way they do? What explains the fact that they have this order and not another? Eternalists have answered such questions in various ways, but as plausible an answer as any is that *they just do* come ordered this way, and there's an end 'ont. It's a brute, contingent fact, on this view, that concrete times come temporally ordered as they do.

Likewise for the presentist: as plausible an answer as any to the penultimate paragraph's questions is that it's a brute, contingent fact that the abstract times come temporally ordered as they do. Explanation has to come to an end somewhere, and it's not unreasonable to suppose that it bottoms out in the contingent fact that certain times are earlier than certain other times.)

4.3 Characterizing Being Present, Being Past, and Being Future

We can now see how the presentist can do without primitive pastness, presentness, and futurity. She need simply take the foregoing *earlier than* relation as primitive, and say that a time is *past* $\text{iff}_{\text{df.}}$ it is earlier than the present time, that a time is *future* $\text{iff}_{\text{df.}}$ it is later than the present time, and that the present time $\text{is}_{\text{df.}}$ whatever time happens to be true. In general, she'll say that a proposition *p* is past (present, future) iff *p* is *included* in a past (present, future) time, where *p* is included in *t* iff *t* entails *p*.[25] On this picture, Bigelow's property *being an x such that dinosaurs roamed x* turns out to be something more like: *being an x such that the proposition that dinosaurs roam x is included in an earlier time.*

5. Objections and Replies

The dialectical situation, then, is this. The grounding objection depends on the supervenience principle and the temporal recombination principle:

> **(SP)** For any proposition *p* and worlds *w* and *w**, if *p* is true in *w* and not in *w**, then (a) according to *w*, something exists that doesn't exist in *w** (or vice versa), or (b) according to *w*, some objects instantiate a fundamental property or relation that they don't instantiate in *w** (or vice versa),

And

> **(TR)** Things could have been just as they are at present (same things in existence, same fundamental properties and relations instantiated) and the past have been different

respectively. The Lucretian objects to the latter on the grounds that present things have properties like *being formerly roamed by dinosaurs*. Sider objects that no serious ontology will have truck with such properties. Primitive properties like *being formerly fat*, *being possibly bald*, *being brave should such-and-such a circumstance arise*—all are dubious because all are irreducibly hypothetical, where a property is hypothetical, roughly, when it points beyond its instances to how things could be, would be, were, or will be. Serious ontologies invoke only categorical or occurrent properties, properties that characterize how a thing actually is. I suggested a way of understanding the hypothetical/categorical distinction on which properties like *being formerly fat* and *being possibly bald* are hypothetical, but analyzable in terms of non-hypothetical properties like *being past* and *being possible*. Says Sider: the latter properties are no better off than their fancier compeers. I replied that this isn't obvious, then considered an objection to the effect that the Lucretian needs at least some tensed properties to be fundamental—to carve at the joints, to ground qualitative similarity—but that none of the tensed properties she postulates seem well-suited for the fundamental property role. I replied that her property *being past* isn't obviously unsuited for that role, but that nothing crucial turns on it because she can analyze this and like properties in terms of the B-theoretic *earlier than* relation (or an analogue thereof), a relation that *is* plausibly thought of as fundamental.

I conclude by considering a few objections.

Objection: your *earlier than* relation is mysterious; you haven't explained what it is for one time to bear this relation to another. I reply that no theory is without primitives; this relation is one of my theory's primitives. Besides, it's not as if I said nothing to explain the relation. I explained several points of analogy between it and the *earlier than* relation invoked by many eternalists.

Objection: this *earlier than* relation of yours is no better than primitive pastness, presentness, and futurity. Just as one can just *see* that these properties are dubious ontological posits, so too, one can just see that your ordering relation on abstract times is unacceptable. I reply that *I* can't see any such thing and I doubt anyone else can either. Why should my *earlier than* relation be any more dubious than the eternalist's B-theoretic *earlier than* relation on times, events, and the like? True, my relation links abstract times; hers links concrete times. But why think that a demerit for my relation?

Objection: your theory costs in primitives; we eternalists don't need this primitive, ersatz temporal ordering relation on abstract times. True enough, eternalists don't need this relation. But they need their own *earlier than* relation. And, it looks as if their relation will either be primitive or defined in terms of primitive temporal notions. (The project of defining temporal notions in terms of non-temporal—e.g., causal—notions is notoriously difficult.) If there's cost to be paid in primitives here, it's a cost which looks to be exacted of both presentism and eternalism.

Objection: unlike the B-theoretic eternalist's *earlier than* relation, your ersatz *earlier than* relation on abstract times is not plausibly thought of as fundamental, as carving the world at the joints. Consequently, it can't do the work you need it to do.

In reply, why think this? Why would my *earlier than* relation on abstract times be any less suited to the fundamental relation role than the eternalist's *earlier than* relation on concrete times? I can't see any reason why it would be.

I'm out of ideas now about what might be wrong with my *earlier than* relation or my definition of pastness and futurity in terms of it. It looks to me like the Lucretian needn't take properties like *being past* as primitive; the work done in a presentistic metaphysic by primitive tensed properties like *being past*, *being present*, and *being future* can be done equally well by the B-theoretic *earlier than* relation. Since this relation seems well-suited to the role of fundamental relation, I conclude that the Lucretian has an attractive reply to the above worry that the primitive properties and relations she postulates don't seem very plausible candidates for the role of fundamental property/relation.

In summary: The grounding objection poses no real threat to presentism. The most powerful form of the objection relies on the supervenience principle, the principle that there could be no difference in what is true without there also being a difference in what things exist and which fundamental properties and relations they instantiate. The presentist can accept this principle with equanimity. The grounding objection also relies on the temporal recombination principle, the claim that things could have been just as they are at present and the past have been different. The presentist can reject this thesis by endorsing Lucretianism. We examined the worry that doing so commits her to objectionable primitives. I think it's clear that this worry is unfounded.[26]

Notes

1. More exactly, presentism is the claim that it is always the case that, quantifying unrestrictedly, for every x, x is present. For more on defining presentism, see, e.g., Crisp 2004a, 2004b; Hinchliff 2000; Ludlow 2004; Markosian 2004; Rea 2003; Sider 2001; and Zimmerman 1998.

2. See, for example, Sider 2001, chap. 2. For discussion, see Bigelow 1996, Keller 2004, and Rea 2003.

3.Cf. John Fox (1987, 189): "By the truthmaker axiom I mean the axiom that for every truth there is a truthmaker; by a truthmaker for A, I mean something whose very existence entails A."

4. See, e.g., Lewis 1992.

5.Fundamental properties and relations: i.e., perfectly natural properties and relations—the ones that "carve at the joints." For a classic discussion, see Lewis 1983.

6. See Lewis 2001, 609ff., and Bigelow 1988, 133. For illuminating discussion, see Keller 2004.

7.Cf. Lewis 2001, 612.

8.Note well: if truth and falsity are fundamental properties, (TR) requires slight reformulation. For suppose truth and falsity are among the fundamental properties. Then, for the presentist anyway, things couldn't have been just as they presently are and the past different because among the present things are propositions about the past—e.g., dinosaurs roamed the earth—and, by hypothesis, among their fundamental properties are properties like being true. Surely things couldn't have been presently such that this very proposition has the truth-value it does and dinosaurs not have roamed the earth! But (TR) and the grounding objection are easily reinstated. If truth and falsity are fundamental properties, simply adjust (TR) to read "things could have been just as they are at present (same things presently in existence, same fundamental properties and relations presently instantiated—ignoring the present distribution of truth, falsity and fundamental properties that entail them) and the past have been different," where a property F entails a property G iff, necessarily, something has F only if it has G.

9. For discussion, see Bigelow 1996, 44-47.

10. Two properties F and G are equivalent iff, necessarily, something instantiates F iff it instantiates G.

11. I've assumed throughout the above discussion that sentences like "x is round" and "x has roundness," may be used interchangeably and that quantification into predicate positions is unproblematic. Some will balk at these assumptions, but nothing crucial turns on them. The above discussion could be reformulated (at some cost in concision) without them.

12. See his 2001, 40-41.

13. Such a reply captures the spirit if not the letter of his remarks in 2001, 41.

14. Lewis 1983.

15. Thanks to an anonymous referee for raising this objection.

16. See Chisholm and Zimmerman 1997.

17. Chisholm 1979.

18. Zalta 1987.

19. See, e.g., Prior and Fine 1977, and Davidson 2003 and 2004.

20. Cf. Prior and Fine 1977, 121.

21. For discussion, see inter alia Adams 1981, Bealer 1998, Plantinga 1983, and Davidson 2000.

22. For discussion of the distinction between truth in and truth at a world, see inter alia Adams 1981, Davidson 2000, Prior and Fine 1977 and 1981, Fitch 1996, and Turner 2005.

23. Perhaps the "others are later" part isn't as prosaic. Believers in a Broad/Tooley style growing block universe (Broad 1923; Tooley 1997) won't find that part as prosaic.

24. Thanks to an anonymous referee for posing this question.

25. See Crisp 2003 and Davidson 2004: 19 for a related approach.

26. Thanks to Matt Davidson, Brian Kierland, Trenton Merricks, Alvin Plantinga, Mike Rea, James Rissler, Ted Sider, Gregg Ten Elshof, and participants in the 2002 Metaphysical Mayhem at Syracuse University for helpful comments and/or conversation about previous drafts of this paper.

References

Adams, R. M. "Actualism and Thisness," *Synthese* 49 (1981): 3-41.

Armstrong, D. M. *A World of States of Affairs*. Cambridge: Cambridge University Press, 1997.

Bealer, G. *Concept and Quality*. Oxford: Clarendon Press, 1982.

———. "Universals," *The Journal of Philosophy* 90 (1993): 5-32.

———. "Propositions," *Mind* 107 (1998): 1-32.

Bigelow, J. *The Reality of Numbers: A Physicalist's Philosophy of Mathematics* (Oxford:

Oxford University Press, 1988.
———. "Presentism and Properties," *Philosophical Perspectives* 10 (1996): 35-52.
Broad, C. D. *Scientific Thought*. London: Keegan Paul, 1923.
———. *Examination of McTaggart's Philosophy*. Cambridge: Cambridge University Press, 1938.
Chisholm, R. M. "Objects and Persons: Revisions and Replies," *Grazer Philosophische Studien*, 7/8 (1979): 317-61.
Chisholm, R. M., and D. Zimmerman. "Theology and Tense," *Noûs* 31 (1997): 262-65.
Crisp, T. "Presentism." Pp. 211-45 in *The Oxford Handbook of Metaphysics*, edited by M. D. Loux and D. Zimmerman. Oxford: Oxford University Press, 2003.
———. "On Presentism and Triviality." Pp. 15-20 in *Oxford Studies in Metaphysics*, edited by D. Zimmerman. Oxford: Oxford University Press, 2004a. Chapter 7 in this volume.
———. "Reply to Ludlow." Pp. 37-46 in *Oxford Studies in Metaphysics*, edited by D. Zimmernan. Oxford: Oxford University Press, 2004b.
Davidson, M. "Direct Reference and Singular Propositions," *American Philosophical Quarterly* 37 (2000): 285-300.
———. "Presentism and the Non-Present," *Philosophical Studies* 113 (2003): 77-92. Chapter 11 in this volume.
———. Critical Notice of Sider 2001. *Philosophical Books* 45 (2004): 17-33
Fine, K. "First Order Modal Theories," *Studia Logica* 39 (1981): 159-202.
Fine, K., and A. Prior. *Worlds, Times, and Selves*. Amherst: University of Massachusetts Press, 1977.
Fitch, G. "In Defense of Aristotelian Actualism," *Philosophical Perspectives* 10 (1996): 53-71.
Fox, J. F. "Truthmaker," *Australasian Journal of Philosophy* 65 (1987): 188-207.
Friedman, M. *Foundations of Space-Time Theories: Relativistic Physics and Philosophy of Science*. Princeton, N. J.: Princeton University Press, 1983.
Hinchliff, M. "A Defense of Presentism in a Relativistic Setting," *Philosophy of Science* 67 (Proceedings) (2000): S575-S586. Chapter 14 in this volume.
Keller, S. "Presentism and Truthmaking." Pp. 83-106 in *Oxford Studies in Metaphysics*, edited by D. Zimmerman. Oxford: Oxford University Press, 2004. Chapter 15 in this volume.
Lewis, D. "New Work for a Theory of Universals," *Australasian Journal of Philosophy* 61 (1983): 343-77.
———. "Review of D. M. Armstrong, A Combinatorial Theory of Possibility," *Australasian Journal of Philosophy* 70 (1992): 211-24.
———. "Truthmaking and Different-Making," *Noûs* 35 (2001): 602-15.
Lucretius Carus, Titus. *On the Nature of the Universe*. Translated by R. E. Latham, revised with an introduction and notes by J. Godwin. London, New York, Victoria, Toronto, Auckland: Penguin Books, 1994.
Ludlow, P. "Presentism, Triviality and the Varieties of Tensism." Pp. 21-36 in *Oxford Studies in Metaphysics*, edited by D. Zimmerman. Oxford: Oxford University Press, 2004.
Markosian, N. "A Defense of Presentism." Pp. 47-82 in *Oxford Studies in Metaphysics*, edited by D. Zimmerman. Oxford: Oxford University Press, 2004.
Mulligan, K., P. Simons, and B. Smith. "Truth-Makers," *Philosophy and Phenomenological Research* 44 (1984): 287-321.
Plantinga, A. "On Existentialism," *Philosophical Studies* 44 (1983): 1-20.

Plantinga, A., and P. Grim. "Truth, Omniscience, and Cantorian Arguments: An Exchange," *Philosophical Studies* 71 (1993): 267-306.

Prior, A. N., and K. Fine. *Worlds, Times and Selves*. Amherst: University of Massachusetts Press, 1977.

Rea, M. "Four-Dimensionalism." Pp. 246-80 in *The Oxford Handbook of Metaphysics*, edited by M. J. Loux and D. Zimmerman. Oxford: Oxford University Press, 2003.

Restall, G. "Truthmakers, Entailment and Necessity," *Australasian Journal of Philosophy* 74 (1996): 331-40.

Sider, T. *Four-Dimensionalism: An Ontology of Persistence and Time*. Oxford: Clarendon Press, 2001.

——. "Reductive Modality." Pp. 180-210 in *The Oxford Handbook of Metaphysics*, edited by M. J. Loux and D. Zimmerman. Oxford: Oxford University Press, 2003.

Tooley, M. *Time, Tense & Causation*. Oxford: Oxford University Press, 1997.

Turner, J. "Strong and Weak Possibility," *Philosophical Studies* 125 (2005): 191-217.

Zalta, E. N. "On the Structural Similarities between Worlds and Times," *Philosophical Studies* 51 (1987): 213-39.

Zimmerman, D. "Temporary Intrinsics and Presentism." Pp. 206-19 in *Metaphysics: The Big Questions*, edited by P. v. Inwagen and D. Zimmerman. Malden, Mass: Blackwell Publishers, 1998. Chapter 17 in this volume.

Part V

Presentism and Persistence

Introduction: Presentism and Persistence

L. Nathan Oaklander

The problem of time is not only interesting in its own right, but it is a topic that is central to many issues in other areas of philosophy such as the philosophy of science, philosophy of religion, and metaphysics to name a few. The final two essays, by Dean Zimmerman and Trenton Merricks, connect time with a fundamental problem of metaphysics, namely, the problem of change or David Lewis' problem of temporary intrinsics (Lewis 1986: 202-4). Zimmerman attempts to defend the presentist solution to this problem against Lewis' rejection of presentism. In the course of his defense he explains the difference between the serious tenser (propositions change their truth value in virtue of what is true *now*) and the detenser (propositions have an unchanging truth value). He also explains the method of paraphrase as the basis of determining ontological commitment, and avoiding non-presentist commitments to undesirable entities. Merricks, on the other hand, does not attempt to defend presentism against one of its main rivals, what he calls "indexicalism," the view that "all times are on an ontological par, and the present time is special only because it is the time at which we (or our current time-slices, current utterances, current actions or current thoughts, etc.) are located; the present time, according to this position, is merely *this* time" (chapter 18, p. 319). Rather, he argues for the difference between these two views by drawing attention to the different implications each has regarding their response to the problem of temporal intrinsics.

The problem of temporary intrinsics is a version of the problem of change. Change involves one and the same thing having incompatible properties at different times. An apple changes if, say, it is first green and then red. The problem of change involves giving some account of time and some account of identity that allows for change to be possible. In his compact discussion Lewis (1986, pp. 202-4) considers three accounts of change and rejects two of them. On one account, which he calls "endurantist," there is a single thing, a continuant that is wholly present and remains literally the same and wholly present though time and change. On the endurantist view, one and the same apple is green at one time and red at another time. Where "a" is the apple, "t_1" and 't_2" are times, and "G" and "R" are the (relational) properties of *being green at* and *being red at* respectively, such an account can be symbolized as G (a, t_1) & R (a, t_2).

Lewis objects, and Zimmerman and Merricks agree, that though such a move eliminates any incompatible properties problem in the changing apple, it turns the monadic, non-relational properties of *being red* and *being green* into relations between an apple and a time. This is phenomenologically false, and metaphysically suspect. There is clearly a difference between properties such as *being red, being round*, and *being middle C*, that are not relations, and properties such as *being taller than, being larger than* and *being the brother of*, which clearly are relations. Moreover, although Lewis doesn't mention it, the endurantist analysis commits one to the existence of moments of absolute time as one term of the relation.

The second account that Lewis considers and rejects is presentism. According to presentism, the problem of change is avoided by rejecting the premise upon which it is based, namely, that one and the same thing *has* incompatible properties. For the presentist, the only properties that a thing *has* are those it *presently* has. Since the presentist maintains that only present things exist, he maintains that the only properties a thing has are those it presently has. Thus, for the presentist the problem of temporary intrinsics, or how an object can have incompatible intrinsic properties does not arise since there are no times other than the present at which an object can have incompatible properties. It is true that an object *had* properties other than those it presently has, and *will have* properties other than those it now has, but it is not the case that an object can *have*, that is, *now* possess or exemplify incompatible properties.

Lewis finds the presentist solution to the problem of temporary intrinsics unacceptable because it is too fantastic to believe. For if only the present exists, then nobody has a future or a past since there are no times other than the present, and that belief "goes against what we all believe" (1986, 204). I shall discuss Zimmerman's defense of presentism against Lewis' objections shortly, but first to the solution Lewis accepts.

According to Lewis' preferred solution there are no enduring entities, or continuants that are wholly present at each moment of their existence. Rather, entities persist through time by perduring. A perduring entity is one that has different temporal parts for each of the different times at which it exists. When an object changes it has different temporal parts that exemplify different monadic properties, and while temporal parts are numerically different from each other, they are so related to each other to be called parts of one and the same thing.

Zimmerman favors the presentist solution to the problem. Like many of the previous presentists we have considered, Zimmerman understands that view as "a thesis about the range of things to which one should be 'ontologically committed'" (p. 308). On the presentist view, the range of ontological commitment is limited to what exists at present. "The 'once was' no longer exists and the 'will be' does not yet exist" (p. 308). It appears, however, that we are committed to entities that the presentist wants to disavow since we make judgments about the past and the future. According to the truthmaker principle, if a statement is true then there must be something in the world that makes it true. The problem of truthmaker for the presentist is that the statement

> (5) There was a person who is not one of the people who presently exist (or, more simply, George Washington did exist, but does not presently exist)

seems to require the existence of a past person, George Washington. However, the presentist is bothered by the past existence of George Washington since she worries how there can *be* something that does not exist.

To avoid ontological commitment to undesirable entities Zimmerman employs the method of paraphrase. This method involves taking statements that seem to have unwanted ontological implications and giving them philosophical glosses or paraphrases which seem to capture the truth in question, "while avoiding the implication that the troublesome things exist" (p. 310). Thus, regarding (5) Zimmerman suggests it be paraphrased as the proposition that there is an individual essence not now exemplified that once was exemplified and was then the essence of that person. Unfortunately, Zimmerman's paraphrase doesn't help avoid reference to the existence of a past individual, since "was once exemplified" raises the question of the semantic correlate of "was." This is problematic since "was once exemplified" seems to make reference to a past time at which the individual essence is exemplified.

This problem reemerges in Zimmerman's response to Lewis' objection to presentism that we are now ready to consider. Just as Zimmerman argues a wholesale acceptance of the truthmaker principle may warrant unacceptable ontological commitments that can be avoided through suitable paraphrases, Zimmerman claims that a similar strategy can avoid Lewis' objection that presentism violates our intuitions about persistence. Lewis argues that presentism denies what we all believe, namely, that things persist through change, that is, have different and incompatible properties at different times. The presentist solution is that nothing *has* incompatible properties since it is never the case that, say, I have the property *being straight* and have the property of *being bent.* Rather, I was straight and I will be straight again, but there are no past or futures times at which I am straight. But then, Lewis argues, presentism must reject the truism that things persist through change, which Zimmerman summarizes as:

> (PC) Three are (at least) two different times; one at which I am bent, another at which I am straight.

Although this commonsense belief is not to be denied, it does not wear its ontological commitment on its sleeve. Indeed, according to Zimmerman (PC) does not imply a commitment to past and future times at all. "When I ask myself whether I think that my childhood exists, or the time of my death . . . [t]he answer comes back a resounding No." Thus, he sets himself the task of paraphrasing (PC) in such a way that it captures what is meant but does not involve direct reference to other times. The analysis or paraphrase he gives is as follows:

> Either I was bent and would become or had previously been straight, or I was straight and would become or had previously been bent, or I will be bent and will have been or be about to become straight, or I will be straight and will have been or be about to become bent. (p. 313)

Of course the problem here is the same as plagued his attempt to paraphrase claims about past individuals, namely, that the appeal to the grammatical tenses (for example, had previously been bent, or I will be bent) leaves unanalyzed precisely what needs to be analyzed and therefore begs the question of whether or not his alleged paraphrase actually does away with what presentists believe are unacceptable ontological commitments.

As I indicated at the outset, Merricks is less concerned with defending presentism than with spelling out the implication of it and indexicalism (Merrick's term for eternalism or the B-theory) as responses to Lewis' problem of temporary intrinsics. His main thesis is that

(1) Presentism entails that there are no perduring, four-dimensional objects; and
(2) Indexicalism entails that there are no enduring, three-dimensional objects.

He concludes by demonstrating, so he believes, that a single world cannot contain both temporal extended perduring entities (objects or events) and three dimensional entities (objects or events).

The argument for (1) is based on the seemingly obvious claim that

(3) An object cannot have another object as a part if that other object does not exist.

Given (3) and presentism, it follows that the only parts a thing can have are those that it presently has, since the only parts that exist are those that presently exist. Since not all of a four-dimensional object's parts exist at the present time, it must have distinct temporal parts that exist at distinct times. However, since the presentist maintains that the only times are present, it follows that "not all of a perduring object's parts could exist at the single time at which it is present" (p. 320). Thus, if presentism is true then there are no perduring objects. In slightly different terms, since not all the parts of an object which lasts over time exist at a single time, presentism is incompatible with perdurance, the view that things persist by having different temporal properties at different times.

For a presentist, an object endures by being wholly present at each time at which it exists. To say that an object is "wholly present" at a time is to say that all the parts an object has, and all the properties it has, exist at each time that the object exists. This might appear to present a problem since an object can change its parts over time and it can have different properties at different times. How then can all the parts and properties an object has be at every time an enduring object exists? In other words, if an object is wholly present at t if and only if it has at t all the parts and properties it ever will have, then it appears to follow that an object has different properties and parts at the same time, indeed, at every time, and that is absurd. While one may question the notion of an object's being "wholly present," Merricks does not take that tack. Rather, his solution to the problem is presentism. An object has only those properties and parts that it has at the present time. Thus, presentism and endurantism are compatible.

The combination of indexicalism and endurantism is however, impossible, according to Merricks. According to endurantism,

> (4) It is possible that an enduring object O is F, and that at some time in the past O was ¬F.

For the presentist there is no contradiction here since the only properties an object has are those which it presently has. For the indexicalist, however, all times exist equally, so we can conclude that if a single object is F at one time, and is ¬F at another (and the object is wholly present at both times), then that object both is and is not F" which is contradictory. Thus, if one is an indexicalist, then one must be a perdurantist and claim that objects change their properties by having different temporal parts which have different properties at different times. Thus, (2) is also true.

In the concluding section Merricks argues that in addition to the incompatibilities demonstrated so far there is the further and more general consequence, namely, that perduring entities and enduring entities are incompatible. The argument is quite straightforward although the conclusion is not intuitive. Given that (1) and (2) are true, Merricks demonstrates that presentism is incompatible with all perduring entities (objects and events) and that indexicalism is incompatible with all enduring entities. Since either presentism or perdurantism are true it follows that either there are no perduring entities or there are no enduring entities, according to Merricks.

There are many aspects of Merricks' reasoning that can be criticized. Some who have opposed presentism have argued that endurantism and indexicalism (see Mellor 1998 and Wahlberg 2009a,b) and continuants and temporal parts (Tegtmeier 2008) are compatible. Others have maintained that presentism and four-dimensionalism are compatible (Brogaard 2000). Merricks also has his defenders (Benovsky 2007 and 2009). The reader who is interested in these issues may wish to consult the writings of these authors further.

References and Further Readings

Benovsky, J. "On Presentist Perdurantism," *SATS: Nordic Journal of Philosophy* 8 (2007): 79–88.

———. "Presentism and Persistence," *Pacific Philosophical Quarterly* 90 (2009): 291-309.

Brogaard, B. "Presentist Four-Dimensionalism," *Monist* 83 (2000): 341-56.

Craig, W. L. "McTaggart's Paradox and the Problem of Temporary Intrinsics," *Analysis* 58 (1998): 122–27.

Carter, W. S. , and H. S. Hestevold. "On Passage and Persistence," *American Philosophical Quarterly* 31 (1994): 269–84.

Geach, P. "Some Problems about Time." Pp. 302-18 in *Logic Matters*, by P. Geach. Berkeley and Los Angeles: University of California Press, 1972.

Haslanger, S. "Endurance and Temporary Intrinsics,"*Analysis* 49 (1989): 119–25.

Hinchliff, M. "The Puzzle of Change," *Philosophical Perspectives*10 (1996) ("Metaphysics"): 119–36.

Lewis, D. *On the Plurality of Worlds*. Oxford: Blackwell Publishing, 1986.

Mellor, D. H. *Real Time II*. London: Routledge, 1998.

Merricks, T. "Endurance and Indiscernibility," *Journal of Philosophy* 91 (1994): 165–84.

———. "On the Incompatibility of Enduring and Perduring Entities," *Mind* 104 (1995): 523–31.

———. "Persistence, Parts, and Presentism,"*Noûs* 33 (1999): 421–38.

Oaklander, L. N. "Craig on McTaggart's Paradox and the Problem of Temporary Intrinsics," *Analysis* 59 (1999): 314–18.

Tegtmeier, E. "Persistence." Pp. 185–98 in *Persistence*, edited by C. Kanzian. Frankfurt: Ontos Verlag, 2008.

Wahlberg, T. H. "The Problem(s) of Change Revisited," *Dialectica*, 61, 2 (2007): 265–74. Reprinted in *Objects in Time: Studies of Persistence in B-time*. Ph.D. Thesis, Department of Philosophy, Lund University, 2009a.

———. "Endurance Per Se in B-time," *Metaphysica* 10 (2009): 175–83. Reprinted in *Objects in Time: Studies of Persistence in B-time*. Ph.D. Thesis, Department of Philosophy, Lund University, 2009b.

Zimmerman, D. W. "Persistence and Presentism,"*Philosophical Papers* 25 (1996): 115–26.

———. "Temporary Intrinsics and Presentism." Pp. 206-19 in *Metaphysics: The Big Questions*, edited by P. van Inwagen and D. Zimmerman. Malden, Mass: Blackwell Publishers, 1998. Chapter 17 in this volume.

17
Temporary Intrinsics and Presentism

Dean Zimmerman

David Lewis develops something like an antinomy concerning change which he calls "the problem of temporary intrinsics." The resolution of this puzzle provides his primary motivation for the acceptance of a metaphysics of temporal parts.[1] Lewis's own discussion is extremely compressed, showing up as a digression in a book about modality. So I shall set forth in some detail what I take to be his line of reasoning before suggesting that, at least for those philosophers who take seriously the distinction between past, present, and future, the argument poses no special threat.

The Structure of Lewis's argument

Lewis's argument for temporal parts has the following structure. He offers reasons to deny that "the only intrinsic properties of a thing are those it has at the present moment"[2]—reasons, that is, for rejecting the "second solution" he considers. But if, in addition to the intrinsic properties I have now, I also have the intrinsic properties I have at other times, then I will end up having pairs like *being bent* and *being straight*—pairs that are, in some sense, incompatible. The challenge is then to answer the question: How can I have a pair of incompatible properties?[3] Lewis thinks there are only two possible ways to answer this question. The first is unacceptable, and the second leads to the doctrine of temporal parts:

> (1) My being both bent and straight is like my son's being both tall and short—tall for a two-year-old, say, but short by comparison to most people. This strategy for dealing with apparent contradiction construes the seemingly incompatible properties as really relations to other things (in the case of tall and short, relations to different comparison classes). The version of this strategy that Lewis considers for temporary intrinsics is his "first solution": that shapes and other seemingly intrinsic properties "are disguised relations, which an enduring thing may bear to times."[4] There is

> no more difficulty in standing in the *bent-at* relation to one time and the *straight-at* relation to another than there is in bearing the *tall-for-a* relation to two-year-olds and the *short-for-a* relation to the citizens of the United States as a whole. But Lewis doesn't like this solution; he thinks it is tantamount to the rejection of intrinsic properties altogether.
>
> (2) There's only one way left, says Lewis, to make the apparent contradiction go away while retaining the incompatibility of *being bent* and *being straight*; and that is to treat it as we do the case of the road that is both bumpy and smooth. How can a road be both? Easily: by having one part that is bumpy and another that is smooth. So, analogously, the only way for me to be both bent and straight is for me to have a part that is bent and a part that is straight. But these cannot be ordinary spatial parts of me, like an arm or a hand. The bent "part" of me is exactly my size and shape, with arms, legs, torso, and head; and likewise for the straight "part" of me. And, like the different spatial parts of the road, these different parts of me must be distinct one from another. So I emerge as a whole spread out along the temporal dimension with different (temporal) parts for the different times I occupy, much as the road is a whole spread out along the spatial dimension with different (spatial) parts for the different places it occupies.

I am willing to grant Lewis's assertion that, once someone admits that I have more properties than just those I have now, she must choose between alternatives (1) and (2). And perhaps it is true that (1) eliminates temporary intrinsics altogether. At the very least, it eliminates temporary *monadic* properties ("one-place" properties, properties that are not relations); and it's easy to see why someone might think that *really* intrinsic properties should be monadic.[5] What I want to question instead is the very first move: Why suppose that I must have more than just the properties I have now?

Serious Tensers and Presentists

Before looking at Lewis's answer, I want to make clear what view Lewis is targeting: namely, "presentism." A closely related position is that of one who "takes tense seriously." As shall appear, one can't very well be a presentist without taking tense seriously, although it's possible to do the reverse.

When a philosopher says, "The only properties I have are those I have now," it is tempting to respond by saying: This thesis is either an uninteresting, tautologous truth; or an obvious falsehood. If the first occurrence of "have" is in the present tense, then the assertion is equivalent to "The only properties I have now are those I have now." Who could disagree? But how dull! On the other hand, suppose this "have" is an instance of what philosophers sometimes call a "tenseless" verb. To say that I (tenselessly) have some property, for instance that I (tenselessly) am straight, is to say something more or less equivalent to this: I either was straight, or I am straight, or I will be straight. But "The only properties I (tenselessly) have are those I have now" is true only if either I never change or I exist for but an instant. Taken, then, in the only way in which it can be true (i.e., with the first "have" in the present tense), the claim seems too trivial to be the focus of a substantive philosophical debate.

I am convinced that there *is* an important disagreement between those who take tense seriously and those who don't. Precisely what the disagreement boils down to will depend to some extent upon metaphysical theses about what kinds of things are, in the first instance, true and false. Here is one example; but I believe that nothing much hinges on accepting just this view about the most fundamental bearers of truth. Suppose you think that the sentences we write down and utter are true or false in virtue of their expressing *propositions* that are true or false in some more basic sense. A proposition is something that can be expressed in many different ways; it can be believed by one person and disbelieved by another; and, at least in the case of a proposition that isn't about a particular sentence or thought, it would have existed and been either true or false even in the absence of all sentences or thoughts. This familiar conception of the ultimate bearers of truth and falsehood[6] can be conjoined with a tensed or a tenseless theory about the nature of the proposition. On a tensed construal, a proposition's being true is not typically a once-and-for-all thing. The sentence "I am bent" could now be used by me to express a true proposition; but the proposition in question hasn't always been true, and it won't continue to be true for very long. A tenseless account of propositions, on the other hand, takes them to be like statements made using tenseless verbs: each is either always true, or never true.

The competition between the tensed and tenseless approaches to the fundamental bearers of truth gives rise to a familiar dispute over the importance of "tense logic." Logic is all about describing the most general patterns of truth-preserving inference. If the things that are true and false can be true but *have been false*, or be *about to become false*, then some of the patterns of inference logicians should be interested in will involve temporal notions. On the tensed conception of truths, it is a question of logic whether, for example, the proposition: It will be the case that I am bent, implies the proposition: It was the case that it will be the case that I am bent. Thus relations like being true simultaneously, and being true earlier or later than, will turn out to be, at least in part, logical notions.[7] On the other hand, those who take truth-bearers to correspond to tenseless statements will regard this as a blunder: temporal relations are for science and (perhaps) metaphysics to explore; but they are not part of the subject matter of logic.[8]

The philosopher who takes a tensed approach to the bearers of truth regards each of them as making a claim about what is the case *now*. Of course some propositions are eternally true: in other words, there are propositions which, either necessarily, or as a matter of contingent fact, have always been true and will always be true. That two and two make four is an example of the first sort. And historical propositions expressed by tenseless statements, such as my utterance in a lecture of "Plato believes in universals," are examples of the latter sort. But the proponent of tensed truth-bearers will insist that the true proposition expressed is composed of tensed propositions; it's a disjunction of three propositions: Either Plato (now) believes in universals, or he did, or he will.[9] This is a truth, but it's made out of three other propositions, only one of which is true, and each of which concerns what is now the case.

I shall call a philosopher who takes this sort of position a "serious tenser." Those who do not take tense seriously include D. H. Mellor, J. J. C. Smart, and others who defend what is sometimes called the "new tenseless theory of time": the thesis that the meaning of every tensed utterance can be captured by stating, in a tenseless language of eternal truths, the conditions in which the utterance would be true.[10] It should be evident that there can be a real dispute between the serious tenser and someone who rejects this view.[11]

Many serious tensers are also *presentists*. The presentist says: The only things that exist are those that exist at present. The "once was" no longer exists and the "will be" doesn't exist yet. But the proponents of presentism are also confronted with a skeptical challenge to the significance of their thesis. Is the first occurrence of "exist" in the presentist's assertion a tensed one? Then the presentist is simply making a fuss over a pointless tautology: "The only things that exist now (i.e. at present) are those that exist at present." Who denies this? Or is "exist" here a tenseless verb, equivalent to "existed or exists now or will exist"? But then it's an implausible metaphysical thesis: the claim that everything exists at all times, that nothing can have a less than eternal history. So either presentism is a boring truth, or an interesting falsehood.

Presentism is neither; it is a substantive thesis, and one that is not equivalent to the claim that everything exists eternally. Just as the serious tenser thinks there is, at bottom, only one kind of truth, and that is "truth-now," so the presentist thinks there is only one largest class of all real things, and this class contains nothing that lies wholly in the past or future.[12] Presentism is, in fact, a thesis about the range of things to which one should be "ontologically committed."

Philosophers are always looking out for the ontological commitments of their views—where someone is ontologically committed to a certain kind of thing just in case something she believes implies that something of that kind exists. There are many perfectly sensible truths which, on the surface, seem to require the existence of highly problematic entities—*entia non grata*, as it were. Consider, for example, the following:

(1) Jeeves was nonplused by the dearth of champagne in the ice box.
(2) Moriarty is the most well-known criminal in detective fiction.
(3) Courage is a virtue displayed by many people.
(4) The children Roger and Margarett Zimmerman would have had, had they met and married sooner, are my merely possible siblings.

On the face of it, these are statements about such things as dearths, fictional characters, characteristics that may be possessed by many people, and merely possible persons. One might think that it could be inferred from them that: there is at least one dearth, there are some fictional criminals, there is something displayed by every courageous person, there are merely possible people. But each of these statements can seem hard to swallow for one reason or another:

A dearth of champagne isn't a kind of *thing*, a sort of invisible anti-champagne located where the champagne should be. To say that Jeeves was nonplused by the dearth of champagne is simply to say that there was no champagne in the ice box, and that he was taken aback by the situation.

Nor are there some criminals (among the least dangerous of criminals) who are fictional. Fictional characters are not an odd group of people who, for some reason, we cannot meet in the way we meet other people, but can only get to know through stories. Statements about, say, Moriarty must really be elliptical for claims about the stories Arthur Conan Doyle wrote that had the name "Moriarty" in them.[13]

It might seem less problematic to suppose that there are some things called "virtues," of which courage is one. But if courage is something that can be displayed (or possessed or exemplified) by many different people at once, then some puzzling questions immediately arise. For how could anything be displayed in many different places at once, except by having a part displayed in each of those places and only there? Those philosophers particularly perplexed by this question (called *nominalists*) claim that (3) doesn't imply that there is one thing possessed by all the courageous people. Some nominalists would say that each of the courageous people has his or her own particular instance of courageousness (in D. C. Williams's terminology, a courageousness "trope"[14]), and that statements about courage are really about the big group or heap of all these instances.

For present purposes, the final case is the most illuminating. Do we really want to say that there are some merely possible people? That some people are tall, some are short, and some are nonexistent—the limiting case of diminutive stature, as it were? Philosophers who answer, No, are called "modal actualists": they hold that there are no nonactual things. But then how to make sense of (4)? In some sense, it must be possible for my parents to have had more children than they did, children not identical with any actually existing persons. One strategy for doing justice to this fact while avoiding ontological commitment to merely possible persons is to posit individual essences for nonexistent individuals, and then to construe talk about nonactuals as really talk about these essences. Then (4) becomes the claim that there are unexemplified individual essences, each of which would be the essence of a person if it were exemplified.[15] Another is to say that what (4) really comes to is the claim that it's possible that there be something that is a child of Roger and Margarett and is not identical with Jones, Robinson, . . . or any of the other actual people. This is an assertion about the possible truth of a certain proposition (that there be something that is a child and is not identical with. . . .); the proposition itself isn't about any particular nonactual thing; and it's not equivalent to the claim that there is something that is a possible person and is not identical with. . . .[16]

These are some typical attempts to avoid ontological commitment to undesirable entities. Statements which, on the surface, seem to imply that there are certain problematic entities are given philosophical glosses or paraphrases which seem to capture the truth in question while avoiding the implication that the troublesome things exist. The presentist is engaged in precisely the same sort of enterprise. But the truths that bother her are of this sort:

(5) There was a person who is not one of the people who presently exist.
(6) There will be a person who is not one of the people who presently exist.

The presentist is a "temporal actualist"—she is troubled by the fact that (5) and (6) seem to imply that there are some people who do not now exist, just as the modal actualist is bothered by the fact that (4) seems to imply that there are some people who do not actually exist. How can there *be* something that no longer exists, or that hasn't existed yet, she wonders? And so the presentist tries to show that the truth of (5) and (6) doesn't really conflict with her thesis that no nonpresent things exist.

One way of trying to show this would be to make use of individual essences again: (5) becomes the proposition that there is an individual essence not now exemplified that was once exemplified, and was then the essence of a person; and analogously for (6). Another is to insist that the truth of (5) implies only that it was the case that there is someone not identical with Jones, Robinson, . . . or any other presently existing person; but not that there is someone who used to exist and is not identical with Jones, Robinson, . . . And likewise for (6).[17]

How is presentism related to taking tense seriously? The presentist must, I think, be a serious tenser. At the very least, tenseless statements that ostensibly require ontological commitment to past and future things must be treated as equivalent to tensed truths that do not. And the presentist could not very well regard all the fundamental truth-bearers as eternally true, corresponding to tenseless statements. According to her, one of the truths is that wholly future things, like my first grandchild, do not exist—and such truths had better be susceptible to change. On the other hand, the serious tenser need not be a presentist. Quentin Smith, for example, is a non-presentist serious tenser.[18] According to Smith, fundamental truths are all tensed; but past and future individuals and events, although no longer present, nonetheless exist. Ostensible ontological commitment to such things cannot, on Smith's view, be paraphrased away.

But the combination of rejecting presentism while taking tense seriously is an unstable one. For the primary motivation for treating the fundamental truth-bearers as mutable and true *now* is the desire to do justice to the feeling that what's in the past is over and done with, and that what's in the future only matters because it will eventually be present. This is the source of the importance Prior attaches to the exclamation "Thank goodness that's over!"[19] If yesterday's headache still exists, and remains as painful as ever, then why should I be relieved now? Would the mere fact that it's no longer present justify this attitude? Most serious tensers, including Smith himself, will agree that it would not. And so, to render reasonable our special concern for the present, Smith strips past and future events of all their interesting intrinsic properties. For instance, yesterday's headache, although it exists, is no longer painful. It has a past-oriented property, *having been painful*—a sort of backwards-looking relation to the property *being painful.* But it is not painful now, and that's why it no longer concerns us.[20]

Although this view makes sense of our relief when pain is past, I find it unappealing in the extreme. The past and future events and objects it posits are too ghostly to be real. A painful headache cannot exist without being painful; a tanker explosion cannot exist without being violent and loud; Plato cannot exist while having neither body nor soul. What's left of these past and future things and events is too thin: yesterday's headache is still an event, but it isn't painful or throbbing or much of anything else; Plato is still a substance, I suppose, but he doesn't talk or think or walk or sleep or have any spatial location. Neither Plato nor headache has any of the ordinary intrinsic properties it displayed while present. Smith's efforts to preserve the intuition behind "Thank goodness that's over!" while rejecting presentism are, I judge, unsuccessful. Past and future things become nearly-bare particulars, unreal echoes of their once or future selves. The serious tenser is much better off without them.

Why Does Lewis Reject Presentism?

Now the serious tenser says that it is simply not true that I have the property *being straight* if I am bent now. I was straight, and will be again; but I am not now, and so there is no problem of my having incompatible intrinsic properties. Of course philosophers are free to invent a tenseless language in which "I am straight" is true just in case I either am now or was or will be straight. Who can stop philosophers from inventing peculiar ways of speaking? But the bare fact that one can talk this way doesn't create any problem about my having incompatible properties.

What is Lewis's response to this serious tenser solution of the problem of temporary intrinsics? He seems to suppose (reasonably, I think) that someone who takes this line must be a presentist. But, by Lewis's lights, presentism is too incredible to be believed. Presentism "rejects endurance; because it rejects persistence altogether"; and it "goes against what we all believe" by implying that "there are no other times." "No man, unless it be at the moment of his execution, believes that he has no future; still less does anyone believe that he has no past." And yet, says Lewis, the presentist denies these obvious facts.[21]

This string of claims represents what might be called the "no persistence objection" to presentism. Lewis takes it that the following thesis of "Persistence through Change" is obviously true:

(PC) There are (at least) two different times; one at which I am bent, another at which I am straight.[22]

Lewis thinks that (PC) is a simple expression of my belief that I persist through changes in my posture: there are times when I'm bent and times when I'm straight. The presentist is committed to the nonexistence of all times but one, the present. (PC) says there is more than one time; so presentism and (PC) are incompatible.

The serious tenser dissolution of the problem of temporary intrinsics given at the beginning of this section does not require the truth of presentism; a non-presentist serious tenser like Smith has little to fear from Lewis's argument. But Smith's combination of views has turned out to be unacceptable; and so the tensed response to the problem of temporary intrinsics stands or falls with presentism.

In order for Lewis's argument to have any teeth, (PC) must have two features: (i) it must be something we all, on reflection, believe; and (ii) it must require ontological commitment to the existence of more than one time. To be something commonly believed, (PC) must correspond to the humdrum assertion that I am bent at some times and straight at others. The question is whether this belief in my persistence through change—and the similar belief had by anyone who can remember changing posture—implies that there exist more times than the present.

If the statements used to express ordinary beliefs could be counted on to wear their ontological commitments on their sleeves, then an affirmative answer would be justified. But virtually everyone must allow that many statements expressing commonsensical beliefs do *not* wear their ontological commitments on their sleeves. It would be just like Bertie Wooster to respond to Jeeves's report about the dearth of champagne in the ice box by saying: "Well, at least there's *something* in the ice box." The source of the joke here would be that, generally speaking, from the fact that there's a such-and-such in the ice box, it follows that there is something in the ice box. But when the "such-and-such" is a *dearth* of something, it doesn't follow. Why? Because the assertion that there's a dearth of something is just a fancy (and old-fashioned) way of saying that there isn't any of that something—and that's compatible with there being *nothing at all* in the ice box.

Compare (PC) with a precisely parallel case involving ontological commitment to nonactuals. I suppose that most of us believe that we could have been put in situations that would have resulted in our lives going differently than they have in fact gone. There are certain possible experiences and events which, had they happened, would have prevented me from becoming a philosopher. But does this statement commit me to the existence of nonactual experiences and events? I should think not.

A few people have believed in the existence of alternative universes, just as real and concrete as this one, but with things going differently in them—worlds in which, for instance, the Axis powers win World War II, and the U.S. is partitioned between Germany and Japan.[23] David Lewis, in fact, believes in the literal existence of alternative universes, just as concrete as our actual world, in which every possible way things could go actually plays itself out.[24] But Lewis is one of the exceptions that prove the rule. The rest of us cannot bring ourselves to believe that there is such an event as the Axis powers' winning the war, an event with which, fortunately, we are not space-time neighbors. It's not that we ordinarily ignore these nonactual events because they are "far" from us, unreachable from our world. Rather, we think they simply are not.

How do we know that we aren't, implicitly, committed to the existence of such merely possible events? Well, we just ask ourselves whether we think they exist—whether we think that there are such things, whether we think we stand in real relationships with them. The answer comes back a resounding No. And then, if we are philosophers, we go about the business of finding plausible paraphrases for our beliefs ostensibly about nonactual possibilities—paraphrases that seem to us to capture more or less what we believed all along, but which do not even appear to imply that there are situations involving me that don't occur, or whole worlds full of people and events that are not actual. If it were to become clear that there is no way to do this, then perhaps we would feel forced to reconsider our judgment that our beliefs about alternate possibilities do not implicitly commit us to the existence of such things. But that's not usually the way things go in philosophy: there's usually more than one way to skin a philosophical cat; usually several competing approaches to a given philosophical problem emerge as favorites, with much to be said for and against each of them. And so it is here: there are ever so many fairly plausible projects underway for paraphrasing away ostensible commitment to nonactual things and situations, each with its own advantages and disadvantages, and few confront such grave obstacles as to suggest that they are absolute dead-ends.

The presentist believes that the situation is precisely parallel when it comes to my belief that there are times at which I'm bent and times at which I'm straight. Does this commit me to the existence of times other than the present? Well, when I ask myself whether I think that my childhood exists, or the time of my death, the snows of yesteryear, or the light of other days, the answer comes back a resounding No. Is it just that I feel that past and future things and events can be regarded as nonexistent because they are "temporally far" from me? I think not—the past is no more, and the future is not yet, in the strictest sense. And so those who share this judgment begin the work of philosophical paraphrase, trying to find plausible construals of statements like (5), (6), and (PC) that capture what is meant but do not involve direct reference to other times, nonpresent individuals, and events. So, for instance, (PC) can be taken as a tenseless statement expressing a disjunction of tensed propositions: Either I was bent and would become or had previously been straight, or I was straight and would become or had previously been bent, or I will be bent and will have been or be about to become straight, or I will be straight and will have been or be about to become bent. Surely this tensed disjunction is true if (PC) is true; furthermore, it contains no mention of anything like a non-present time. So, given the presentist's desire to avoid ontological commitment to nonpresent times, this tensed statement provides a perfectly sensible paraphrase of my conviction that I can persist through change of shape.

Furthermore, it is not as if Lewis himself allows (PC) to stand as it is, with no paraphrastic gloss. After all, he thinks that I am bent at one time and straight at another only in virtue of the fact that I have temporal parts located at these times, one of which is bent, the other straight. So "there is a time at which I am bent," as it occurs in (PC), receives the paraphrase "there is a time at which I have a temporal part that is bent." Lewis salvages our common conviction that we persist through change by introducing the uncommon notion of a temporal part. But if his temporal-parts reading of (PC) captures enough of our pretheoretical convictions to be acceptable, then surely he must allow the presentist similar leeway in her attempt to affirm persistence through change while avoiding talk of non-present times.[25]

The large-scale project of paraphrasing truths ostensibly about nonpresent times and things is as complex and difficult as the counterpart project concerning nonactuals. Ways must be found to capture all truths about past and future things without the appearance of ontological commitment to such things.[26] Presentists must, for example, find a way to understand statements ostensibly about relations that hold between presently existing things and things in the past and future. Causation is one instance of this problem: the causal relation holds between events; but no relation can hold between a present event and some future or past event, since such events do not exist. Must the presentist then conclude that no event in the present can be caused by anything earlier, or cause anything later?[27] Such difficulties must be overcome for presentism to remain plausible.

And there are familiar chestnuts bedeviling *anyone* (presentist or not) who takes tense seriously, such as McTaggart's paradox[28] and the puzzle about the rate at which the present "moves." Is this rate one minute per minute? It couldn't very well move any faster! And yet this doesn't sound like a proper rate at all.[29] Perhaps most worrisome is that positing facts about what is present *absolutely* (and not merely about what is "present relative to me" or "present relative to my inertial frame") seems inconsistent with a well-confirmed scientific theory: special relativity.[30] But, as indicated in the notes to this and the previous paragraph, these are problems which presentists and others who take tense seriously have tried to address. Have the solutions been satisfactory? Perhaps not in every case. But rejecting presentism on the basis of such problems would require careful exploration of these debates—debates which have nothing to do with the problem of temporary intrinsics per se. Furthermore, there's reason to be hopeful that they will be resolved in the presentist's favor—or at least that they will not be resolved decisively in favor of her opponents. After all, as John Bigelow points out [chapter 8 in this volume—*Editors*], presentism was accepted everywhere by nearly everyone until a mere hundred or so years ago.[31] A thesis with a track record like that shouldn't be expected to go down without a fight.

So far as I know, all presentists (and almost all who take tense seriously) reject the doctrine of temporal parts; indeed, Prior, Geach, and Chisholm have been among its most vocal opponents.[32] What I have tried to show is that the part of Lewis's argument aimed at these philosophers requires considerable buttressing before it will convince. In particular, we need a reason to think that some truths ostensibly about nonpresent things cannot be given plausible paraphrases that eschew commitment to such things. So far as I can see, there isn't any reason to think this is so. At any rate, Lewis hasn't (yet) given us one.

Notes

1. Cf. *On the Plurality of Worlds* (Oxford: Blackwell, 1986), pp. 202-4; the relevant passage is reprinted as the previous chapter of this volume, and also (as "The Problem of

Temporary Intrinsics: An Excerpt From *On the Plurality of Worlds*") in *Metaphysics: The Big Questions*, pp. 204-6.

2. Cf. Lewis, *Plurality*, p. 205.

3. It might be replied that there is no problem with having both if the verb "having" is taken *tenselessly*—that is, in such a way that "I have both" is equivalent to something like: "I had, now have, or will have the one; and I had, now have, or will have the other." But then we should want to know why these properties deserve the label "incompatible." How do they differ from a pair of compatible intrinsics, like *being red* and *being round*?

4. Lewis, *Plurality*, p. 204.

5. One might, however, attempt to treat intrinsic properties as monadic while the *having* of them is a relation between a thing, a property, and a time. See, for example, Peter van Inwagen, "Four-dimensional Objects," *Noûs* 24 (1990), pp. 245-55; Sally Haslanger, "Endurance and Temporary Intrinsics," *Analysis* 49 (1989), pp. 119-25.

6. It can be found in Bolzano, Frege, Church, Chisholm, and Plantinga, to name but a few.

7. This point of view is exemplified by A. N. Prior and Peter Geach. See Prior, "Changes in Events and Changes in Things," in his *Papers on Time and Tense* (Oxford: Clarendon Press, 1968), pp. 1-14 [chapter 3]; Prior, "The Notion of the Present," *Studium Generale* 23 (1970), pp. 245-48; and Geach, "Some Problems about Time," reprinted in his *Logic Matters* (Berkeley and Los Angeles: University of California Press, 1972), pp. 302-18.

8. See Gerald Massey, "Tense Logic! Why Bother?" *Noûs* 3 (1969), pp. 17-32.

9. If the tenseless verb used in my lecture were the ordinary historical present tense, the proposition in question would lack the final conjunct; only the more arcane tenseless verb introduced by philosophers is used to express disjunctive propositions with disjuncts concerning the future.

10. Mellor, *Real Time* (Cambridge: Cambridge University Press, 1981), ch. 5; J. J. C. Smart, "Time and Becoming," reprinted in his *Essays Metaphysical and Moral* (Oxford: Blackwell, 1987). For further defenders of the new tenseless theory, see the contributions of L. Nathan Oaklander, Michelle Beer, and Clifford Williams to Part 1 of *The New Theory of Time*, ed. by Oaklander and Quentin Smith (New Haven, Conn.: Yale University Press, 1994); for criticisms, see Quentin Smith's contributions to Part 1 of the same volume, and his *Language and Time* (New York: Oxford University Press, 1993).

11. Some have held that there are both tenselessly true propositions and "tensedly true" propositions, and that the former are not equivalent to disjunctions of the latter. But this is a minority opinion; most who take at least *some* fundamental truth-bearers to be mutable regard all truth as tensed truth. For some reasons to do so, see Roderick M. Chisholm and Dean W. Zimmerman, "Tense and Theology," *Noûs* 31 (1997), pp. 262-65.

12. For a paradigmatic statement of this position, see Prior, "The Notion of the Present."

13. On some problems for carrying out this project, see Peter van Inwagen, "Creatures of Fiction," *American Philosophical Quarterly* 14 (1977), pp. 299-308; "Fiction and Metaphysics," *Philosophy and Literature* 7 (1983), pp. 67-77; and "Pretense and Paraphrase," in Peter J. McCormick (ed.), *The Reasons of Art* (Ottawa: University of Ottawa Press, 1985), pp. 414-22.

14. See Williams, "On The Elements of Being," *Review of Metaphysics* 7 (1953), pp. 3-18 and 171-92.

15. Compare Alvin Plantinga, "Actualism and Possible Worlds," *Theoria* 42 (1976), pp. 139-60; reprinted in *The Possible and the Actual*, ed. by Michael J. Loux (Ithaca: Cornell University Press, 1979).

16. See Prior, *Papers on Time and Tense*, pp. 142-43; and Kit Fine's Postscript to Prior and Fine, *Worlds, Times and Selves* (London: Duckworth, 1977).

17. For a general discussion of the treatment of past and future individuals in tense logic, see Prior, *Past, Present, and Future* (Oxford: Clarendon Press, 1967), ch. 8. See also Chisholm, "Referring to Things That No Longer Exist," *Philosophical Perspectives*, vol. 4 (Action Theory and Philosophy of Mind) (1990), pp. 545-56.

18. See Quentin Smith, *Language and Time* (New York: Oxford University Press, 1993); see esp. ch. 5.

19. See Prior, "Some Free Thoughts about Time," reprinted in Zimmerman and van Inwagen, *Metaphysics: The Big Questions*, pp. 104-7; and his "Thank Goodness That's Over," reprinted in Prior, *Papers in Logic and Ethics* (London: Duckworth, 1976).

20. Incidentally, Smith's approach to past and future events and things provides him with the means to define "being present"—something he claims cannot be done. Just take all the kinds of intrinsic properties which a contingent thing cannot have when it is wholly past or future; and then say a thing is present just in case it is either a necessary thing (and so must always be present), or it is a contingent thing that has properties belonging to this special class.

21. Lewis, *On the Plurality of Worlds*, p. 204.

22. This thesis, and its name, are taken from personal correspondence with Lewis, and used with his permission.

23. Such is the world of the "alternate history" novel, *The Man in the High Castle*, by Philip K. Dick. Dick became convinced that such alternate streams of history are not mere fictions, but that they are real; he claimed to have been able to "recall" events from lives lived in other worlds.

24. For Lewis's reasons for believing in concrete worlds besides this one, see his *On the Plurality of Worlds* (for the senses in which his worlds are *concrete*, see section 1.7 of the book). I should point out that, unlike Dick, Lewis's reasons are purely theoretical and a priori, not empirical.

25. Trenton Merricks pointed this out to me, in conversation; and Sally Haslanger said it (long before that conversation) in "Endurance and Temporary Intrinsics," *Analysis* 49 (1989), pp. 119-25; see especially pp. 119-20.

26. For some presentist responses to the problem, cf. Prior, *Past, Present and Future*, ch. 8 ("Time and Existence"); Prior, *Papers on Time and Tense*, ch. 8 ("Time, Existence, and Identity"); Chisholm, "Referring to Things That No Longer Exist"; and John Bigelow, "Presentism and Properties," in *Philosophical Perspectives*, 10 (1996), pp. 35-52.

27. John Bigelow and I have offered, independently, very similar solutions to this problem. Cf. the final section of my "Chisholm and the Essences of Events," in *The Philosophy of Roderick M. Chisholm* (The Library of Living Philosophers), ed. by Lewis Edwin Hahn (Chicago: Open Court, 1997); and Bigelow's "Presentism and Properties," p. 47.

28. Cf. McTaggart, *The Nature of Existence: Vol. 2* (Cambridge, U.K.: Cambridge University Press, 1927), pp. 9-22 [see chapter 2]; and, for a response which, by my lights, completely dissolves this "paradox," cf. C. D. Broad, *Examination of McTaggart's Philosophy*, Vol. II, Part I (Cambridge: Cambridge University Press, 1938), pp. 309-17. (McTaggart's argument and Broad's response are reprinted in van Inwagen and Zimmerman, *Metaphysics: The Big Questions*, pp. 67-79.)

29. J. J. C. Smart raises the puzzle about the rate of passage in *Philosophy and Scientific Realism* (London: Routledge & Kegan Paul, 1963), p. 136. Presentists can hope that Ned Markosian has settled the problem for good and all in his "How Fast Does Time Pass,?" *Philosophy and Phenomenological Research* 53 (1993), pp. 829-44.

30. Prior's description of the problem and his response may be found in "Some Free Thinking About Time." Cf. also Geach, "Some Problems about Time." More recent

treatments may be found in Quentin Smith, *Language and Time*, ch. 7. For one scientist who thinks that Prior may have been right about the prematurity of giving up on the notion of absolute simultaneity, cf. J. S. Bell, *Speakable and Unspeakable in Quantum Mechanics* (Cambridge: Cambridge University Press, 1987), p. 77; and Bell's remarks in *The Ghost in the Atom*, ed. by P. C. W. Davies and J. R. Brown (Cambridge: Cambridge University Press, 1986), esp. pp. 48-51.

31. Bigelow, "Presentism and Properties," pp. 35-36.

32. For characteristic rejections of temporal parts, cf. Prior, "Some Free Thinking about Time"; Geach, "Some Problems about Time"; and Chisholm, *Person and Object: A Metaphysical Study* (La Salle, Ill.: Open Court, 1976), Appendix A, pp. 138-44. For a serious tenser who accepts temporal parts, see Quentin Smith, "Personal Identity and Time," *Philosophia* 22 (1993), pp. 155-67.

References

Bell, J. S. *Speakable and Unspeakable in Quantum Mechanics*. Cambridge: Cambridge University Press, 1987.

Bigelow, J. "Presentism and Properties," *Philosophical Perspectives* 10 (1996): 35-52. Chapter 8 in this volume.

Broad, C. D. *Examination of McTaggart's Philosophy*, Vol. II, Part I. Cambridge: Cambridge University Press, 1938.

Chisholm, R. *Person and Object: A Metaphysical Study*. La Salle, Ill.: Open Court, 1976.

———. "Referring to Things That No Longer Exist," *Philosophical Perspectives*, Vol. 4 (Action Theory and Philosophy of Mind) (1990): 545-56

Chisholm, R. M., and D. Zimmerman. "Tense and Theology," *Noûs* 31 (1997): 262-65.

Davies, P. C. W., and J. R. Brown, eds. *The Ghost in the Atom*. Cambridge: Cambridge University Press, 1986.

Dick, P. K. *The Man in the High Castle*. New York: Vintage Books, 1962.

Geach, P. "Some Problems about Time." Pp. 302-318 in *Logic Matters*, by P. Geach. Berkeley and Los Angeles: University of California Press, 1972.

Haslanger, S. "Endurance and Temporary Intrinsics," *Analysis* 49 (1989): 119-25.

Lewis, D. *On the Plurality of Worlds*. Oxford: Blackwell, 1986.

Markosian, N. "How Fast Does Time Pass?" *Philosophy and Phenomenological Research* 53 (1993): 829-44.

Massey, G. "Tense Logic! Why Bother?" *Noûs* 3 (1969): 17-32.

McTaggart, J. M. E. *The Nature of Existence: Vol. 2*. Cambridge, U.K.: Cambridge University Press, 1927.

Mellor, D. H. *Real Time*. Cambridge: Cambridge University Press, 1981.

Oaklander, L. N., and Q. Smith, eds. *The New Theory of Time*. New Haven, Conn.: Yale University Press, 1994.

Plantinga, A. "Actualism and Possible Worlds," *Theoria* 42 (1976): 139-60; in *The Possible and the Actual*, edited by M. J. Loux. Ithaca, N. Y.: Cornell University Press, 1979.

Prior, A. *Past, Present, and Future*. Oxford: Clarendon Press, 1967.

———. "Changes in Events and Changes in Things." Pp. 1-14 in *Papers on Time and Tense*, by A. Prior. Oxford: Clarendon Press, 1968.

———. *Papers on Time and Tense*. Oxford: Clarendon Press, 1968.

———. "The Notion of the Present," *Studium Generale* 23 (1970): 245-48.

———. "Thank Goodness That's Over." Pp. 78-84 in *Papers in Logic and Ethics*, by A. Prior. London: Duckworth, 1976.

———. "Some Free Thinking about Time." Pp. 47-51 in *Logic and Reality: Essays on the Legacy of Arthur Prior*, edited by J. Copeland. Oxford: Clarendon Press, 1996.

Prior, A., and K. Fine. *Worlds, Times and Selves*. London: Duckworth, 1977.

Smart, J. J. C. *Philosophy and Scientific Realism*. London: Routledge & Kegan Paul, 1963.

———. "Time and Becoming." Pp. 78-90 in *Essays Metaphysical and Moral*, by J. J. C. Smart. Oxford: Blackwell, 1987.

Smith, Q. *Language and Time* (New York: Oxford University Press, 1993).

———. "Personal Identity and Time," *Philosophia* 22 (1993): 155-67.

Van Inwagen, P. "Creatures of Fiction," *American Philosophical Quarterly* 14 (1977): 299-308.

———. "Fiction and Metaphysics," *Philosophy and Literature* 7 (1983): 67-77.

———. "Pretense and Paraphrase." Pp. 414-22 in *The Reasons of Art*, edited by P. J. McCormick. Ottawa: University of Ottawa Press, 1985.

———. "Four-dimensional Objects," *Noûs* 24 (1990): 245-55.

Williams, D. C. "On the Elements of Being," *Review of Metaphysics* 7 (1953): 3-18 and 171-92.

Zimmerman, D. "Chisholm and the Essences of Events." Pp. 79-93 in *The Philosophy of Roderick M. Chisholm*, edited by Lewis Edwin Hahn. Chicago: Open Court, 1997.

Zimmerman, D., and P. Van Inwagen, eds. *Metaphysics: Big Questions*. Cambridge, Mass.: Basil Blackwell, 1998.

18
On the Incompatibility of Enduring and Perduring Entities

Trenton Merricks

Presentism is the doctrine that the present time is ontologically privileged. According to the presentist, all that exists, exists at the present time; and an object has only those properties it exemplifies at the present time. Those who reject presentism hold that all times are on an ontological par, and the present time is special only because it is the time at which we (or our current time-slices, current utterances, current actions or current thoughts, etc.) are located; the present time, according to this position, is merely this time. This place (the place in which I now sit) is simply one place among many; likewise, says one who denies presentism, the present time is simply one time among many. I shall call the view opposed to presentism, since it treats being present as simply being this time, "the indexical view of the present" or "indexicalism" for short.[1]

One might be suspicious that presentism, as I have described it, is trivially true, and thus my attempt to describe a substantive philosophical thesis has failed. For one might think that simply knowing English and its use of tense is enough to tell us that an object has all and only those properties it has at the present moment. Since no one moved by this concern will be helped by my saying that presentism is the doctrine that all that really exists is what exists at the present time and all the properties an object really has (I mean really has) are the ones it has at the present time, I shall try a different approach entirely. I shall ignore this objection and go on to argue that each theory of the present rules out a different account of persistence. Seeing the sort of controversial inferences each account sanctions in the arguments below should clarify the ways in which each is a substantive philosophical thesis.

I

I will argue that:

(1) Presentism entails that there are no perduring, four-dimensional objects; and
(2) Indexicalism entails that there are no enduring, three-dimensional objects.

I will conclude by demonstrating that the conjunction of (1) and (2) leads to an even more striking claim—the claim that a single world cannot contain both temporally extended, perduring events and three-dimensional, enduring objects.

In order to see why (1) is true, we must first consider the following:

(3) An object cannot have another object as a part if that other object does not exist.

(3) states that one object cannot have a second object as a part if there is no second object. (3) is also entailed by the thesis that being apart cannot be exemplified by what does not exist. (3) is, I think, unimpeachable. After all, who could deny that, for example, the question of whether the universe has among its parts massive crystalline spheres which contain the heavenly bodies is closed once it is shown that no such spheres exist?[2]

A four-dimensional object is composed of "temporal parts," some of which are three-dimensional and exist but a moment.[3] A four-dimensional object lasts over time by way of having distinct temporal parts existing at distinct times.[4] This sort of lasting or persisting is called "perduring." So the central thesis of perdurance or four-dimensionalism is that objects which last over time have temporal parts which exist at many different times and that not all of their parts exist at any single time; thus, not all of a perduring object's parts could exist at the single time which is present. If presentism is true, then those parts of an object which do not exist at the present time do not exist at all. So if presentism is true, a perduring object has some parts—the vast majority of its parts, in fact—which do not exist. But this possibility has been ruled out by (3). We can, therefore, conclude that if presentism is true, there are no perduring objects. (If indexicalism is true, it doesn't follow from the fact that a four-dimensional object has parts which fail to exist at the present time that those parts fail to exist; this is just the sort of inference that indexicalism denies.)

That objects last over time by perduring is, of course, only one of two competing views. Its competitor is that objects have no temporal parts and last, not by perduring, but by enduring. Enduring objects lack temporal extent and have three dimensions instead of four. If a three-dimensional enduring object lasts from one time to another, then there is a three-dimensional object existing at one of those times which is literally identical with a three-dimensional object existing at the other. Although presentism combined with perdurance runs afoul of (3), no similar problems need afflict the combination of presentism and endurance. For the endurantist claims that at any single time at which an object exists, all of that object's parts exist at that time; this claim is captured by the slogan that an enduring object is "wholly present" at each time at which it exists. The endurantist might claim, of course, that an object can change parts, and that what was once a part of an enduring object no longer exists. But presentism combined with (3) is consistent with an enduring object's having had something as a part which no longer exists. This is because, according to presentism, it does not follow from the fact that O had p as a part at some time in the past that O has p as a part. Of course, the perdurantist cannot deny that a perduring object has parts which do not exist at the present time. To deny this is simply to deny that a perduring object has temporal parts; it is to renounce perdurance.[5]

So we can see that (1)—the claim that presentism entails that there are no perduring objects—is true. The inconsistency of perdurance with presentism is not a product of some revisable or tangential feature of perdurance, but rather follows from the very nature of perdurance—the claim that not all the parts of an object which lasts over time exist at a single time.

The first step in arguing that indexicalism entails that there are no enduring objects (i.e., that (2) is true) is to note that, obviously, objects undergo change. It is no part of the doctrine that objects endure to deny this truism, so any endurantist should accept the following:

(4) It is possible that an enduring object O is F, and that at some time in the past O was ¬F.

The endurantist must understand (4) as stating that the object that existed at a previous time and was ¬F is identical with that object which exists at the present time and is F. But then, it has often been objected, it follows that that object both is F and is ¬F.[6] The obvious response to this objection is that it rests on a fallacious inference from O's failing to exemplify F at some time other than the present to O's failing to exemplify F. The inference is fallacious because, so this response goes, O exemplifies only those properties that it has at the present time. So if presentism is true, we can see that the endurantist can easily avoid contradiction in the face of change.

Suppose, however, one rejects presentism for indexicalism. The "obvious response" that keeps (4) from leading the endurantist into contradiction is no longer available. Given indexicalism, we can conclude that if a single object is F at one time, and is ¬F at another, then that object both is and is not F. But that is, of course, contradictory. Granting the obvious, that objects undergo change, we can see that endurance and indexicalism are inconsistent. We can thus see that (2) is true. This argument for (2) is the rather familiar argument that change, endurance, and the indiscemibility of identicals lead to absurdity. My only addition is to make explicit the role played by indexicalism.[7]

The perdurantist is not forced into absurdities by indexicalism and the possibility of change. The claim that a perduring object is F at the present time but is ¬F at some past time amounts to the claim that O's present temporal part is F, and that it has a past temporal part that is ¬F. This leads to nothing contradictory.[8] This way of making change consistent with indexicalism trades on the fact that a perduring object may have a property at a particular time in virtue of having a temporal part which has that property at that time. Because the endurantist denies that objects have temporal parts, she must deny that an object has a property in virtue of its temporal part having that property. Therefore, the way in which the perdurantist reconciles indexicalism and change is not open to the endurantist.

Some have tried, however, to reconcile endurance and indexicalism. Not, of course, by making use of the perdurantist's strategy, but instead by claiming that all properties that an enduring object seems to gain or lose are really either disguised relations to times or time indexed. So, according to the defender of this position, no enduring object is ever, strictly speaking, simply red. Rather, it stands in the being red at relation to a certain time t, or it exemplifies the time indexed property being red at time t.[9] There is, of course no contradiction in saying a single object exemplifies both being red at time t and not being red at t*, or in saying that a single object stands in the being red at relation to one time but not to another.

But it is *not* the case that all of the properties that an object seems to gain or lose are really relations to times or time indexed.[10] A short list of those properties which are not—they are known as "temporary intrinsics"—includes shape, color, size, and mass. I defend this claim elsewhere (Merricks 1994). Rather than rehearse those arguments here, let me simply note that if they are successful, then we can embrace temporary intrinsics and, with them, (2).

Before leaving the argument for (2) based on change, I would like to look at a species of this argument which trades on a central claim of endurance: that an enduring object is wholly present at each time at which it exists. If an enduring object exists at time t and fills place P, then all of its parts are located within place P at time t. It is possible, however, for an enduring object to move; it is possible that at one time t such an object be in place P, and that at another time t* it be in place P* (which does not overlap P). If we assume that indexicalism is true, it seems we can conclude that an enduring object can have all of its parts in one place, P, and also have all of its parts in a distinct, non-overlapping place, P*. This absurdity follows from the possibility of motion combined with indexicalism and the view that an object is wholly present at each time at which it exists. So if indexicalism is true, we should reject the claim that an object is wholly present at each time at which it exists; that is, if indexicalism is true, we should reject endurance. Of course, nothing untoward follows from the motion of a perduring object if indexicalism is true; for a perduring object's being in a place at a time implies not that all of its parts are in that place at that time, but rather that one of its temporal parts is in that place at that time. So a perduring object could be in both place P and P*; this just means it could have one part in P and a distinct part in P*.

We can see that perduring objects are not compatible with a presentist theory of time, and also that indexicalism is inconsistent with endurance. While this conclusion, that (1) and (2) are true, is one that many would not be surprised to discover, I will argue in the next section that it leads directly to a conclusion that is very striking indeed.[11]

II

Note that the reasons for endorsing (1) and (2) are not specific to enduring and perduring objects. The arguments here are general enough to include not only objects, but events (and any other kind of entity that might have either three or four dimensions). Suppose someone held that events endure; suppose that is, that someone held that an event was wholly present at each time at which it existed.[12] We can quickly see that enduring events are inconsistent with indexicalism by adapting one of the arguments for (2). Suppose a particular event, for example, a moving feast is wholly present, one day, in place P; unsurprisingly, the feast moves and the next day is in a non-overlapping place, P*. If the moving feast endures, then when it is in P* it is wholly in P*, and, therefore, has no part in P; therefore, if the feast endures and indexicalism is true, then all of the feast's parts are in place P, and none of the feast's parts are in place P.[13] This is unacceptable. It is easy to see that the more general objection in terms of change and indexicalism could also be adapted to show that changing events, as well as changing objects, cannot endure if the indexical view of the present is true. After all, an event can no more both be F and ¬F than can an object.[14] So we can conclude:

(2*) Indexicalism entails that there are no enduring entities.

"Entities" here and below is meant to be broad enough to include, at least, both objects and events.

Suppose that presentism is true, and all that exists exists at the present time. Assume for reductio that an event is four-dimensional and has many (temporal) parts which do not exist at the present time. Add that something cannot be a part (including a temporal part) if it does not exist. From all this we get the very sort of contradiction that resulted from trying to combine presentism and perduring objects. So we can conclude:

(1*) Presentism entails that there are no perduring entities.

We now have all the resources necessary to prove that there are not (and cannot be) both three-dimensional and four-dimensional entities.

(1*) Presentism entails that there are no perduring entities.

(2*) Indexicalism entails that there are no enduring entities.

(5) Either presentism or indexicalism is true.[15]

Therefore,

(6) Either there are no perduring entities or there are no enduring entities

Therefore,

(7) It is not the case that there are both perduring entities and enduring

entities.

It is, I think, intuitively very plausible that there could be enduring entities such as, for instance, baseballs alongside perduring events, such as baseball games. And it seems intuitively plausible that there could be a three-dimensional object that has a four-dimensional history. But if (7) is true, these intuitions have been shown to be misguided.

Not only is an ontology that combines three-dimensional and four-dimensional entities intuitively appealing, it is also one that some philosophers explicitly defend. Among them are: David Wiggins (1980, p. 25, n. 12) and Lawrence Lombard (1986, pp. 127-31), both of whom claim that objects such as cats and dogs endure, but events perdure. These philosophers are not idiosyncratic in their defense of perduring events—that events perdure has achieved the status of philosophical orthodoxy, even among those who eschew four-dimensionalism for objects. So it should be clear that (7) has serious consequences for ontology in general, and, in particular, for which theories of events are consistent with an ontology of three-dimensional objects.

Finally, I would like to direct attention to (7)'s impact on the debates surrounding personal identity. If one claims that persons are perduring entities composed of "person-stages," then, since perduring persons are inconsistent with enduring objects, one is committed to a thoroughgoing four-dimensionalism. Conversely, if one thinks that at least some objects, such as bodies or houses or metaphysical simples, endure, then one may not endorse the fashionable view that persons are composed of person-stages. The question of personal identity over time, i.e., personal persistence, cannot be approached independently of the question of persistence of objects and events in general.[16]

Notes

1. Presentists are also known as those who "take tense seriously" or believe in "temporal becoming." Indexicalists are also known as those who think time is "static," think of time as "spacelike," or "do not take tense seriously."

2. (3) is entailed by the position known as "serious actualism," according to which an object can exemplify a property only in worlds in which it exists. Some have objected to serious actualism by claiming that an object exemplifies non-existence in worlds in which it doesn't exist. The merits of this as a counterexample to serious actualism aside, it should be noted that no one has objected to serious actualism by maintaining that being a part can be exemplified by an object in worlds in which it doesn't exist.

3. Some of a four-dimensional object's temporal parts also have temporal extent, and these temporal parts therefore have temporal parts of their own. Some perdurantists, such as Whitehead, think that all of an object's parts have some duration—none last for but an instant—and thus all parts of an object are four-dimensional. In this paper I assume perduring objects would have some three-dimensional, instantaneous parts, but nothing rides on this. Below I argue that presentism combined with perdurance commits one to the view that perduring objects have parts that do not exist, and the argument for this would work even if one thought that all temporal parts had some duration. In fact, the defense of (1) might be even

easier if one took a view like Whitehead's: assuming that the present time has no temporal duration, if all objects and parts do have temporal duration, then no objects or parts exist at the present time. The relevance of these remarks will be clear below. (See Whitehead 1920, p. 56.) Mark Heller (1990, pp. 4-6) also discusses reasons the perdurantist might have for denying that a perduring object has three-dimensional temporal parts.

4. My use of "persists," "perdures," and "endures" follows the convention introduced by Mark Johnston and David Lewis (see Lewis 1986, p. 202). I will use "perduring" and "four-dimensional" (and their cognates) interchangeably; likewise with the expressions "persisting," "three-dimensional," and "enduring." However, these expressions are not equivalent. Perhaps there could be souls "spread out in time"; if so, it would be possible that an object perdure without having spatial dimensions, and thus without having four dimensions. Likewise, there might be an extensionless point that endures but, since it lacks extension, does not have three dimensions. Conversely, physicists might discover (or it might be possible) that space has more than three dimensions, in which case enduring physical entities would have as many dimensions as space, and perduring entities would have one more than that.

5. Suppose the perdurantist were to follow the endurantist's lead here, and say that all the parts of a perduring object, O, exist at the present time, because although O had some past temporal part as a part, O now does not have that temporal part as a part. Suppose, that is, that the perdurantist says that the only parts O has (as opposed to the parts it had or will have) are the ones that exist at the present time. Then it would follow that O has only those parts that exist now, that is, that O has all and only the parts its present time-slice has. But then O collapses into its present slice—they have, for instance, exactly the same parts—and O therefore turns out to be a non-persisting temporal part, and not a persisting object. Or conversely, if the perdurantist demands that, no matter what, O be a persisting entity, we get the conclusion that a three-dimensional thing—a thing with parts that exist at only one time—persists. (This collapse of what persists into what exists at the moment is, of course, exactly what the endurantist wants.)

6. This objection has been raised and defended by, among others, D. M. Armstrong (1980, pp. 68-69), Michael Jubien (1993, pp. 24-27), and David Lewis (1986, pp. 202-4).

7. In defense of the claim that my "addition" to the argument only makes explicit what it already contains, note that Lewis explicitly rejects presentism in his defense of the argument. See Lewis's attack on the view that "the only intrinsic properties of a thing are those it has at the present moment" (Lewis 1986, p. 204).

8 In some cases of change of a perduring object, it may not be the instantaneous time-slice that does the work, but rather one of the perduring object's larger, temporally extended parts. So one might say that O is now exactly ten days old, but tomorrow will not be, and that this means that the temporally extended part of O which starts at the time of O's birth and ends now lasts exactly ten days, but not so for the part that stretches from birth to tomorrow.

9. Or, according to "adverbialism," it exemplifies being red in a t-ly manner. For a clear articulation and defense of adverbialism, see Haslanger (1989).

10. This point has been emphasized by Lewis (1986, p. 204).

11. For another and more extended defense of (1) and (2) I recommend Carter and Hestevold (1994).

12. An odd supposition perhaps, since, so far as I know, no one explicitly defends such a claim. There is, however, no reason in principle why such a claim could not be defended. Suppose events are property exemplifications (such as O's being red). Why couldn't O's being red be wholly present at more than one time?

13. The notion of an event undergoing the sort of change that seems to result in contradiction might seem a little implausible; one might insist that objects move, but that events do not. See for example Lombard's defense of this claim (1986, pp. 127-30). But this is a sensi-

ble claim only if one assumes (as Lombard does) that events do not endure, but instead have temporal parts. If events endure, then they should be able to undergo the sort of change that objects do—that is, the sort of change that raises problems when combined with indexicalism.

14. The susceptibility of enduring events in an indexical world to this last, most general, problem depends on the fact that events have temporary intrinsic properties (not so for the more limited claim about events and their parts).

15. I assert this without argument. As I have presented the doctrines here, presentism and indexicalism are so general as to pretty clearly exhaust the options. (5) might even follow from each of the following tautologies: either the present time is ontologically privileged or it is not, and either all times are on an ontological par or they are not. At any rate, if anyone has a real alternative to either presentism or indexicalism, then the argument of this paper can be understood as arguing for either (7) or the denial of (5). That conclusion would still be interesting.

16. Thanks to Marian David, Anthony Ellis, Jaegwon Kim, Eugene Mills, Alvin Plantinga, Philip Quinn, Michael Rea, Peter Vallentyne, James van Cleve, and Dean Zimmerman for helpful comments on this paper.

References

Armstrong, D. M. "Identity through Time." Pp. 67-78 in *Time and Cause: Essays Presented to Richard Taylor*, edited by P. v. Inwagen. Dordrecht: Reidel, 1980.

Carter, W. S., and H. S. Hestevold. "On Passage and Persistence," *American Philosophical Quarterly* 31, no. 4 (1994): 269-84.

Haslanger, S. "Endurance and Temporary Intrinsics," *Analysis* 49, no. 3 (1989): 119-25.

Heller, M. *The Ontology of Physical Objects*. Cambridge: Cambridge University Press, 1990.

Jubien, M. *Ontology, Modality, and the Fallacy of Reference*. Cambridge: Cambridge University Press, 1993.

Lewis, D. *On The Plurality of Worlds*. Oxford: Basil Blackwell, 1986.

Lombard, L. *Events: A Metaphysical Study*. London: Routledge & Keegan Paul, 1986.

Merricks, T. "Endurance and Indiscernibility," *Journal of Philosophy* 91 (1994): 165-84.

Van Inwagen, P. *Material Beings*. Ithaca, N.Y.: Cornell University Press, 1990.

Whitehead, A. N. *The Concept of Nature*. Cambridge: Cambridge University Press, 1920.

Wiggins, D. *Sameness and Substance*. Cambridge: Harvard University Press, 1980.

Index

A-B theory, 239
absolute future, 233
absolute moments, 18
absolute pastness, 233, 241
absolute present, 233
absolute simultaneity, 233, 240, 246
actual history of the world 151, 290
Alston, William P., 64-65
A-properties, 1-2, 4, 11, 15, 17-19,21-23, 86, 256, 285
Aquinas, Thomas, 66-67, 70n39, 70n41, 71n44
A-relation(s), 83-84, 86
argument from causation, 129-132, 136
Aristotle, 132, 133, 232
Armstrong, D.M.A.,113n5, 142, 146,272nn8-9, 325n6
A-series, 1-2, 16,-19, 36-43, 43n2, 43n3, 75-76, 80-82, 86
atemporal entities, 163-164, 177, 180n5
atemporal facts, 170-171, 174, 179
a-temporal things, 94, 110
A-theoretical ontology, 59
A-theoretic becoming, 21, 59, 81
A-theory, 2-3, 11-12, 17, 22, 24, 59, 61,80-82, 84, 239
Augustine, 4, 12-15, 17, 21, 27-34, 47-48, 50, 259-260

Benovsky, J., 303
Bergson, H., 130-132
Bigelow, John, 4, 14, 99, 106, 113n2, 117-119,158n6, 180n1, 192, 207, 247n1,253, 254, 256, 273n25, 277, 278, 279, 281, 314, 316n27
Bourne, Craig P., 4, 14, 118-121, 159n13
branching future possibilities, 133
branching view, 150, 153-157
Brentano, F., 127
Broad, C. D., 47, 59-60, 63-64, 66, 69n22, 113n6, 130, 132, 158, 181n24
Brogaard, B., 303, 193n2
B-series, 2, 16-18, 21, 36-38, 41, 43n2, 44, 58, 76
B-theoretic eternalist, 287
B-theoretic ontology, 22, 24, 57
B-theory see also eternalist, 2-3, 11-12
Butterfield, Jeremy, 218

Callender, Craig, 100, 113n1, 168n6, 238, 239
causation
argument from, 129-132, 136; causal determinism,145, causal links, 147, causal relation, 129,130,132, 135, 208, 210, 216, 314; causal theory of reference, 130
change, 15, 18-19, 23, 36-37, 39, 46, 52-53, 76, 163, 299
Chisholm, Roderick M., 197-198, 266, 273n21, 273nn24-25, 286, 315n11, 316n27
Christensen, Ferrel, 59, 61, 69n14, 72n52
compound substance, 36
consciousness, 38
contingent statement, 153, 234, 235n1
contingent truths, 142, 257, 278
continuant(s), 299, 303
correspondence theory of truth, 143
Cosmic Microwave Background Radiation (CMBR), 242
Craig, William Lane, 12, 21-23, 24, 25,

75-86, 87n3, 141, 143, 144, 158, 180n1, 181n19, 183n55, 224
Crisp, Thomas M., 4, 14, 95--96, 20, 219n2, 251, 254, 255, 256
C-series, 76

Davidson, Matthew, 173-174, 180n1, 182n26, 182n32, 191-193
De Clerck, Rafael, 193, 193n1
deflationary theories of truth, 143
demonstratives and pronouns, 175
dependent existence, 134
de re beliefs, 216
descriptivism, 63
Dick, Philip K., 316n23
directly referential term, 174-175
direct reference theories, 176
Dummett, M., 107n2, 145
duration, 28

endurantism, endurantist, 299, 303, 321, 322, 325n5
enduring entities, 300, 324
enduring objects, 320, 321, 322
ersatz actualism, 104
ersatz B-series, 255, 256, 290
ersatz precedence, 120
ersatz theory, 96, 107n7
ersatzism, 96, 104,149
eternalism, 2-3, 96, 112, 113, 118, 120, 195, 239, 255, 289,
eternalists, 291, 292, 238
event(s), 20-21, 36-37, 42, 45-46, 52,
existence, 22, 57, 62-63, 65-67
existential quantification solution, 200-201, 203-204, 206
existential quantifier, 103
existentialism, 287, 288
exists simpliciter, 94, 100

Fine, Kit, 287
Fiocco, Oreste, 14, 121-122
four dimensionalism, 125, 136, 252, 260,263,264, 266, 268, 269, 271n3, 272n5,303, 325,320, 323, 324
four-dimensional objects, 302, 320, 324
Fox, John F., 293n3
Freddoso, A. J., 81
Frege, G., 135
future, 27-30, 36, 38, 41-43, 196, 256
future contingents, 153, 233
Futurity, 3, 60, 254, 256. *See also* A-properti

Gale, Richard, 84
Geach, P. T., 314, 315
general fact, 21, 53
Godfrey-Smith, William, 242, 244
Grim, Patrick, 286
grounding objection, 277-293, 293n8

haecceities, 274n29, 274n32
Haslanger, Sally, 315n5, 315n25
Hegel, 35, 137
Heller, Mark, 325n3
hidden logical form solution, 173-174, 200, 203
Hinchliff, M, 113n, 180n1, 163n54, 273n15
homonyms, 211
Humean theory, 145, 146, 147
hypothetical/categorical distinction, 283

Ibn Sina, 66, 70n41
immediate awareness, 30
indeterminate truth value, 153
indexicalism, see also B-theory, eternalism, 202, 299, 302, 303, 319, 320-323, 324n1, 326n13, 326n15
individual essences, 105, 201, 202
individual facts, 53
instants, 219n2, 219n3
intentional claims, 206
intentionality, 13, 17, 21, 127, 130
inter-temporal sentences, 192
IS, 285

Johnston, Mark, 325n4

.ant, Immanuel, 35, 62-63, 65
.eller, Simon, 180n1, 181n22, 181n23, 251-254
:ripke, Saul A., 103, 176, 183n45, 183n46, 211

eibniz, G. W., 136, 274n26
e Poidevin, R., 143, 145, 158n1
evinson, 59
ewis, David, 4, 104, 106, 149, 153, 159n7, 181n19, 181n22, 195, 220n16, 268, 272n7, 272n10, 273n23, 278, 299, 300, 301, 305-306, 311-314, 316n24, 325n4, 325n7
ewisian concretist, 195
›gical constructions, 143
›gical operators, 80
ombard, Lawrence, 113n1, 324, 326n13

orentz, H. A., 224n1
orentz-Fitzgerald contraction, 246
owe, E. J., 69n10
ucretianism, 254, 279, 280, 284,
ucretius, 117, 133-135, 279-280
udlow, P., 141, 145, 219n2,
ukasiewicz, J. 145

Iarkosian, Ned, 99, 113n2, 165-167, 175, 180n1, 181n12,181n22, 182n44, 207, 211, 316n29,
Iaudlin, Timothy, 246
IcCall, Storrs, 59, 133, 158n1,
IcTaggart, J. E. M., 4, 14-19, 24, 57-59, 75-77, 81, 86, 87n4, 136, 237,
IcTaggart's paradox, 4, 11-12, 20, 24, 59, 75, 77, 79-81, 167-168, 181, 219n2, 254, 314
Ieinong, Alexius, 127, 273n15
Iellor, D. H., 81,82, 158n1, 180n1, 219n2, 303, 308
ıemory, 29-30, 34, 36,
Ierricks, Trenton, 4, 94, 113n2, 122n1, 180n1, 299, 302, 303
ıethod of paraphrase, 301
Ieyer, Ulrich, 4, 94-97

Minkowski, H., 244
Minkowskian interpretation, 178-179
modal actualists, 309, 310
modal operators, 61, 104, 107n6
modal primitivism, 104, 107n6
modal realism, 61, 96, 104, 126, 148,
Molinists, 281
moments, 37, 41, 43, 170
Moore, G. E., 63-64, 66, 71n46,
Moorean fact, 112

names, 202
  ambiguous, 211; proper names, 143, 211; token, 211
negation operator, 62
negative existential, 260, 261
negative paraphrase strategy, 198-200, 203
neo-Lorentzian interpretation, 178, 182, 246
nestling temporal operators, 219n2
Newtonian physics, 289
nominalists, 309
nonactuals, 314
non-existent entities, 162-252
non-existent objects, 273
non-present things, 191-192
non-referring singular terms, 176

Oaklander, L. Nathan, 5n1, 23-25, 85, 86n5, 158n1, 158n6, 167-168, 181n19,
open future or block theory, 2
overlap strategy, 197-198, 203

Parsons, T., 272n4, 273n15
particular fact, 21
Passmore, 48
past, 27, 30, 36, 38, 41-43, 196, 219n7, 256, past entities, 193; past facts, 141; pastness, 59-60, 254, 256
Perdurance, 302, 320. *See also* four-dimensionalism

perdurantism, 272n4, 303, 321, 324n3, 325n5,
perduring entities, 300, 301, 303, 323, 324

perduring events, 320, 324
perduring objects, 302, 320, 321, 322, 323, 324, 325n8
personal identity, 324
Plantinga, Alvin, 62-64, 69n22, 70n33, 143, 181, 286, 287
Plato, 137
possibilism, 96, 136
possible worlds, 97n2, 104-105, 107nn6-107n8, 148, 195, 238
predicates, 22, 282
A-predicates, 22; monadic existence, 103, quasi predicate, 60; tenseless predicate, 232
prediction, 29
present, 27-30, 35, 38, 41-43, 196
absolute, 233; non-present things, 191-192; present tense facts, 24, 80; 147-155; present tense statements, 233; present tensed truths, 264; specious, 48
presentism,
actualist, 165-166; anti-existentialist, 173; atomic, 253, 269-270; cone, 242-245; degree, 169; ersatzer, 97, 105, 120, 148, 158, 159n12; eternal, 182n2; flowing, 168; haecceitist, 253, 267-269; indexical, 180n2; Lucretian, 280, 282, 283,284, 285, 291-293; Meinongian, 252, 262-263, 273; ontology, 62, 67-68; point, 241-242; primitive, 120; 169-170, 180n2; Priorian, 157; reductive, 120, 145-148; serious, 202- 203; transient, 161-180; 183n57; truthmaker-denying, 262, 265; truthmaker-preserving, 269, 270
presentness, 22-23, 57, 59-62, 67-68, 87n3, 256
principle of achronality, 243, 244
principle of bivalence, 145
principle of instantiation, 146
Prior, Arthur N., 19-21, 59-62, 72n52, 75, 80, 97n1, 102, 117, 132, 142, 143, 148, 181n24, 182n25, 207, 208, 210, 211,245, 246, 272n14, 314, 317n30, 317n32
privileged observers, 229, 231, 233, 234
properties see also A-properties
categorical, 282-283; hypothetical, 280-283; intrinsic, 306, 326n14; modal, 281; monadic, 24, 300; non-dispositional, 280; non-fundamental, 278;non-relational, 19; past-tensed, 280; past and future-tensed, 253, 254, 266, 267; present future-tensed, 118; present past-tensed world, 118; primitive, 283, 284; relational, 60, 201-204 239; temporal, 16-17, 22, 163-164, 168; tensed, 117-119, 122n1, 254, 281, 284-285, 292
propositions, 135, 307
characterizing, 63-64; embedded, 148; false relational proposition, 198; future tense, 152, negative existential, 63; past- and future-tensed proposition, 142; past-present-future-tensed proposition, 144; past-tensed proposition, 151; present tense, 132, 142, 148-149, 151; propositional function, 71; propositional operators,61; Russellian, 182n42; singular, 175,182; tensed, 142, 185, 285 313; tenseless, 255, 285, 286; true propositions, 132, 134, 135, 315n11; unembedded, 148-149, 151
Putnam, Hilary, 223, 240, 241

quasi-subjects, 61
Quine, W. V., 132, 133, 135, 273n16, 281

recursion theorist, 216
redundancy theory of the present, 142
referent, 101, 176
relations
accessibility, 156; analogical earlier than, 287; A-relations, 83- 84,86; argument from,126-127, 133, 135, 136; a-temporal, 218, 219n5; B-relations, 2, 15-17, 22, 76, 82-86; causal, 129, 130, 132, 135, 208-210, 216, 314; comparative, 208-210; cross-temporal, 171-174, 319n2; cross-time, 207-219; disguised,

305; earlier than (E relation), 293; earlier/later than, 2; ersatz B-relations, 255-256; eternalist's, 289; existence entailing, 126-131; false relational proposition, 198; greater-than,156; instantiation, 285; intentional, 193, 199, 208, 215-217; 220n12; inter-temporal, 192-193; logical, 181n8; necessitation, 146; non-intentional, 193; non-relational properties, 19; objection from, 117; perceptual, 216; problem of, 196-197, 204; relational fact, 1; relational properties, 60, 201-204, 239; relational truth, 265; semantic, 210-214; simultaneity, 228, 230, 231; temporal, 239, 255, 307; tensed, 193n2; tenseless, 193n2; transitive, 202; transitive assymetrical, 35; triadic, 202
:ussell, Bertrand, 15, 38, 67, 71n45, 71n46, 132, 133,159n12, 211, 212

ainsbury, R. M., 176
almon, N., 181n24
aunders, Simon, 99, 177, 180n1
avitt, Steve, 99, 183n54, 238, 240, 241, 242, 243, 244, 245
chaffer, Jerome, 70n27
chlesinger, G. N., 59, 158n1
emantics, 61
erious tenser, 299, 308, 310-312
extus Empiricus, 130, 132
hoemaker, S. 1
ider, Theodore, 94, 99, 106, 113n1, 176, 178,182n25,183n45, 205, 207, 211, 280, 281, 283, 284, 291, 292
klar, Lawrence, 178, 240
mart, J.J.C., 271n3, 308, 316n29
mith, Quentin, 59, 61-62, 67, 71n47, 87n5, 158n1, 169-170, 181n9, 224, 310, 316n20
Soames, Scott, 214
Sosa, Ernest, 67, 71n50
Special Theory of Relativity (STR), 158n3, 177-179,183n55, 223, 228, 230, 231, 232, 237, 238, 240, 247
specious present, 48
Spinoza, B., 35
Stein, Howard, 183n54, 241
Stoics, 117, 130, 132-135,
substances, 134
succession, 16,18, 25, 223,
supervenience principle, 251, 256, 279

Tegtmeier, E, 303
temporal actualist, 310
temporal becoming, 11-14, 16-19, 22, 24, 75, 80-81, 85, 240, 324n1
temporal being, 68
temporal change. *See* temporal becoming
temporal distance, 255, 289
temporal entity, 163-164, 179, 180n5
temporal existence, 23, 68, 100-101
temporal objects, 85
temporal parts, 167, 260, 300, 302, 303, 305, 313, 314, 320,324n3
temporal passage, 20. *See also* temporal becoming
temporal priority, 299
temporal reality, 121, 162-167, 169-172, 174, 176-177, 181n12
temporal recombination, 279
temporal things, 113
temporary intrinsics, 167-168, 181n21, 219, 299, 300, 302, 305-314, 322
tense(d)
detenser, 239, 299; determinations, 60-61; disjunction, 313; expression, 169; facts, 144,180n2; indexical tense operator, 238; language, 11; new tenseless theory of time, 308; operators, 61, 102-103, 105; past-and-future tense facts, 24; past-and future tensed statements,146,158n6, 259; past-and-future tense truths, 252, 254, 260-263, 266, 270, 271, 273n19, past- or future- tensed statements, 252;

past tensed statements, 3, 142, 147,155, 158; present tense facts, 24, 80; 147, 155; present tense statements, 233; present tensed truths, 264; theory (view) of time, 102-104, 157, 158n3, 207; truths, 3, 119, 170, 310, 315n11;
tenseless
notion of existence, 234-235; predicates, 232; statements, 307, 310; theory of time, 102-104, 157,158; verb, 306, 315n3, 315n9
things, 52, 170
three-dimensionalism, 272n5, 302, 324
time, 12-13, 27, 31-33, 38, 40, 45, 255
absolute, 37; A-theoretic view of, 287; A-time, 19, 76; B-time, 76; Continuous time, 156; cross-time, 207-219; Einstein-Minkowski space-time, 224, 240-241; ersatz times, 120, 149; four-dimensional space time, 260; local proper times, 245; movement of, 44n2; new tenseless theory, 308; passage of, 11, 45-46; physical time, 235; real time series, 149; timeless existence, 23, 68; timeless universal, 68; time series, 38, 40
Tooley, Michael, 5n1, 113n6, 129-130, 132, 158, 180n1, 219n2, 219n7, 246, 272n3
triviality argument, objection, 93-94, 96, 109-111
truth
truth at a world, 287, 288, truth simpliciter, 150; truth-at-a-time,150; truth-for-me, 235n truthmaker, 251-252, 256, 260-265, 271, 272; truthmaker axiom, 277; truthmaker objection, 168, 171;truthmaker principle, 300, 301;ttruthmaker(s), 3,5, 22, 24, 128, 141, 142, 148, 158, 170; truthmaking objection, problem, 191, 260, 271; truth-value links problem, 147, 157

Van Inwagen, P., 314n5, 314n13, 315n5

Wahlberg, T. H., 303
Watts, Isaac, 45
Whitehead, A. N., 325n3
Wiggins, David, 324
Williamson, Timothy, 217-218

Zalta, Edward, 286
Zimmerman, Dean W., 4, 100, 105, 113n2, 122n1, 207, 299, 300, 301, 315n11,

About the Editors

Ernâni Magalhães is adjunct professor of philosophy at West Virginia University. He has published articles in the *Australasian Journal of Philosophy* and *Philosophia.* He also co-edited a volume of *Philosophia* (38) on the centennial of McTaggart's argument for the unreality of time.

L. Nathan Oaklander, David M. French Distinguished Professor of Philosophy, has taught at the University of Michigan-Flint since 1972. Among his numerous books is an edited four volume major work, *The Philosophy of Time, Critical Concepts in Philosophy* (Routledge, 2008), *The Ontology of Time* (Prometheus, 2004), and *Existentialist Philosophy: Second Edition* (Prentice Hall, 1996).

www.ingramcontent.com/pod-product-compliance
Lightning Source LLC
Chambersburg PA
CBHW031415220926
51612CB00004B/40
9780739145005